Thermodynamics and Kinetics
of Biological Processes

Thermodynamics and Kinetics of Biological Processes

Editors
Ingolf Lamprecht · A. I. Zotin

W DE G

Walter de Gruyter · Berlin · New York 1983

Editors
Prof. Dr. Ingolf Lamprecht
Institute of Biophysics
Free University Berlin
D-1000 Berlin 33

Prof. Dr. A. Zotin
Institute of Developmental Biology
Academy of Science of the U.S.S.R.
Moscow

Library of Congress Cataloging in Publication Data

Thermodynamics and kinetics of biological processes.
Translated from the Russian.
„This monograph is, in fact, a second volume to the book Thermodynamics of biological processes published by Walter de Gruyter (Berlin, West) in 1978"–Pref.
Bibliography: p.
Includes index.
1. Biophysics. 2. Thermodynamics. 3. Ecology. 4. Adaptation (Biology) 5. Ontogeny. I. Lamprecht, Ingolf, 1933– . II. Zotin, A. I. (Aleksandr Il'ich), 1926– . [DNLM: 1. Biophysics. 2. Kinetics. 3. Thermodynamics. QT 34 T319]
QH505.T393 1982 574.19'16 82-17279
ISBN 3-11-008200-4

CIP-Kurztitelaufnahme der Deutschen Bibliothek

Thermodynamics and kinetics of biological processes / ed. Ingolf Lamprecht ; A. I. Zotin. – Berlin ; New York : de Gruyter, 1982.
ISBN 3-11-008200-4
NE: Lamprecht, Ingolf [Hrsg.]

 Printing: Georg Wagner, Nördlingen. – Binding: Dieter Mikolai, Berlin. Printed in Germany.

This monograph deals with thermodynamic and kinetic approaches to differing biological problems (ecology and populations, physiological reactions of organisms to external factors and adaptation, kinetics of ontogenesis).

The main difficulty in considering these problems in terms of thermodynamics is concerned with the absence of a commonly accepted theory of nonlinear, irreversible processes. Therefore a large section is devoted to the discussion of current problems in statistical and phenomenological thermodynamics. Leading physicists and mathematicians were asked to contribute to the monograph. This book, therefore, deals with contemporary problems of the thermodynamics of irreversible processes.

This monograph will be useful for biophysicists, physiologists, developmental biologists, and for physicists and mathematicians interested in modern problems of thermodynamics, in the thermodynamics of biological processes and in mathematical biology.

We dedicate this book to the memory of our friend Dr. Anatoli Kotomin who met his death as a result of an accident during his expedition to the Far East. Dr. Kotomin was born in 1943 and graduated from the Biological Faculty of the Moscow University in 1965. He got his degree of Candidate of Biological Sciences in 1969 for a dissertation entitled "Use of physical and chemical methods of inactivation of the nuclei for studying their morphogenetic function in development".

Between 1969 and 1974 Dr. Kotomin worked at the Laboratory of Biochemical Embryology, Institute of Developmental Biology (IDB) of the Academy of Sciences of the USSR (Moscow). During this period his studies were concerned with RNA transport in fish oocytes and embryos, molecular mechanisms of formation of mitochondria and the interregulation of mitochondrial and cytoplasmic protein synthesis.

From 1974 Dr. Kotomin worked at the Laboratory of Developmental Biology and began his studies on the energetics and thermodynamics of development and growth. His accidental death unfortunately stopped these studies.

All those who knew Anatoli Kotomin during his lifetime will always remember his frankness and eagerness, his deep interest in the problems of science and his sincere attitude towards all people.

I. Lamprecht A. I. Zotin

Preface

This monograph is, in fact, a second volume to the book "Thermodynamics of biological processes" published by Walter de Gruyter (Berlin, West) in 1978.

From the point of view of irreversible thermodynamics the behaviour of any system can be established and described if the value and the behaviour of dissipative functions of the system are known. As has been shown in the previous book, four types of phenomena determining the specific dissipative functions of living organisms can be described in the most general form as: 1) basal metabolism, 2) the constitutive approach of living systems to the final stationary state in the course of development, growth and aging, 3) inducible-impulse reactions connected with a short-term reaction to external or internal factors, and 4) inducible-adaptive processes connected with long-term variations of external or internal conditions, i. e. with adaptation of living organisms. The monograph of 1978 was devoted mainly to constitutive processes associated with developmental biology. The present monograph is devoted to inducible processes associated with the reaction of living systems to external actions and the adaptation of organisms to varying conditions in the environment. Continuity of this and the previous monograph is obvious since both treat living processes from a thermodynamic point of view.

A thermodynamic treatment of biological processes is complicated because the thermodynamic theory of nonlinear, irreversible processes has not yet been sufficiently worked out. Therefore, in the present monograph attention is paid mainly to a discussion of some contemporary problems in the statistical and phenomenological thermodynamics of nonequilibrium processes whereas biological problems take a more modest place.

Leading research workers in the field of modern irreversible thermodynamics have contributed to this book. It is natural, therefore, that a wide range of opinions and approaches in such an unsettled field of knowledge as irreversible, nonlinear thermodynamics is presented here. We hope that this will show, to a wide range of biologists, how difficult the problem of a thermodynamical description of biological phenomena is, and will warn them against too trusting a view on the various thermodynamical theories which are often actively advocated by different authors (of course, this includes some of the present authors). Besides, we believe that the comparison of the first chapters which are devoted to thermodynamical and mathematical problems with the chapters describing, for the most part, biological problems will clearly show to the reader what a deep divide lies between biology and thermodynamics: the level of theoretical investigations is much higher in thermodynamics than in biology. We think it instructive to display this difference in the levels of investigations to make it clear that modern thermodynamics has achieved much, whilst the thermodynamic interpretation of biological processes cannot be built on the basis of the achievements of the past century, a feature which is often found though in biological research.

Some remarks should be made about the basic idea supported, perhaps by not all of the present authors, but which is the practical basis and reason for the book. Onsager (1931 b) came to the conclusion that irreversible processes that proceed in thermodynamic systems may be compared with fluctuation relaxation processes. This idea was advocated still more explicitly by Gurov (1978) who suggested a suitable term of "large-scale fluctuation", which he compared with irreversible processes. We ususally distinguish between spontaneous fluctuations and those induced by external forces. It is obvious that when fluctuations and irreversible processes are compared, we mean, as a rule, induced fluctuations, which collapse as soon as the external forces stop acting, and thus the system returns to the initial equilibrium or stationary state. Hence, any transition process in a thermodynamic system is either the emergence or relaxation of fluctuations. That is why the first problems to be considered in the book are those of statistical thermodynamics which are, to a great extent, concerned with fluctuation theory, and then the secondary problems of phenomenological thermodynamics, the kinetics of transition processes and biological phenomena.

The appearance of this book is due to the generous support of the director of the Institute of Developmental Biology Prof. T. M. Turpayev.

In the preparation of the book important help was rendered by G. V. Dontsova, G. S. Lozovskaya, I. S. Nikolskaya, S. N. Malyghin and particularly by E. V. Presnov, who participated in the discussion of the plan and structure of the book and also checked and corrected the bibliography. We are grateful to all of them as well as to the authors of each contribution to the book for their willingness to colaborate in such a complex field as the thermodynamics of biological processes.

The English translation was thoroughly checked by Dr. A. E. Beezer, London, to whom we are indebted very much.

Berlin, October 1982

I. Lamprecht
A. I. Zotin

Contents

V. Ecology and Populations

VI. Reaction of Organisms to External Factors

I. General Problems of Statistical Nonequilibrium Thermodynamics

Statistical Thermodynamics of Nonlinear Irreversible Processes

F. M. Kuni, L. Z. Adjemian, A. P. Grinin, T. Yu. Novozhilova,
B. A. Storonkin

The thermodynamics of irreversible processes, as a science of general laws of energy, momentum and mass transfer, has appeared and developed mainly on the basis of phenomenological considerations of matter properties. The great achievements of nonequilibrium thermodynamics that made it one of the fundamental features of modern natural sciences are first of all due to the heuristic strength of the phenomenological approach. In the thermodynamics of irreversible processes, though, there are a number of important problems which seem to be, in principle, inaccessible to the phenomenological approach.

Firstly, there is the basic problem of irreversibility. Within the phenomenological approach irreversibility is introduced as a postulate. But can one formulate those conditions responsible for dissipation? The conditions under which irreversibility would, not only be consistent with, but follow from, the reversible character of the law of motion of individual particles in a thermodynamic system.

Secondly, there is the problem of substantiations of a thermodynamic description and defining the limits of its applicability. If a thermodynamic description proves insufficient and requires extension then the problem, naturally, develops into a more general one: how to choose additional variables so that they are independent of the thermodynamic ones. For example, what part of the tension tensor, independent of thermodynamic parameters should be introduced to describe highly viscous liquids (Kuni, 1974)? The problem becomes still greater when one has to transfer from the ordinary thermodynamics of nonlinear irreversible processes to an extreme nonequilibrium thermodynamics. To clarify this we should note that in

Thermodynamics and Kinetics of Biological Processes

ordinary nonequilibrium thermodynamics, based on the local equilibrium hypothesis, the very relations between dissipative flows (heat, viscous, diffusive) and the corresponding thermodynamic forces are linear. Nonlinearity in the thermodynamic equations is due to the dependence of kinetic coefficients on thermodynamic parameters and to the transformation from a coordinate system moving with a local velocity (wherein dissipative flows are determined) to a fixed (laboratory) system. The treatment of nonlinearity, purely local in the first case and purely convective in the second, is not difficult, and is described below. If the gradients are large and, therefore, dispersion is essential, the nonlinear processes never approach equilibrium. In this case, the local equilibrium hypothesis becomes meaningless, and besides thermodynamic quantities it becomes necessary also to consider the averaged products of the microscopic analogs of the thermodynamic quantities. How can one establish the very structure of equations of nonequilibrium thermodynamics in which (equations (1), (2), (3)) the kinetic centres alone contain the correlation effects of the differences between averaged products and the products of averaged values?

Thirdly, there is the problem of an ambiguity in the definition of a nonequilibrium thermodynamic ensemble entropy and the even greater problem of whether or not the entropy concept exists at all for systems far from equilibrium. It is not necessary to explain the importance of this problem which is being widely discussed by physicists concerned with the thermodynamics of irreversible processes.

And finally, there is, of course, the problem of calculating kinetic coefficients and defining kinetic centres which are also present in nonequilibrium thermodynamic equations. In a simpler and more realistic form the problem can be formulated as an urgent question of the long-term behaviour of kinetic centres. Without solving this question it is, of course, impossible to discuss seriously a local-in-time form of nonequilibrium thermodynamic equations, and the possibility that these equations may consider time dispersion.

In such a brief outline, we have, of course, made simplifications and have been probably somewhat subjective in the division into separate problems. But we may state quite categorically that taken together they demonstrate, objectively, the actual situation.

These problems become particularly urgent in view of the potential application of nonequilibrium thermodynamics to objects qualitatively more complicated than physico-chemical ones. Such objects are biological and biophysical systems requiring not only descriptive, but also quantitative investigation.

Evidently, all these problems cannot be solved without reference to the fundamental laws of the interaction and motion of particles which form thermodynamic assemblies. We do not intend, of course, to take into account all the details of particle interaction and motion and even of the kind of mechanics, classical or quantum, by which this interaction or motion is described. If all this were necessary, no thermodynamic theory as a general science of energy, momentum and mass transfer would exist at all. We mean identification of some more general conditions of the interaction and motion mechanism as a whole, i. e. the type of conditions describing the initial state, the existence of a time hierarchy, rather rapidly weakening correlations, and a thermodynamically limiting transition.

The condition of the initial state is, of course, beyond the framework of purely dynamic concepts. It is, in effect, a probability-statistical hypothesis. The fact that such a hypothesis is needed and dynamism alone is not enough seems natural since thermodynamics, as distinct from mechanics, deals with qualitatively new laws inherent in the entire ensemble of an enormous amount of particles. It also seems natural that the probability-statistical hypothesis is met only once, at inition. The initial condition is defined by dynamics only.

Thus we see that the solution of the problems formulated is within the competence of nonequilibrium statistical mechanics, which combines probability and dynamic concepts of the properties of bodies composed of a large number of separate particles. Although statistical mechanics has for a long time been applied as an auxiliary means for the phenomenological approach (e. g. in the derivation of reciprocal relations by Onsager), it is only fairly recently that nonequilibrium statistical mechanics has taken root in the thermodynamics of irreversible processes. Gradually a new science, the statistical thermodynamics of irreversible processes, has developed.

The formulation of nonequilibrium statistical thermodynamics was decisively stimulated by earlier work of Bogolyubov, van Hove and Prigogine, in which the foundations of a dynamic consideration of nonequilibrium systems in general were laid down. The main merit of the dynamic approach was that within it the statistical hypotheses concerning the macroscopical system were minimized, and concentrated in the initial state only in the form of the initial condition of the correlation weakening. Further evolution of the system was determined by dynamic laws only and did not need the chaotic *shakings* which were undertaken earlier (e. g. in well-known works by Pauli) as an artificial means for reaching dissipative behaviour in the system.

The approach suggested by nonequilibrium statistical mechanics is microscopic. It is consistent and at the same time dynamic in the above-mentioned sense. A complete realization of this approach in principle automatically solves all the problems formulated above previously inaccessible to the phenomenological approach. This confirms the fact that it is just a microscopic approach that is aimed at solving these problems.

In reality, however, we are still far from obtaining a complete solution. Our aim now is to show the present possibilities of a statistical thermodynamic treatment of irreversible processes and also to illustrate the additional problems which it presents and which earlier were completely hidden in the phenomenological approach.

First we shall show how the theoretical technique of ordinary nonequilibrium thermodynamics is constructed (Mori, 1965; Zubarev, 1974).

The bases for consideration are the mechanical conservation laws in the local form

$$\frac{\partial \hat{a}(t)}{\partial t} = - \nabla_\beta \hat{j}_\beta \tag{1}$$

Here t is time, ∇ is the gradient operator, Greek indices characterize Cartesian components; $\hat{a}$ and $\hat{j}_\beta$ are the densities of the conserved quantities (energy, momentum and mass of the components) and the densities of the corresponding flows (energy, momentum and mass flows); "ˆ" indicates that these are microscopic (and not averaged) values; for convenience these values are determined from averaged values in the final equilibrium state. In order to keep our calculations simple we understand $\hat{a}$ and $\hat{j}_\beta$ as column vectors, some elements of which are given by a discrete index which characterizes the conserved quantity (energy, momentum and the mass of certain components) and by a continuous index characterizing the space coordinate of the observation point. We should also note that concrete mathematical expressions for microscopic densities of the conserved quantities and their flows in terms of the coordinates and momenta of separate particles of the system (which microscopic quantities always depend on) are well known. For the conserved quantities they follow directly from their meaning. For flows they are so chosen that the conservation laws automatically follow from the dynamic equations of motion.

As is seen from the conservation laws, the less the space gradients are, the slower the densities of the conserved quantities change in time. Under conditions close to homogeneity, the energy, momentum and mass transfer processes proceed, therefore, more slowly than all the other possible processes in the system approaching the equilibrium state. This means the existence of a time hierarchy. In this case, the densities of the conserved quantities are thought of as quasi-integrals of motion. These are just the conditions to make a thermodynamic description possible.

If we suppose, as an approximation, that as compared to thermodynamic processes all the other processes are instantaneous, the local equilibrium state will be established in the system for each value of the density of the conserved quantity. The distribution function of such a local equilibrium state is determined as the maximum in the information state entropy at a given mean value of the densities of the conserved quantities

$$\hat{\varrho}_\ell(t) = \hat{\varrho}_0 \frac{e^{-\hat{a}^+ \cdot f(t)}}{\langle e^{-\hat{a}^+ f(t)} \rangle} \tag{2}$$

Here $\hat{\varrho}_0$ is a final equilibrium distribution function, $f(t)$ is a column vector of the thermodynamic parameters (more precisely, their deviation from equilibrium values) "+" designates Hermitian conjugation, "." implies convolution, i. e. internal multiplication (multiplication of the row obtained by Hermitian conjugation by the column, the brackets "< >" mean equilibrium averaging with the function $\hat{\varrho}_0$. The values of the thermodynamic parameters are determined from the condition used in entropy maximization, that averaging of microscopic densities of the conserved quantities with local equilibrium distribution functions yields their actual mean values. These values may, in principle, change, even over microscopic distances. The procedure formulated for determination of thermodynamic parameters will be valid in this case, too. The concept of a local equilibrium distribution function is, thus, beyond the framework of the local equilibrium hypothesis. The latter is obtained from a local equilibrium distribution, under the limitation of small gradients, when the values of the thermodynamic parameters at the observation point can be determined from the space integral (which is present in the convolution designated by ".". For linear processes (from methodical consideration this case is the first to be discussed) the above-mentioned procedure makes it possible to obtain easily explicit expressions for thermodynamic parameters

$$f(t) = - \langle \hat{a}\hat{a}^+ \rangle^{-1} \cdot a(t) \tag{3}$$

Here the correlator $\langle \hat{a}\, \hat{a}^+ \rangle$, wherein under the average symbol we have a direct product of the column vector $\hat{a}$, may be thought of as a finite rank square matrix in discrete indices and as an infinite rank one in continuous indices. A matrix inverse of the selected vector is indicated symbolically by the power -1. The internal multiplication, designated by ".", should be understood as a multiplication of the inverse matrix by the column vector $a(t)$ of the mean densities of the conserved quantities.

In effect, all the other processes concerned in the approach to equilibrium are, of course, not instantaneous and an incomplete equilibrium state is not, practically, realized. This is emphasized explicitly by the fact that the true nonequilibrium distribution function $\hat{\rho}(t)$ differs from the local equilibrium function $\hat{\rho}_\ell(t)$:

$$\hat{\rho}(t) = \hat{\rho}_0 \hat{\Psi}(t); \quad \hat{\rho}_\ell(t) = \hat{\rho}_0 \hat{\Psi}_\ell(t). \tag{4}$$

This difference is conveniently (after separating a usual equilibrium function) described in terms of the quantity

$$\hat{\Psi}(t) = \hat{\Psi}_\ell(t) + \Delta\hat{\Psi}(t). \tag{5}$$

We should particularly emphasize that in spite of the fact that the nonequilibrium correction $\Delta\hat{\Psi}(t)$ is non zero, the local equilibrium function still yields, by definition, true mean values for the densities of the conserved quantities. For other microscopic quantities, say flows, this property is lost. The calculation of a nonequilibrium correction to a local equilibrium distribution function thus becomes an urgent problem.

The actual proximity of the local equilibrium distribution to the true one makes it possible, though, to consider a local equilibrium function as the initial condition. Indeed, if the initial moment is chosen in an infinite past, the dynamics of the system will restore the distorted correlations, and the system will correct itself. The probability-statistical hypothesis formulated as a condition of the correlation weakening in the inital state makes it possible to approach a reduced description of the system on the level of thermodynamic quantities.

The equation for the nonequilibrium correction to the local equilibrium distribution function has the form

$$\frac{\partial \Delta\hat{\Psi}(t)}{\partial t} \quad - i L \Delta\hat{\Psi}(t) - i L \hat{\Psi}_\ell(t) - \frac{\partial \hat{\Psi}_\ell(t)}{\partial t} \tag{6}$$

This equation follows directly from the fundamental principles of many-particle system dynamics whose evolution is determined by the Liouville operator L . Owing to the initial condition of the correlation weakening

$$\Delta\hat{\Psi}(t_0) = 0 \qquad (t_0 \to -\infty) \tag{7}$$

the form of the given functional dependence of the microscopic quantity $\Delta\hat{\Psi}$ of an enormous number of coordinates and the momenta of all the particles in the system is determined by a much smaller but indeterminate number of variables. It is determined by thermodynamic parameters and by their time derivatives (last term of equation (6)). But transfer to a thermodynamic description requires a still larger reduction in the number of independent variables. It also requires an elimination of the time derivatives of the thermodynamic parameters.

For the solution of this problem it proves efficient to consider the use of functional spaces and to use the projection operator method suggested by Zwanzig (1960) and applied by Mori (1965). The operators, $\mathcal{P}_a$, of projection of conserved quantities on microscopic densities $\hat{a}$ are defined in the following way

$$\mathcal{P}_a \hat{G} = \langle \hat{G}\hat{a}^+ \rangle \cdot \langle \hat{a}\hat{a}^+ \rangle^{-1} \hat{a} . \tag{8}$$

Here G is an arbitrary microscopic quantity taken from the Hilbert space of such quantities, i. e. from the space of the functions that depend on coordinates and momenta of the particles of the system. This definition corresponds to the choice of the correlation of two microscopic quantities as being their scalar product. This is the choice that enables us to represent the local equilibrium distribution function as a projection of the nonequilibrium function

$$\hat{\Psi}_l(t) = 1 + \mathcal{P}_a \hat{\Psi}(t) . \tag{9}$$

This device solves the problem of eliminating the time derivatives of thermodynamic parameters.

The solution of equation (6) for the nonequilibrium correction $\Delta\hat{\Psi}(t)$ is

$$\Delta\hat{\Psi}(t) = \int_0^\infty dt' \, \hat{I}^+_\gamma(-t') \nabla_\gamma f(t-t') . \tag{10}$$

The important result, that time derivatives were eliminated from this equation by the projecting operator method, is clearly seen from the fact that the microscopic quantity $\hat{I}_\gamma(t)$ does not contain projection on the quasi-integrals of motion, i. e. the densities of the conserved quantities (such projections are automatically cut off by applying the operator $(1 - \mathcal{P}_a)$

$$\hat{I}_\gamma(t) = e^{it(1-\mathcal{P}_a)L}(1-\mathcal{P}_a)\hat{j}_\gamma . \tag{11}$$

It is essential that this orthogonality does not change with time. The dependence of this correction to the local equilibrium function for macroscopic quantities is realized only through the thermodynamic parameters and, more precisely, only through their gradients. Thus a reduced description in terms of thermodynamic parameters and the setting up of a statistic ensemble corresponding to this description has been achieved. The appearance of memory as an integral over the retardation time, t', is apparently a consequence of this reduced description.

Now, it is not difficult to find mean values of the microscopic flow densities:

$$\dot{j}_\beta = \dot{j}^{(l)}_\beta + I_\beta(t), \tag{12}$$

where

$$\dot{j}^{(l)}_\beta(t) = \langle \hat{j}_\beta \rangle^t_l , \qquad I_\beta(t) = \langle \hat{I}_\beta \rangle^t \tag{13}$$

$$I_\beta(t) = \int_0^\infty dt' \langle I_\beta(t') \hat{I}^+_\gamma \rangle \cdot \nabla_\gamma f(t-t') . \tag{14}$$

The brackets < > and < > imply averaging at a time t over local equilibrium and nonequilibrium ensembles. The first component in equation (12) represents, evidently, a flow in a local equilibrium state. It may be found from the known local equilibrium distribution function and is of a purely statistical nature. The nonequilibrium contributions are, therefore, entirely included in the second component; which is a dissipative heat flow (3), a viscous and a diffusion flow. That the dissipative flow is represented as an averaged value of the quantity $\hat{I}_\beta$ makes is possible to term the latter a microscopic dissipative flow. An explicit expression for a mean dissipative flow is given in equation (14). This expression is an integral over the retardation time. In addition, it is a space integral and a sum over the discrete indices, which is obvious if one remembers that "." symbolizes a convolution operation. The role of the kinetic centre under gradients of thermodynamic parameters is represented by a microscopic dissipative flow correlator. Thus for kinetic centres expressions are obtained which, in principle, can be used for real calculations. It is essential that these expressions, and the dissipative flows determined by them, can be non zero only under the condition that a thermodynamic limiting transition to infinitely large dimensions of the system proceeds whilst the initial moment tends to an infinite past. From the physical point of view such a sequence of limiting transitions is necessary for excluding Poincaré cycles, i. e. making the reverse rates of the system vanish.

The kinetic centre regularization performed by subtracting the thermodynamic quasi-integrals of motion from the correlating dissipative flow permits us to leave in the kinetic centres only the contributions from processes approaching equilibrium more rapidly than the thermodynamic ones. Time hierarchy in this case guarantees a rapid decrease in kinetic centres as the retardation time increases as compared with the change of thermodynamic parameters. A rapid memory weakening is thus a mathematical expression of time hierarchy.

A rapid weakening of memory indicates evidently that thermodynamic processes have a quasi-local-in-time (quasi-Markov) character. This important consequence of the existence of time hierarchy in its turn confirms the considerations used above in basing the correlation weakening condition on the initial state. At the same time it means a practical independence of the expressions obtained for mean dissipation flows and kinetic centres on the above assumption and on a strictly defined closed system during an evolution from the initial state. In effect, kinetical centres have the chance to damp for the times during which an isolation of a quasi-closed system may brake (if there is a contact with the thermostat) - and this is anyway the case after a thermodynamic limiting process.

Now we may appraise the true worth of what seem to be the quite abstract ideas of functional spaces. It is just these ideas that permit the introduction of a concept of dependence or independence, of different microscopic quantities, and permit discussion of the contributions of one quantity to the others, and, what is most important, we can now eliminate the contributions from slow motions and thus establish a hierarchy of quantities relative to the rate of their change in time. As has already been mentioned, the phenomenological approach operating as it does previously with averaged quantities is absolutely devoid of these possibilities.

The expression obtained for mean flows has a quasi-local structure not only in time but also in space. The value of the flow depends on thermodynamic parameters only in the immediate neighbourhood of the observation point of the flow. The radius of this area coincides with the intermolecular correlation radius and usually has negligible dimensions of the order of mean intermolecular distance as compared to microscopic scales (we do not mean special situations in e. g. considerably rarefied or close-to-critical states where the correlation radius may be rather large).

A quasi-local space-time character of flows permits the application of a local equilibrium hypothesis which is a basic approximation for the usual thermodynamic treatment of irreversible processes. We shall show how, within this hypothesis, one can extend the results obtained above to nonlinear processes (Kuni, Storonkin, 1971, 1972 a, b, 1973; Storonkin, Kuni, 1973). At each space-time point imagine a hypothetical equilibrium thermodynamic system, such that the true mean densities of the conserved quantities of the real system coincide at a given point with the corresponding equilibrium values in the hypothetical system. The velocity of such a system is evidently equal to the local mass velocity. True thermodynamic parameters, which are strictly determined by a local equilibrium distribution, will also coincide with the corresponding values for the hypothetical equilibrium system. In the framework of quasi-local equilibrium hypothesis the local values of mean densities of the conserved quantities and thermodynamic parameters are thus interconnected by the usual equilibrium thermodynamic equations.

The local structure of mean flows should evidently be conserved in a nonlinear theory, too: their values in some space-time point are to depend on the state of the system in its close neighbourhood only. But it is in a nonlinear theory only that by its state this neighbourhood is close not to the final state of complete equilibrium but to the hypothetical equilibrium system introduced at a given point. In the coordinate system fixed to the hypothetical equilibrium system we shall therefore have, for mean flows, certain expressions analogous to the previous ones

$$j'_\beta(t) = \langle \hat{j}_\beta \rangle_{\vec{r},t} + \int_0^\infty dt' \langle \hat{I}_\beta(t') \hat{I}_\gamma^+ \rangle_{\vec{r},t} \cdot \nabla_\gamma f(t-t') \quad (15)$$

The prime implies here that the flow pertains to a coordinate system moving relative to the laboratory one; the suffixes $\vec{r}$ and t stress that averaging is carried out not over the finite state distribution function but over that of the hypothetical equilibrium system referred to the point $\vec{r}$, t at which the mean flow is calculated. In accordance with this the kinetic centres (dissipative flow correlators) now also depend on this point. Within the local equilibrium hypothesis the gradients of the thermodynamic parameters f should also be referred to this point. In practical application they may be expressed in the normal way through gradients of temperature, mass velocity and chemical potentials. In the final equations one should put the mass velocity equal to zero. The remaining integration over the retardation time and space coordinates will then be referred to the kinetic centres only and will give the kinetic coefficients.

Having found expressions for flows in a moving coordinate system we can, with the aid of known transformation formulae, easily obtain expressions for the flows in a fixed (laboratory) coordinate system. A subsequent substitution of these expressions into the averaged conservation laws completes the system of master equations of the level of thermodynamic description. So we have finished constructing the system for the study of the statistical thermodynamics of nonlinear, irreversible processes. One could build its quantum analogue in a similarly simple way.

The presence of closed expressions for kinetic centres in statistical mechanics is a reliable foundation not only for their numerical calculation (on a computer or by model simplifications) but also for their general application in theoretical research.

In simpler situations one can also succeed in solving the fundamental problem of irreversibility. For example, this can be done using Markov's approximation, i. e. by neglecting retardation. In a linear theory the result is as follows. All the hydrodynamic modes, as a superposition of which one can represent the general solution of nonequilibrium thermodynamics, damp in time. If a system is capable of collective motions the damping may be of an oscillating character. The frequency of natural oscillations in the same as the very possibility of collective motions is entirely determined by local equilibrium contributions to the flow terms. The result of a nonlinear theory is more modest. Of course, no simple solutions exist here and, therefore, one may draw a conclusion concerning only the existence of necessity for dissipation inequalities in kinetic coefficients. The necessity of a preliminary thermodynamic limiting transition manifests itself clearly in the problem of irreversibility (corresponding to instantaneous change of the initial moment to an infinite past). This is just the transition that makes it possible to obtain a continuum in spectral representations of correlators and to draw the above-mentioned conclusions.

The expressions for kinetic centres given by statistical thermodynamics are also successfully used in describing their time-asymptotic behaviour. As the system approaches a homogeneous state, when its thermodynamic parameters remain essentially constant over time, the kinetic centres keep on slowing down but not according to an exponential law, as was supposed earlier in the phenomenological approach, but according to a much slower power law with the power $(-\frac{3}{2})$ (Dufty, 1971; Ernst et al., 1971; Pomeau, 1971; Dorfman, Cohen, 1973). This confirms the validity of Markov's approximation, since the integral of $t'^{3/2}$ in the kinetic coefficient converges, albeit very slowly, at long times t'. Any time-moment expansions are, here, out of the question. This is the answer to another question formulated above, which is inaccessible to the phenomenological approach.

We should note now, that the methods of nonequilibrium, statistical mechanics are already applied to more complicated physical problems, which overstep the framework of the usual statistical thermodynamics of nonlinear irreversible processes. In the first place these concern the problems mentioned at the very beginning where the thermodynamic description alone is insufficient and the description must include additionally some other long term quantities, namely, tension tensor, moment of momentum, etc. The capacity of the microscopic approach to isolate from the additional quantities, the contributions independent of thermodynamic ones makes it possible to move further in solving thes problems.

Another interesting area of applications is nonequilibrium thermodynamic systems within which both extreme nonlinearity and large dispersion are central features. The first papers concerned with this problem took into account nonlinear effects that overstepped the framework of the local equilibrium hypothesis. The transformation properties of generalized kinetic centres were investigated and the analogues of Curie relations for them were established (Kuni, Storonkin, 1971, 1972 a, b, 1973; Storonkin, Kuni, 1973; Zubarev, 1974). The basic problem was again the kinetic centre regularization: requiring subtraction from them of secular slow motion contributions which must be wholly expressed in terms of quantities which form the basis of the description of the system (Kawasaki, 1970; Kuni et al., 1973 a, c, d, 1974; Adjemian et al., 1974 a, b; Kuni, 1974). This problem turned out to be closely connected with the elimination from kinetic centres of unconnected parts violating the space and time correlation weakening (Kuni et al., 1973 a, 1975; Kuni et al., 1973 d; Adjemian et al., 1974 b,c). A complete solution of the problem required an introduction of the so called *noncumulative projecting operators* which step-by-step separate all disconnected parts from kinetic centres (Adjemian, Kuni, 1975). This formulation of nonequilibrium statistical thermodynamic equations in terms of connected diagrams provided a structure which would fit the general criterion of dissipative systems formulated by van Hove in the quantum theory of balanced equations (in the form of the well-known diagonal singularity condition). Van Hove divided all the problems of theoretical physics into two problem types - dissipative and scattering ones - which depended on whether or not this requirement can be satisfied. The essence of this condition is that among an innumerable number of all possible transitions in a system with an infinitely large number of degrees of freedom it selects a comparatively small part of them but excluding the one which leads the system away from the chosen level of its reduced description.

This condition applied to the thermodynamic problems under consideration implies that no such excitation re-pumping can be realized in the course of the system's evolution which would lead to the creation of a region of extreme inhomogeneity. In the theory presented above this condition was fulfilled automatically, and we did not note it. In the theory of extreme nonequilibrium thermodynamic systems this condition becomes non-trivial. The derivation of equations whose structure meets this condition becomes also non-trivial. In the introduction we already mentioned briefly that in the phenomenological approach it is impossible to take into account all the correlation effects of the difference between mean products of the microscopic densities of conserved quantities and the products of their mean values in the kinetic centres. These correlation effects correspond to the equilibrium-approaching processes becoming more rapid than thermodynamic ones (in the nonlinearity order in question).

We are now in a position to formulate all the conditions responsible for dissipation. These are the initial condition of correlations weakening, the existence of time hierarchy, the diagonal singularity condition, and finally, the thermodynamic limiting transition which, in effect, must appear in the construction of the theory, i. e. before the starting point has moved away to an infinite past.

The methods worked out for the derivation of nonequilibrium thermodynamic equations made it impossible to investigate more completely the time asymptotic behaviour of the kinetic centres where, in addition to the contributions found by the Landau-Plachek method, the contributions to the kinetics of hydrodynamic modes (due to connected diagrams) have been taken into account. More accurate results have confirmed the appearance of the term $t^{3/2}$ in the defining equations but have also led to renormalization of the coefficient found at the asymptotic (Adjemian et al., 1975; Grinin, 1975, 1976). The influence of external fields, including those non-stationary in time, upon extreme nonequilibrium systems have been investigated, too (Kuni et al., 1973 b, 1977).

Only one of the questions posed in the introduction remains unanswered. This is the problem of ambiguity in definition of entropy. In a more detailed formulation of the usual statistical thermodynamics of nonlinear, irreversible processes and, in the above mentioned extensions, including extreme nonequilibrium systems, the concept of entropy was met in the derivation of the local equilibrium distribution function only. Entropy was understood there as the informational entropy of the whole microscopic state, i. e. it appeared as an integral over all the microscopic variables of the system (coordinates and momenta of its particles). Under the integral sign there was a complete distribution function (depending on the given variables) multiplied by the logarithm of this funciton. The unambiguous definition of this entropy, as well as the expression for the local equilibrium distribution function obtained with the aid of this entropy, is beyond doubt. This is the only expression effectively required in the theory presented above - and in its generalizations. It has been used for the proper introduction of thermodynamic parameters and for the establishment of their connection with the mean densities of the conserved quantities. Even in extreme nonequilibrium situations this connection is analogous to the normal equilibrium picture and differs only by the replacement of discrete thermodynamic characteristics by field characteristics that depend on the space point at a given moment. The entropy concept necessary in the phenomenological approach (for correct choice of flows and thermodynamic forces and, therefore, as a means of closure of the set of nonequilibrium thermodynamic equations) has been met neither in the theory presented nor in its generalizations. This shows that, in principle, such a concept is not necessary in deriving equations for the densities of conserved quantities which are the characteristics manifesting themselves directly in experiment. At the same time this concept may be extremely useful for identification of the existence of dissipative character in nonequilibrium theory equations since it does not require a preliminary solution of the equations.

Thermodynamical Approach to the Analysis of Statistical Properties of Nonlinear Dynamic Systems

A. A. Platonov

From the macroscopic point of view, nonequilibrium thermodynamic systems described with reference to fluctuations, are simply a specific class among a variety of systems considered in the statistical theory of control (Pugachev et al., 1974; Kazakov, 1975), often called the theory of dynamic systems (Krasovsky, 1974).

The specificity of thermodynamic systems consists in the connection of their macroscopic description with the characteristics of micromotion and manifests itself in a rather strict regulation of the properties of the macroprocesses developing in the system.

Similarity exists between the mathematical models of macro-processes in thermodynamic systems (e. g. Langevin's equations and their generalization) and processes in dynamic systems considered in statistical control theory. The latter, naturally, are not obliged to possess all the properties of multiple microprocesses in the thermodynamic systems. Nevertheless, proceeding from this similarity, or more precisely, from the similarity between the formalism of the Markovian theory of dynamic systems and fluctuation-dissipation thermodynamics in Markov's version (Stratonovich, 1967, 1969, 1970 a), it is of great practical and theoretical interest to consider the application of the methods and the results of the latter to the analysis of arbitrary Markovian dynamic systems. Note, that the inverse problem seems to be no less interesting.

One of the fundamental results of fluctuation-dissipation thermodynamics consist in the derivation for all thermodynamic systems, of the general relations which connect equilibrium, nonequilibrium fluctuations of the para-

Thermodynamics and Kinetics of Biological Processes

meters, and the parameters characterizing the dissipative properties of the systems or their reaction to external forces (Stratonovich, 1967, 1969, 1970 a; Efremov, 1968, 1972).

An important feature of the Markovian version of the theory is that in the derivation of these relations one does not require consideration of a model of the micro-structure of the system and of the mechanism of micro-interactions. The latter attaches more generality to the results and makes it possible to expect that the number of dynamic systems for which the thermodynamic relations hold will appear to be large.

The derivation of fluctuation-dissipation relations in thermodynamic is based, first of all, on the time reversibility of thermodynamic parameter fluctuations in a stationary state. Investigation of their efficiency, in Markovian systems of an arbitrary physical nature, requires a specification of the concept of time reversibility of the processes in these systems. To this end, we shall first derive the equations which make it possible to describe transition statistics in a *reverse* time.

Kinetic Equations for Transition Probability Density in Reverse Time

Let $x(t) = \{x_1(t), \dots, x_n(t)\}$ be a vector Markovian process. At $t \geqslant t_0$, i. e. in *normal* time, the transition probability density satisfies the direct and conjugate kinetic equations

$$\frac{\partial p(x,t/x_0,t_0)}{\partial t} = L\left(x, \frac{\partial}{\partial x}, t\right) p(x,t/x_0,t_0), \tag{1}$$

$$\frac{\partial p(x,t/x_0,t_0)}{\partial t_0} = M\left(x_0, \frac{\partial}{\partial x_0}, t_0\right) p(x,t/x_0,t_0), \tag{2}$$

where the kinetic operators L and M are defined as follows:

$$L(x, \frac{\partial}{\partial x}, t) = \sum_{n=1}^{\infty} \frac{(-1)^n}{n!} \frac{\partial}{\partial x_{\alpha_1}} \cdots \frac{\partial}{\partial x_{\alpha_n}} K_{\alpha_1 \ldots \alpha_n}(x,t) \tag{3}$$

$$M(x, \frac{\partial}{\partial x}, t) = -\sum_{n=1}^{\infty} \frac{1}{n!} K_{\alpha_1 \ldots \alpha_n}(x,t) \frac{\partial}{\partial x_{\alpha_1}} \cdots \frac{\partial}{\partial x_{\alpha_n}} . \tag{4}$$

The coefficients of intensity $K_{\alpha_1 \ldots \alpha_n}$ appearing in equations (3) and (4) are given by the relations

$$K_{\alpha_1 \ldots \alpha_n}(x,t) = \lim_{\tau \to 0} \frac{1}{\tau} \langle \prod_{i=1}^{n} [x_{\alpha_i}(t+\tau) - x_{\alpha_i}(t)] / x(t) \rangle \tag{5}$$

In addition, the one-dimensional distribution of the parameters $p(x,t)$ satisfies the equation

$$\frac{\partial p(x,t)}{\partial t} = L(x, \frac{\partial}{\partial x}, t)\, p(x,t), \tag{6}$$

for the solution of which one needs to know the form of the initial distribution

$$p(x,t)\big|_{t=t_0} = W_0(x)$$

(If $W_0(x) = \delta(x - x_0)$, equation (6) coincides with (1)).

Note the significance of the condition $t \geqslant t_0$ for equations (3), (4). At $t < t_0$ they are no longer meaningful and valid. However, the Markovian process is known to keep the Markovian properties also in *future* $\to$ *past* transitions, i. e. in a reverse time. Kinetic equations must, exist therefore, which determine the statistics of transitions in a reverse time, i. e. equations for the transition probability density $p(x,t/x_1,t_1)$ at $t \leq t_1$.

The derivation of such equations is rather simple. Differentiation of the usual probability identity with respect to t_1 , leads to

$$\frac{\partial}{\partial t_1} \{ p(x,t)\, p(x_1,t_1/x,t) = p(x_1,t_1)\, p(x,t/x_1,t_1) \} \tag{7}$$

Making use of equations (1), (2) and (6), we can rewrite (7) in the form

$$p(x_1,t_1)\frac{\partial}{\partial t_1} p(x,t/x_1,t_1) = p(x,t)L(x_1,\frac{\partial}{\partial x_1},t_1)\, p(x_1,t_1/x,t) - \tag{8}$$
$$- p(x,t/x_1,t_1)L(x_1,\frac{\partial}{\partial x_1},t_1)\, p(x_1,t_1).$$

Taking into account the explicit form of the operator L , and the fact that

$$p(x,t)L(x_1,\frac{\partial}{\partial x_1},t_1)p(x_1,t_1/x,t) = L(x_1,\frac{\partial}{\partial x_1},t_1)\{p(x_1,t_1)p(x,t/x_1,t_1)\}$$

and making simple transformations, we can represent the right-hand side of the equality (8) in the form of the sum

$$\sum_{n=1}^{\infty}\frac{(-1)^n}{n!}\left\{\sum_{m=0}^{\infty}\frac{(-1)^m}{m!}\frac{\partial}{\partial x^1_{\beta_1}}\cdots\frac{\partial}{\partial x^1_{\beta_m}}\left[K_{\alpha_1\ldots\alpha_n\beta_1\ldots\beta_m}(x_1,t_1)p(x_1,t_1)\right]\right\}\cdot$$
$$\cdot\frac{\partial}{\partial x^1_{\alpha_1}}\cdots\frac{\partial}{\partial x^1_{\alpha_n}}\, p(x,t/x_1,t_1). \tag{9}$$

If we introduce the designation

$$Q_{\alpha_1\ldots\alpha_n}(x,t) = \frac{1}{p(x,t)}\sum_{m=0}^{\infty}\frac{(-1)^m}{m!}\frac{\partial}{\partial x_{\beta_1}}\cdots\frac{\partial}{\partial x_{\beta_m}}\left[K_{\alpha_1\ldots\alpha_n\beta_1\ldots\beta_m}(x,t)p(x,t)\right] \tag{10}$$

then from (8) and (9), we obtain one of the desired equations for $p(x,t/x_1,t_1)$ at $t \leq t_1$:

$$\frac{\partial p(x,t/x_1,t_1)}{\partial t_1} = \bar{M}(x_1,\frac{\partial}{\partial x_1},t_1)\,p(x,t/x_1,t_1) \tag{11}$$

where the kinetic operator, $\bar{M}$, is of the form

$$M(x,\frac{\partial}{\partial x},t) = \sum_{n=1}^{\infty}\frac{(-1)^n}{n!}Q_{\alpha_1\ldots\alpha_n}(x,t)\frac{\partial}{\partial x_{\alpha_1}}\cdots\frac{\partial}{\partial x_{\alpha_n}} \tag{12}$$

In a similar way we can obtain the second kinetic equation

$$\frac{\partial p(x,t/x_1,t_1)}{\partial t} = \bar{L}(x,\frac{\partial}{\partial x},t)\,p(x,t/x_1,t_1), \tag{13}$$

where

$$\bar{L}(x,\frac{\partial}{\partial x},t) = -\sum_{n=1}^{\infty}\frac{1}{n!}\frac{\partial}{\partial x_{\alpha_1}}\cdots\frac{\partial}{\partial x_{\alpha_n}}Q_{\alpha_1\ldots\alpha_n}(x,t). \tag{14}$$

Equations (11), (13) should correspond with the natural initial (*final*) condition:

$$p(x, t / x_1, t_1)\Big|_{t=t_1} = \delta(x - x_1).$$

Notice that, as distinct from equations (1), (2), equations (11), (13) involve, besides the coefficients of intensity, also an explicit form of the one-dimensional probability distribution density $p(x,t)$, which makes the analysis and the solution of these equations somewhat peculiar.

It is possible to set up, or to define, the form of $p(x,t)$ in different ways subject to specific conditions. For example, for a number of practically interesting problems this distribution is assumed to be stationary and it's form is set up from additional considerations (Boltzmann, Gibbs and other type distributions).

In the general case the form of the distribution $p(x,t)$ may be established by the solution of equation (6) under given initial and boundary conditions.

Equations (11), (13) offer considerable possiblities for interpretation.

Firstly, the transition probablity density, $p(x,t/x_1,t_1)$, may acquire the following meaning: if at a moment of time t_1 an exact value $x_1 = x(t_1)$ of the process is fixed (say, measured), then $p(x,t/x_1,t_1)$ describes the probability distribution density of the values $x = x(t)$ of the process at the time t_1 preceding t , i. e. the values from which a transition to the value x_1 is possible.

Thus, the knowledge of the form of the function $p(x,t/x_1,t_1)$ permits use of the information contained in the measurement of $x_1 = x(t_1)$ for specification of the statistical characteristics of the Markovian process in question at the time $t \leq t_1$.

The second possibility for interpretation is associated with the evolution of the Markovian process in reversed time. If we change the signs at the odd time components of the process at the moment of time $t_1 = t_0$ and consider the process $\varepsilon x(t-\tau)$ ($\varepsilon = +1$ for even time and $\varepsilon = -1$ for odd time components $x(t), \tau \geq 0$), the function $p(\varepsilon x, t_0-\tau/\varepsilon x_0, t_0)$ may be interpreted as the transition probability density in reversed time. Therefore, equations (11), (13), after the substitution of variables

$$x \to \varepsilon x, \quad x_1 \to \varepsilon x_0, \quad t_1 \to t_0, \quad t \to t_0-\tau,$$

are kinetic equations for processes in reversed time.

Note that equations (11), (13) are quite in agreement with the general axiomatic principles and equations of Markovian theory. They seem to have rather wide limits of applicability.

Time Reversibility of Markovian Processes

Time reversibility of random processes is understood as invariance in their statistical properties under time reversal (with a simultaneous inversion of signs at odd time components of the process).

The mathematical formulation of time reversibility conditions both in early papers (Stratonovich, 1967, 1969, 1970 a) and in more recent works (Bochkov, Kuzovlev, 1977; Stratonovich, 1978) is given considerable and unnecessary restrictions on the classes of processes considered to be time reversible. Thus, a rather wide class of reversible processes, which do not meet the requirements established in the above-mentioned papers, is excluded from consideration. Introducing definitions permits a more general consideration of the time reversibility problem.

Definition 1. Let the Markovian process $x(t)$ be reversible at the moment of time t_0, then for any $\tau_2 \geq \tau_1 \geq 0$ the equality

$$p(\varepsilon x_2, t-\tau_2/\varepsilon x_1, t_0-\tau_1) = p(x_2, t_0+\tau/x_1, t_0+\tau_1) \tag{15}$$

holds.

Definition II. We call a Markovian process quite reversible, if equality (15) is valid for any values of t_0 .

From definition II it follows that complete reversibility of Markovian processes is possible only if they are stationary. Definition I naturally extends the concept of time reversibility to non-stationary processes.

Notice that our definitions are connected only with the dynamics of the transitions and do not directly involve the one-dimenional density $p(x,t)$ as in the case, say, in the developments of Bochkov, Kuzovlev (1977) and Stratonovich (1978).

At $\tau_1 = \tau_2$ relation (15) becomes the identity

$$p(\varepsilon x_2, t_0-\tau_2/\varepsilon x_1, t_0-\tau_1)\Big|_{\tau_1=\tau_2} = p(x_2, t_0+\tau_2/x_1, t_0+\tau_1)\Big|_{\tau_1=\tau_2} = \delta(x_2-x_1) . \quad (16)$$

From condition (15) it is easy to obtain the Chapman-Smolukhowski equation for the transition probability density in reversed time: at $\tau_2 \geqslant \tau_0 \geqslant 0$

$$\begin{aligned} &p(\varepsilon x_2, t_0-\tau_2 | \varepsilon x_0, t_0-\tau_0) = \\ &= \int p(\varepsilon x_2, t_0-\tau_2/\varepsilon x_1, t_0-\tau_1) p(\varepsilon x_1, t_0-\tau_1/\varepsilon x_0, t_0-\tau_0)\, dx_1 . \end{aligned} \quad (17)$$

It is readily seen that when (15) holds, the following relation

$$p(\varepsilon x_2, t_0-\tau_2) = \int p(x_2, t_0+\tau_2/x_1, t_0+\tau_1) p(\varepsilon x_1, t_0-\tau_1)\, dx_1$$

is valid, which is in agreement with the ordinary probability identity

$$p(x_2, t_0+\tau_2) = \int p(x_2, t_0+\tau_2/x_1, t_0+\tau_1) p(x_1, t_0+\tau_1)\, dx_1 ,$$

provided only that the equality

$$p(\varepsilon x, t_0-\tau) = p(x, t_0+\tau) \quad (18)$$

is satisfied at all values of t.

Thus, the time symmetry of the statistical characteristics of the transition is responsible for the form of the one-dimensional distribution, and it is not necessary to postulate the fulfilment of relations of the type (18), as performed in the derivations of Bochkov, Kuzovlev (1977), and Stratonovich (1978).

For reversible Markovian processes condition (18) means that the corresponding one-dimensional distribution must be stationary even with respect to odd time components of the vector, x, $p(x,t) \to p_0(x)$ and

$$p_0(\varepsilon x) = p_0(x). \tag{19}$$

On the basis of formula (15) we can find relations which are fulfilled by the coefficients of intensity $K_{\alpha_1 \cdots \alpha_n}(x,t)$ of time-reversible Markovian processes.

We shall show that the fulfilment of the operator equality

$$L(x, \frac{\partial}{\partial x}, t_0+\tau) = -\bar{L}(\varepsilon x, \varepsilon \frac{\partial}{\partial x}, t_0-\tau) \tag{20}$$

is a necessary and sufficient condition for time reversibility of a Markovian process at the point t_0.

Let condition (15) be satisfied. According to (11) the transition probability density $p(\varepsilon x_2, t_0-\tau_2/\varepsilon x_1, t_0-\tau_1)$ at $\tau_2 \geqslant \tau_1$ satisfies the equation

$$\frac{\partial p(\varepsilon x_2, t_0-\tau_2/\varepsilon x_1, t_0-\tau_1)}{\partial \tau_2} = -\bar{L}(\varepsilon x_2, \varepsilon \frac{\partial}{\partial x_2}, t_0-\tau_2) p(\varepsilon x_2, t_0-\tau_2/\varepsilon x_1, t_0-\tau_1). \tag{21}$$

The equation for $p(x_2, t_0+\tau_2/x_1, t_0+\tau_1)$ according to (1) has the form

$$\frac{\partial p(x_2, t_0+\tau_2/x_1, t_0+\tau_1)}{\partial \tau_2} = L(x_2, \frac{\partial}{\partial x_2}, t_0+\tau_2) p(x_2, t_0+\tau_2/x_1, t_0+\tau_1). \tag{22}$$

Fulfilment of condition (15) for all x, τ allows us to write down equation (21) as an equation with respect to $p(x_2, t_0+\tau_2/x, t_0+\tau_1)$:

$$\frac{\partial p(x_2, t_0+\tau_2/x_1, t_0+\tau_1)}{\partial \tau_1} = -\bar{L}(\varepsilon x_2, \varepsilon \frac{\partial}{\partial x_2}, t_0-\tau_2) p(x_2, t_0+\tau_2/\varepsilon x_1, t_0+\tau_1) \tag{23}$$

whose solution will be the same function, $p(x_2, t_0+\tau_2/x_1, t_0+\tau_1)$, as equation (22), provided only that the kinetic operators are equal, i. e. that equation (20) holds.

And, in contrast let the equality (20) be valid for all x, τ. Then, equation (18) can be written in the form

$$\frac{\partial p(\varepsilon x_2, t_0-\tau_2/\varepsilon x_1, t_0-\tau_1)}{\partial \tau_2} = L(x_2, \frac{\partial}{\partial x_2}, t_0+\tau_2) p(\varepsilon x_2, t_0+\tau_2/\varepsilon x_1, t_0+\tau_1). \tag{24}$$

Under the initial condition (16) and at the same boundary conditions as for $p(x_2, t_0+\tau_2/x_1, t_0+\tau_1)$ equation (24) will have the function $p(\varepsilon x_2, t_0-\tau_2/\varepsilon x_1, t_0-\tau_1)$ as its solution, which is exactly coincident with $p(x_2, t_0+\tau_2/x_1, t_0+\tau_1)$.

In its turn the equality (20) holds if, and only if, the coefficients $K_{\alpha_1 \dots \alpha_n}$ of the kinetic operators satisfy the following relations

$$(-1)^n \varepsilon_{\alpha_1} \dots \varepsilon_{\alpha_n} K_{\alpha_1 \dots \alpha_n}(\varepsilon x, t_0+\tau) = Q_{\alpha_1 \dots \alpha_n}(x, t_0-\tau) = \\ = \frac{1}{p(x, t_0-\tau)} \sum_{m=0}^{\infty} \frac{(-1)^m}{m!} \frac{\partial}{\partial x_{\beta_1}} \dots \frac{\partial}{\partial x_{\beta_m}} [K_{\alpha_1 \dots \alpha_n \beta_1 \dots \beta_m}(x, t_0-\tau) p(x, t_0-\tau)] \tag{25}$$

$(n = 0, 1, 2, \dots \infty)$.

It is readily seen that the process is quite reversible in the case of equality in the stationary relations

$$(-1)^n \varepsilon_{\alpha_1 \dots \alpha_n} K_{\alpha_1 \dots \alpha_n}(\varepsilon x) = \frac{1}{p_0(x)} \sum_{m=0}^{\infty} \frac{(-1)^m}{m!} \frac{\partial}{\partial x_{\beta_1}} \dots \frac{\partial}{\partial x_{\beta_m}} [K_{\alpha_1 \dots \alpha_n \beta_1 \dots \beta_m}(x) p_0(x)] \tag{26}$$

a result which coincides with the analogous relations of Stratonovich (1967, 1969, 1970 a, b, c, 1978) but which are obtained here under less stringent limitations.

It should be noted that in case where the coefficients $K_{\alpha_1 \dots \alpha_n}(x, t)$ are time-dependent, the Markovian process may as well be quite reversible. To this end the statistical bonds between the values of the process must necessarily be more dependent on time intervals much less than the characteristic time of the macro-processes. In other words, if the correlation time τ_x of fluctuations of the process $x(t)$, reversible at the point t_0, are much smaller than the time, Δ , within which the transition probability densities change considerably, then the equality (15) will be approximately fulfilled for all t_0 (the so-called quasi-stationary case). This is in spite of the fact that the transition distribution will depend essentially on t_0 .

In the quasi-stationary case the relations between the coefficients of intensity are obtained as the leading terms of the expansion of relations (25) in powers of τ . Time dependence appears on both sides of the equation but becomes local and instantaneous.

Note, finally, that from condition (15), and its consequence (18), there follows the relation

$$p(x_1, t_0 - \tau_1 / x_2, t_0 - \tau_2) = p(\varepsilon x_1, t_0 + \tau_1 / \varepsilon x_2, t_0 + \tau_2) \qquad (\tau_2 \gg \tau_1) \tag{27}$$

which could be used instead of (15) in the definitions of time reversibility I, II.

Thermodynamic Analogies in the Statistic Theory of Nonlinear Dynamic Systems

This topic requires a preliminary brief discussion of the axiomatic principles and the formalism of fluctuation-dissipative thermodynamics.

The derivation of relations of fluctuation-dissipative thermodynamics first of all makes use of the assumption that microscopic motion in the system obeys the laws of mechanics and is, consequently, time reversible. The latter is responsible for time reversibility of fluctuations of the macroscopic parameters in a stationary state.

The introduction of macroscopic parameters is concerned with reduction in the description, i. e. with the transition from a large number of dynamical variables $z = \{q_i, p_i\}_{i=1}^{M}$, where q_i, p_i are particle coordinates and momenta, to a small number of parameters $x_k = x_k[z(t)]$ $(k = 1, .., N \ll M)$, that make up a complete set. In principle, as has been shown e. g. in Stratonovich (1978), proceeding from the laws of motion at the micro-level on the assumption that the time scale of macro-processes considerably exceeds the characteristic time of micro-processes, one can obtain, for the thermodynamic parameters $x(t)$, a set of stochastic differential Langevin-type equations

$$\dot{x}_\alpha = f_\alpha(x) + \xi_\alpha(t) \qquad (\alpha = 1, \dots, N) \tag{28}$$

where $\xi_\alpha(t)$ are delta-correlated stationary random forces. The possibility to progress to equations (28) implies that, under the assumptions made in their derivation, the fluctuations of thermodynamic parameters constitute a reversible Markovian process. This means in its turn that for thermodynamic systems (in the Markovian approximation) relation (26) is valid.

By definition, the first order coefficients of intensity coincide with the functions $f(x)$ in equation (28)

$$K_\alpha(x) = f_\alpha(x) \tag{29}$$

and, thus, at n = 1 from formula (25) there follow the relations connecting the dissipative (relaxation) characteristics of thermodynamic systems - the functions $f(x)$ - with the fluctuational ones, i. e. with the intensities of the fluctuations of random forces $K_{\alpha_1 \dots \alpha_n}(x)$.

Integration of relation (26) over all X, with the weight function $\exp\{xX\}$, leads to the set of exact relations

$$(-1)^n \varepsilon_{\alpha_1 \dots \alpha_n} \mathcal{K}_{\alpha_1 \dots \alpha_n}(\varepsilon X) = \sum_{m=0}^{\infty} \frac{1}{m!} \mathcal{K}_{\alpha_1 \dots \alpha_m \beta_1 \dots \beta_n}(X) X_{\beta_1} \dots X_{\beta_m} . \tag{30}$$

This requires that $p_0(x)$ and its derivatives vanish at the boundaries. The *images* of the coefficients of intensity $\mathcal{K}_{\alpha_1 \dots \alpha_n}(X)$ entering in (30) are determined by the formula

$$\mathcal{K}_{\alpha_1 \dots \alpha_n}(X) = \int K_{\alpha_1 \dots \alpha_n}(x) p_0(x) e^{xX} dx . \tag{31}$$

The equilibrium distribution $p_0(x)$ of the thermodynamic parameters is expressed by the well-known formula

$$p_0(x) = \text{const.} \exp\left\{\frac{1}{k} S(x)\right\} \tag{32}$$

where $S(x)$ is the entropy of the system. Because of the presence of the large parameter describing the number of particles in the system, this distribution has a sharp peak in the vicinity of the equilibrium point. This fact makes it possible to pass over the exact relation (30) to an asymptotic relationship which is valid to high accuracy for thermodynamic systems. To this end an asymptotically valid equality

$$\mathcal{K}_{\alpha_1 \ldots \alpha_n}(X) = \int K_{\alpha_1 \ldots \alpha_n}(x) p_0(x) e^{xX} dx = K_{\alpha_1 \ldots \alpha_n}[x(X)] \tag{33}$$

is used, where $K_{\alpha_1 \ldots \alpha_n}[x(X)]$ are the coefficients of intensity in which the initial arguments are expressed through the conjugate thermodynamic variables X with the aid of the formula

$$x(X) = \left\{x : \frac{\partial \ln p_0(x)}{\partial x_\alpha} = X_\alpha\right\} \quad (\alpha = 1, \ldots, N). \tag{34}$$

The asymptotic relation

$$(-1)^n \varepsilon_{\alpha_1 \ldots \alpha_n} K_{\alpha_1 \ldots \alpha_n}[x(\varepsilon X)] = \sum_{m=0}^{\infty} \frac{1}{m!} K_{\alpha_1 \ldots \alpha_n \beta_1 \ldots \beta_m}[x(X)] X_{\beta_1} \ldots X_{\beta_m} \tag{35}$$

is the basis (see Stratonovich, 1967, 1969, 1970 a, b, c, 1978) for obtaining the groups of linear and nonlinear relations widely used in applications of statistical thermodynamics (including Onsager relations, Markov's version of the fluctuation-dissipative theorem and their nonlinear analogs).

The results obtained prove, in particular, the non-Gaussian character of the random forces $\xi(t)$ when the functions $f(x)$ are nonlinear. This accounts for the errors in repeated attempts to analyse the parameter fluctuations in nonlinear thermodynamic systems with the aid of the Fokker-Planck equation (i. e. in the assumption of the Gaussian character of $\xi(t)$).

Complications in the theoretical calculation of an explicit form of the functions $f_\alpha(x)$ in equation (28) has raised the questions about the phenomenological methods for their determination. In a simple way it is possible to construct by means of experiment, relaxation equations of the form

$$\langle \dot{x}_\alpha \rangle = \varphi_\alpha\{\langle x \rangle\} \qquad (\alpha = 1, \ldots, N) \tag{36}$$

where the functions $\varphi_\alpha(x)$ do not, generally speaking, coincide with $f_\alpha(x)$ from (28). However, *narrowness* of the equilibrium distribution referred to above, as compared to the scale of nonlinearity in the relaxation functions, allows us to consider the functions $f(x)$ as coincident with the empirical functions $\varphi(x)$ without loss of accuracy (see Stratonovich, 1967, 1969, 1970 a, 1978).

Let us now turn to the theory of dynamic systems. One of the most integrated classes of systems with a wide range of application in this theory (Krasovsky, 1974; Pugachev et al., 1974; Kazakov, 1975) are Markovian dynamic systems defined by the stochastic differential equation

$$\dot{x}_\alpha = f_\alpha(x, t) + \xi_\alpha(t) \qquad (\alpha = 1, \ldots, N) \tag{37}$$

where $\xi(t)$ is a delta-correlated random process, which, in most cases, is considered a Gaussian process with a symmetric, and in the general case, nonstationary matrix of intensities.

Comparison of equations (37) and (28) shows a superficial coincidence of the formal description of thermodynamic and Markovian dynamic systems. However, in dynamic systems the form of the functions $f_\alpha(x, t)$ and noise characteristics are constructed independently and are determined by the structure, characteristics of the elements of the system, and by the conditions under which it functions. In thermodynamic systems the form of $f_\alpha(x)$ and the intensity of fluctuations of the Langevin forces are not arbitrary - they are connected, by necessity, through the characteristics of micro-processes. This qualitative difference between the systems should constantly be born in mind whilst investigation of formal analogies between the thermodynamic theory and the theory of dynamic systems proceeds. And also when the methodology of the analysis and the results of one theory are extended to the other.

The limitations imposed on dynamic systems, in the application of thermodynamic relations to them, are clear from the following brief account of the bases of fluctuation-dissipation thermodynamics.

First of all, there is the condition of the time reversibility of those processes proceeding in dynamic systems in a stationary state. The problem of time reversibility in the statistical theory of dynamic systems is elaborated to a very small extent and requires further development. We may follow different directions in such development. It may be the investigation of the physics of processes proceeding in concrete dynamic systems of different classes and the conclusions drawn upon these grounds with regard to their reversibility or irreversibility (see, for example, Stratonovich, 1970 b, c; Stratonovich, Platonov, 1972, 1973).

On the other hand, it may be the search for the conditions under which, for a sufficiently wide range of systems, relation (15), or, those of type (25) and (26) would hold. For example, as is easily seen, the condition of the absence of correlations between even time and odd time components of any normal stationary Markovian process is a sufficient condition for its complete reversibility (in the sense of definition II).

The question of the possibility of introduction into a dynamic system of additional random or regular forces which would transform the initial irreversible process proceeding in the system into a time-reversible one, is of less interest. Some other interesting formulations of the problem are also possible.

Another set of limitations is concerned with the necessity of existence of a steady state in the system, and the corresponding stationary distribution $p(x)$. This problem has been investigated more thoroughly. In particular, relations between the coefficients of a power series expansion of the functions $f(x)$ and the intensities of noise $\xi(t)$ were first obtained by

Krasovsky (1974) for dynamic systems from analysis of the conditions of existence of $p(x)$. These relations are close in character to the thermodynamic ones obtained from (35) at $n=0$. They also appear in the case of irreversibility in the processes within the system when (35) does not hold for all $n \geq 1$.

Finally, if in a stationary, reversible Markovian dynamic (nonlinear) system the stationary distribution has a sharp peak, then for this system not only are the exact relations valid (30) (therefore, Onsager-type relations obtained from (35) and others are also valid).

Note that the establishment of the fact of the existence (as well as the determination of the form) of a stationary distribution for dynamic systems is comparatively simple to realise experimentally.

What has been said above determines the conditions under which a dynamic system may be considered as an analogue of a certain thermodynamic system, and, which permit the use of the results of statistical thermodynamics in the solution of the problems associated with the dynamic system. This makes it possible to simplify considerably the analysis of its properties, and to extend the possibilities of the theory for the investigation of statistical characteristics, effects caused by nonlinearity etc.

In addition, it should be noted that the very fact of time-reversibility in the processes in dynamic systems (in the sense of definitions I or II) contains much additional information. In this case equations (25) or (26) hold, which can serve as the basis for a number of theoretical constructions both simplifying the solution of traditional problems in the dynamic theory, and extending the number of problems that may be solved.

In conclusion we should note that equations (11) and (13) which allow us to find the form of the function $p(x, t/x_1, t_1)$ for $t \leq t_1$, create prerequisites for solution of the problems not associated with time reversibility both in the theory of dynamic systems, and in other theories.

The Problem of Entropy Increase and Stability of Motion in Quantum Mechanics

O. D. Chernavskaya, D. S. Chernavsky

The problem of entropy increase and time irreversibility of processes is closely associated with the problem of transition of a dynamic system into a statistical system. In dynamic systems, whose behaviour is determined by equations of motion and by the intial conditions, the entropy remains zero throughout the process; in statistical systems, according to the second principle of thermodynamics, it increases with time. In classical mechanics, the problem of transition of a dynamic system into a statistical system is solved practically; the main condition for the transition is the instability of the dynamic trajectories of the system (Krylov, 1950; Sinay, 1963). A system becomes ergodic when all its trajectories are unstable. The number of particles in the system plays a secondary role. There exist systems, containing a large amount of particles but which are, nevertheless, stable and, therefore, dynamic systems. An example is the Solar system, which contains a large number of objects but constitute a stochastic system.

On the other hand, there exist systems, containing only one particle, but possessing stochastic features. The most investigated case is Sinay's billiard ball, i. e. a system containing only one particle (sphere) that moves along a plane, limited by internal convex walls (Sinay, 1963).

It was shown, also, that the definition of entropy as a logarithm of phase space, is to be improved. This improvement is essential only for unstable systems and leads, for such systems, to the correct (from the thermodynamic point of view) expression for entropy (Krylov, 1950; Zaslavsky, 1970).

The same problems arise in quantum mechanics. For instance, the entropy, defined according to the normal algorithm, remains constant throughout any

Thermodynamics and Kinetics of Biological Processes

process. This problem has not been solved; some considerations and approaches to its solution are to be found in a number of papers (Zaslavsky, Chirikov, 1971; Chernavsky, Pukhov, 1971; Chernavsky, 1973).

The aim of the present paper is to suggest an approach to the problem formulated above which is, in principle, analogous to the classical approach, i. e. to investigate the influence of the system's stability (or instability) on its evolution. It is suggested, by analogy with classical mechanics, to improve the observed value definition. It will be shown that this improved entropy does increase with time in unstable quantum mechanical systems.

We point out, however, that the notion of instability in quantum mechanics possesses a number of essential peculiarities (as compared with classical mechanics) and will discuss these peculiarities below. We shall consider here autonomic, and limited-in-space systems and, in particular, a quantum system placed in Sinay's billiard ball environment.

Until recently, the problem under consideration was treated as an abstract one. Now it is becoming urgent, particularly in biophysics. Biological systems are far from thermodynamic equilibrium, and at the same time they are not purely dynamic systems (if they were, time-irreversible processes would not proceed). Entropy does increase in biological systems, but only to the extent necessary and sufficient for maintaining life. In other words, one of the tasks of biological systems is to regulate the entropy increase process both in classical and in quantum-mechanical subsystems (since both are important in biology). To understand the principles of regulation of entropy increase it is necessary to have insight into the physical basis of the process. This is precisely the concern of the present paper.

Note that *stochastization* becomes particularly important for the problem of appearance, conservation and transfer of biological information (Prigogine, 1963, Volkenstein, Chernavsky, 1978). Gaining and keeping information are impossible without energy dissipation; on the other hand, entropy increase up to its maximum value means annihilation of the available information.

Formulation of the Problem in Quantum Mechanics

The aim of this section is to consider the transition of a system from a pure state to a mixed state and the process of establishing statistical equilibrium of the system in quantum mechanics. We shall consider autonomic and limited-in-space quantum mechanical systems. In this case the Hamiltonian $\hat{H}(p,x)$ is time-independent and possesses a system of eigenfunctions and eigenvalues E_j , such that

$$\hat{H}\Psi_j(x) = E_j\Psi_j(x). \tag{1}$$

The energy spectrum of the limited system is discrete. However, we shall consider systems in which the characteristic difference of neighbouring levels is small as compared with the mean energy of the system

$$E_j - E_{j-1} \ll \bar{E}.$$

This means that the dimensions, L , of the system are much larger than the particle wave length. Since we are going to be concerned with the time evolution of the system, we consider here nonstationary (and, therefore, nonproper) states.

We shall not require that the number of particles, N , in the system be large since, as has already been mentioned, the condition $N \gg 1$ is neither a necessary, nor a sufficient ergodicity condition (in classical mechanics). Moreover, we shall consider systems that contain only one particle.

In quantum mechanics, the pure state is characterized by the wave function $\Psi(x,t)$. In the general case it is not an eigenfunction and, therefore, depends on time, t. The observed value of the operation $\hat{O}$ (now we do not specify it) in this case is equal to

$$\langle\hat{O}\rangle = \int \Psi^*(\vec{x},t)\hat{O}\Psi(\vec{x},t)d\vec{x} = \sum_{j,k} O_{j,k} C_j^* C_k e^{-i(E_j-E_k)t}, \tag{2}$$

where

$$O_{jk} \equiv \int \Psi_j^*(\vec{x}) \hat{O} \Psi_k(\vec{x}) d\vec{x} \tag{3}$$
$$C_j \equiv \int \Psi_j(x, 0) \Psi_j^*(\vec{x}) d\vec{x}.$$

$\Psi_j(\vec{x})$ is the eigenfunction $H(\vec{p}, \vec{x})$. $\Psi(x, 0)$ is the wave function of the system at the initial moment (Landau, Lifshitz, 1974). The mixed state is characterized by the density matrix $\rho(x, x', t)$ (or by its image $\rho_{jk}(t)$) and the observed value of the operator $\hat{O}$ is given by

$$\langle \hat{O} \rangle = \sum_{j,k} O_{jk} \rho_{kj}(t). \tag{4}$$

In particular, in the limiting case of a statistically determined equilibrium ergodic system, the density matrix is diagonal and (see Landau, Lifshitz, 1976):

$$\langle \hat{O} \rangle = \sum_{j} O_{jj} \rho_{jj}(t). \tag{5}$$

To the pure state there corresponds a factorized density matrix, such that

$$\rho_{jk} = C_j^* C_k e^{-i(E_j - E_k)t} \tag{6}$$

which is essentially non-diagonal.

The definition of entropy in quantum mechanics is equivalent to (see von Neuman, 1939):

$$\langle \hat{S} \rangle = Sp\{\hat{\rho}, \ln \hat{\rho}\} = \sum_{j,k} \rho_{jk} \{\ln \rho\}_{kj}. \tag{7}$$

Although expression (7) resembles (4) in its form, they are essentially different. In equation (4) the matrix, O_{jk} , is independent of the initial state. In other words, any operator $\hat{O}$ may be determined before the initial state is set up and it will have the identical form for all states. At the same time in equation (7) the operator is state-dependent. This means that the entropy operator is some special operator. More precisely, any operator, in its usual sense, may be associated with entropy. The expression (7) should be regarded as an algorithm for calculating the value of entropy.

The expression (7) possesses the following properties: 1) the entropy of any pure state, as is readily seen from (7), is equal to zero; 2) the value of $\langle \hat{S} \rangle$, set up by (7), is time-dependent if the state described by the matrix ρ develops according to the laws of quantum mechanics (Feynman, 1972)

$$\frac{d}{dt} \langle \hat{S} \rangle = [\hat{H}, \hat{S}] = 0, \tag{8}$$

3) the expression (7) gives a correct value of the entropy (equal to the logarithm of the phase volume of the system) in the limiting case when the density matrix is diagonal

$$\langle \hat{S} \rangle = \sum_j \rho_{jj} (\ln \rho)_{jj} . \tag{9}$$

Thus, it is impossible to describe the effect of entropy increase in the framework of quantum mechanics without involving additional conditions. We shall try to show below that the transition from a pure state to a mixed state (and, therefore, the effect of entropy increase) can be described within an improved formalism, which takes into account the stabilization (or instability) properties in the evolution of the quantum-mechanical system.

Before this, we should make a few preliminary remarks:

1. The energy spectrum of a quantum-mechanical system evolving in time (the only type of system where instability may occur) is sufficiently wide; the uncertainty of energy (half-width of the spectrum) is associated with the characteristic time of evolution, Δt, by the uncertainty relation $\Delta E \Delta t \geq \hbar$. This differentiates the quantum-mechanical system from the classical closed system, within which energy is considered to be established with an arbitrary accuracy. This influences the phase space structure: in the classical case, the condition $E = const$ cuts off a narrow region in the phase volume - an energy surface - whose volume contains, however, many

elementary phase cells. In quantum mechanics, the system occupies all the energy levels of the phase space (distribution about layers is determined by the energy spectrum), the volume of each energy layer, if not degenerate, corresponding to the elementary cell volume $(2\pi\hbar)^3$.

2. In general there exist two different types of instabilities: dynamic and parametric ones (Andronov et al., 1959). The former means an increase in reaction to perturbation of the initial conditions (without changing the system parameters).

The parametric instability implies that the behaviour of two systems with identical initial conditions but different (albeit only slight) parameters becomes essentially different with time. In classical systems these two types of instability are, as a rule, associated: the presence of dynamic instability implies the presence of parametric instability and vice versa. In quantum mechanics, the situation is different. There dynamic instability is absent (Wigner, 1970). To clarify this we shall consider two systems differing in their initial conditions

$$\Psi_1(\vec{x},t) = \sum_j C_j^{(1)} \Psi_j(\vec{x}) e^{-iE_j t}$$
$$\Psi_2(\vec{x},t) = \sum_j C_j^{(2)} \Psi_j(\vec{x}) e^{-iE_j t} \quad . \tag{10}$$

At initiation, $t = 0$, their difference, $\delta\Psi$, is

$$\delta\Psi = \Psi_1(\vec{x},0) - \Psi_2(\vec{x},0)$$

and let it be so small that

$$\int |\delta\Psi(\vec{x})|^2 d\vec{x} = \delta \ll 1 \, . \tag{11}$$

This means that the quantity

$$\sum_j |C_j^{(1)} - C_j^{(2)}|^2 = \sum |\delta C_j|^2 = \delta \ll 1 \tag{12}$$

is small.

At any arbitrary moment, $t > 0$,

It is easy to see that, in this case, the measure of the deviation $\int|\delta\Psi(\vec{x},t)|^2 d\vec{x}$ always remains small

$$\delta\Psi(\vec{x},t) \equiv \Psi_1(\vec{x}_1,t) - \Psi_2(\vec{x},t) = \sum (C_j^{(1)} - C_j^{(2)}) \Psi_j(x) e^{-iE_j t} . \tag{13}$$

The physical meaning of the result is simple (it is associated with phase space structure); on change of the intitial conditions only, the position of the levels of the system E_j does not change; only their occupation changes (i. e. the coefficients C_j). In this case, the energy levels are not displaced, and the phase space structure remains the same. When the parameters of the system change, the situation is different: the statement that the measure of deviation, $\int|\delta\Psi(\vec{x},t)|^2 d\vec{x} = \sum_j |\delta C_j|^2 = \delta \ll 1$. remains small is, in this case, irrelevant. We shall therefore, now investigate the parametric instability of the quantum-mechanical system.

3. We have seen above that, in the transition from a pure state to a mixed state (whose limiting case is a thermodynamic equilibrium state), non-diagonal elements of the density matrix disappear. In this connection, the main manifestation of instability in the evolution of a quantum-mechanical system must also be due to the disappearance of the non-diagonal elements ρ_{jk} . In other words, the disappearance of ρ_{jk} $(j \neq k)$ may be regarded as the manifestation of that instability, and the characteristic time, τ , of its decrease with time of instability development.

4. In classical systems, an instability of the particle trajectory entails the necessity of an additional averaging over an ensemble of macroscopically identical systems. This averaging is an additional, not permitted by the postulates of classical mechanics. Note, that being applied to stable systems, this operation does not contradict the results of classical mechanics, i. e. becomes identical with unity (Chernavsky, 1973). An operation of this type should also be permitted in quantum mechanics. Therefore, let us define the density matrix as

$$\tilde{\rho}_{jk} = \frac{1}{n} \sum_{\nu}^{n} \rho_{jk}^{\nu} . \tag{14}$$

Averaging over an ensemble of macroscopically identical systems, whose parameters differ slightly from each other (the measure of difference is discussed below) is introduced; ν is the system's index in the ensemble, $\mathfrak{N}$ is the number of elements in the ensemble.

Similarly, the observed quantities are defined as

$$\langle\tilde{\hat{O}}\rangle \equiv \frac{1}{\mathfrak{N}}\sum_{\nu}^{\mathfrak{N}}\sum_{j,k}\rho_{jk}^{\nu}O_{kj} \equiv \sum_{k,j}\tilde{\rho}_{jk}O_{kj} \tag{15}$$

or, in the case of an initially pure state,

$$\langle\tilde{\hat{O}}\rangle \equiv \frac{1}{\mathfrak{N}}\sum_{\nu}^{\mathfrak{N}}\int\Psi^{*\nu}(\vec{x},t)\hat{O}\Psi^{\nu}(\vec{x},t)\,d\vec{x}. \tag{16}$$

Entropy is defined as

$$\langle\tilde{\hat{S}}\rangle \equiv Sp\{\tilde{\rho}_{jk}(\ln\tilde{\rho})_{kj}\}. \tag{17}$$

The equality (17) is coincidental with (7) in two limiting cases: when the evolution of the system is stable, $\tilde{\rho}_{jk}\to\rho_{jk}$, and, when the density matrix is diagonal (it is seen below that for the diagonal elements we always have $\tilde{\rho}_{jj}=\rho_{jj}$). In unstable systems the value determined by (17) increases in time at the expense of the disappearance of the non-diagonal elements, $\tilde{\rho}_{jk}$ $(j\neq k)$.

The operation of averaging has the following features:
1. If all the states of the ensemble are identical (which takes place e. g. at t = 0), averaging introduces no corrections and is an identity equal to unity operation.
2. If the state is stable (e. g. it is stationary), the corrections introduced by averaging are small with respect to small perturbation parameters. In this case, the operation of averaging may also be regarded as an identity.
3. If the system's evolution is unstable, the operation of averaging after rather long time is not an identity.

Note that the operation of averaging (14) is even more natural in quantum mechanics than in classical mechanics. Indeed, measurement (observation) in quantum mechanics implies the presence of an ensemble of systems, and averaging over this ensemble (Bom, 1963). Parameters of the systems within one ensemble may, in general, differ from one another. Thus, the operation of averaging in quantum mechanics may be considered not as an additional condition, but rather as a realization, in an explicit form, of the axiom referred to above.

The simplest model

To illustrate our statements let us consider a very simple system - a one-dimensional box with infinite walls (Fig. 1). The Hamiltonian of the system has the form

$$H = \frac{p^2}{2} = -\frac{1}{2}\frac{\partial^2}{\partial x^2} . \tag{18}$$

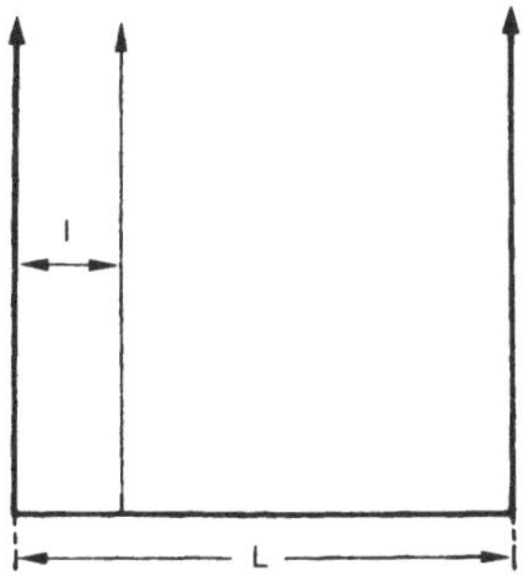

Fig. 1 - Scheme of a one-dimensional box with infinite walls

We have chosen the system of units such that the particle mass m and the Planck constant $\hbar$ are set equal to unity. Length, l , is a dimensioned quantity: energy has the dimension L^{-2} , and time has the dimension of the squared length. The initial pure state is chosen in the form

$$\Psi(x,0) = \begin{cases} \sqrt{2/l}\,\sin\frac{\pi}{l}x & 0 < x < l \\ 0 & \text{else} \end{cases} \tag{19}$$

The basic functions have the form

$$\Psi_m(x) = \sqrt{\frac{2}{L}} \sin\left(\frac{m\pi}{L} x\right). \tag{20}$$

The energy eigenvalues are equal to

$$E_m = m^2 \left(\frac{\pi}{L}\right)^2. \tag{21}$$

The coefficients C_m are given by

$$C_m = \frac{2}{\sqrt{\ell L}} \int_0^{\ell} \sin\left(\frac{\pi}{\ell} x\right) \cdot \sin\left(\frac{m\pi}{L} x\right) dx = \frac{1}{\sqrt{\ell L}} \cdot \frac{2\pi/\ell}{\left(\frac{\pi^2}{\ell^2} - \mu^2\right)} \sin(\mu \ell), \tag{22}$$

where $\mu = \frac{m\pi}{L}$ (see (3)).

From (22) it follows that the coefficients C_m are large at $m \sim \frac{L}{\ell}$ and small at $m \gg \frac{L}{\ell}$ and $m \ll \frac{L}{\ell}$. The characteristic energy in this case is $E_{ch} \cong \frac{\pi^2}{\ell^2}$, and the characteristic velocity, $v_{ch} \cong \frac{\pi}{\ell}$. Thus the collision time, i. e. the time of wave propagation throughout the system, is given by

$$\tau_{ch} = L / v = L . \ell . \tag{23}$$

Let us consider an ensemble of systems that differ from each other by the value of the parameter L:

$$L^{\nu} = L_0 - \lambda_{\nu} \qquad (\lambda_{\nu} \ll L_0).$$

Then

$$E_m^{\nu} = m^2 \left(\frac{\pi}{L^{\nu}}\right)^2 \cong \frac{m^2\pi^2}{L_0^2} + \frac{\lambda_{\nu}}{L_0} \cdot \frac{m^2\pi^2}{L_0^2} = E_m^0 \left(1 + \frac{\lambda_{\nu}}{L_0}\right) \tag{24}$$

and

$$C_m^{\nu} = C_m^0 + \frac{\lambda_{\nu}}{L_0} \cdot \frac{\partial C_m}{\partial L}\Big|_{L_0} + \ldots = C_m^0 + O\left(\frac{\lambda_{\nu}}{L_0}\right).$$

The values of ρ_{mn}^{ν} are $\rho_{mn}^{\nu} = C_m^{*\nu} C_n^{\nu} e^{-i(E_m^{\nu} - E_n^{\nu})t} =$ (24 a)

$$= \rho_{mn}^0 \exp\{-i(\Delta E_m^{\nu} - \Delta E_n^{\nu})t\} + O\left(\frac{\lambda_{\nu}}{L_0}\right) \cong \rho_{mn}^0 \, e^{-i(\Delta E_m^{\nu} - \Delta E_n^{\nu})t}.$$

The averaged matrix elements (with account taken of (24 a) are equal to

$$\tilde{\rho}_{mn} = \frac{1}{\mathfrak{N}} \sum_{\nu=1}^{\mathfrak{N}} C_m^{*\nu} C_n^{\nu} \exp\{-i(E_m^{\nu} - E_n^{\nu})t\} = \qquad (25)$$
$$= \overset{\circ}{\rho}_{mn} \frac{1}{\mathfrak{N}} \sum^{\mathfrak{N}} e^{-i(\Delta E_m^{\nu} - \Delta E_n^{\nu})t} \cong \overset{\circ}{\rho}_{mn} \frac{1}{\mathfrak{N}} \sum_{\nu=1}^{\mathfrak{N}} e^{-i(E_m - E_n)\frac{\lambda_\nu}{L_0} t} .$$

Expansion with respect to the (small) parameter $\frac{\lambda_\nu}{L_0}$ appears in all the expressions, the leading terms alone remaining everywhere except for $\exp\{-i(E_m^{\nu} - E_n^{\nu})t\}$. The quantity $\Delta E \cdot t$ will not be considered small since we are interested in the behaviour of the system at $t \to \infty$.

Assuming that the ensemble of systems is sufficiently large, i. e. $\mathfrak{N} \gg 1$, such that the difference $L_\nu - L_{\nu-1} \to 0$ and $\frac{\lambda_\nu}{L_0} \ll 1$, we can replace the summation in (25) by integration, after which we obtain

$$\tilde{\rho}_{mn} = \overset{\circ}{\rho}_{mn} \frac{1}{\mathfrak{N}} \int \left(\exp\{-i(E_m - E_n)\frac{\lambda_\nu}{L_0} t\}\right) \cdot W(\nu)\, d\lambda_\nu .$$

Here the function $W(\nu)$ is the relative weight of the systems of length λ_ν in the ensemble. We choose $W(\nu)$ in the form

$$W(\nu) = \frac{\Delta}{\Delta^2 + (\lambda_\nu)^2} . \qquad (26)$$

Then, as a result of integration, we have

$$\tilde{\rho}_{mn} = \overset{\circ}{\rho}_{mn} \exp\{-|E_m - E_n| t \frac{\Delta}{L_0}\} = \qquad (27)$$
$$\overset{\circ}{\rho}_{mn} \exp\{-\frac{\pi}{L_0} \frac{\Delta \ell}{L_0} |n^2 - m^2| \frac{t}{\tau}\} .$$

From this it is seen that the non-diagonal terms of the averaged density matrix vanish in the course of time, whereas the diagonal terms do not depend on time. The damping decrement depends on the matrix element position.

Characteristic matrix elements, such that $n \sim \frac{L}{\ell}$; $m \sim \frac{L}{\ell}$; $|n-m| \sim \frac{L}{\ell}$ damp with the characteristic time

$$t_c = \tau \cdot \frac{\ell}{\Delta} . \qquad (28)$$

Elements close to the diagonal, such that $n, m \sim \frac{L}{\ell}$; $|n-m| \sim 1$ damp with time

$$t_d = \tau \frac{L}{\Delta} , \tag{29}$$

i. e. much more slowly.

Matrix elements with small quantum numbers, such that $m \sim n \sim 1$; $|m-n| \sim 1$ damp still more slowly

$$t' = \tau \frac{L^2}{\Delta \ell} . \tag{30}$$

The time, t_c , may be considered to be the characteristic time of transition from a pure state to a mixed state (time of mixing), since the terms ρ_{mn} (corresponding to $m \sim n \sim \frac{L}{\ell}$; $|m-n| - \frac{L}{\ell}$) make the main contribution to the observed values.

The time of mixing coincides with the time of collision, τ , provided that $\Delta \sim \ell$. This means that the system will be effectively mixed only if the change in the parameter L is comparable with the dimension of the packet. If $\Delta \ll \ell$, then $t_c \gg \tau$, and, in this case the system may be considered stable. In this sense the system considered here is analogous to classical one-dimensional systems, which are also stable toward small perturbations. On the other hand, in ordinary systems, where L is macroscopic, while ℓ is microscopic, the fluctuations of parameters $\Delta \sim \ell$ practically always exist. In this case, a pure state will be transformed into a mixed with a characteristic time of the order of the collision time τ . It is readily seen that this will take place in a two-dimensional (and three-dimensional) model with plane walls. Therefore we arrive at the conclusion that in any macroscopic system a pure state is transformed into a mixed state within a time comparable with the collision (or flight) time, τ . This conclusion is in agreement with the sequently quoted statement about *stochastization* of the quantum state in the macroscopic system (although this statement is often advanced without sufficient argument).

Let us now consider an ensemble of particles, within which the initial states, e. g. parameters ℓ , are different. In this case, intrinsic self-energies do not depend on ν , and the coefficients, C_m^ν , are equal to

$$C_m^\nu = C_m^0 + \ell_0 \left(\frac{\lambda_\nu}{\ell_0}\right)\left(\frac{\partial C_m}{\partial \ell}\right)_{\ell_0} + O\left(\frac{\lambda_\nu}{\ell_0}\right). \tag{31}$$

The averaged density matrix elements are given by

$$\begin{aligned} \tilde{\rho}_{mn} \cong \rho_{mn}^0 + \frac{1}{\mathfrak{N}} \sum_1^{\mathfrak{N}} \ell_0 \left\{ C_m^{*0} \left(\frac{\partial C_n}{\partial \ell}\right)\Big|_{\ell_0} + C_n^0 \left(\frac{\partial C_m}{\partial \ell}\right)\Big|_{\ell_0} \right\} \times \\ \times \frac{\lambda_\nu}{\ell_0} \exp\{-i(E_m - E_n)t\} . \end{aligned} \tag{32}$$

The second term in (32) is, in this case, always small compared to the first one (with respect to the small parameter λ_ν / ℓ_0). Therefore, the averaged density matrix coincides (to this accuracy) with the non-averaged result. This is a consequence of the dynamic stability of the quantum-mechanical systems.

Two-Dimensional Model with Convex Walls (Sinay's Billiard)

Let us consider the quantum-mechanical two-dimensional system shown in Fig. 2. Let the Hamiltonian inside the system describe free particles motion.

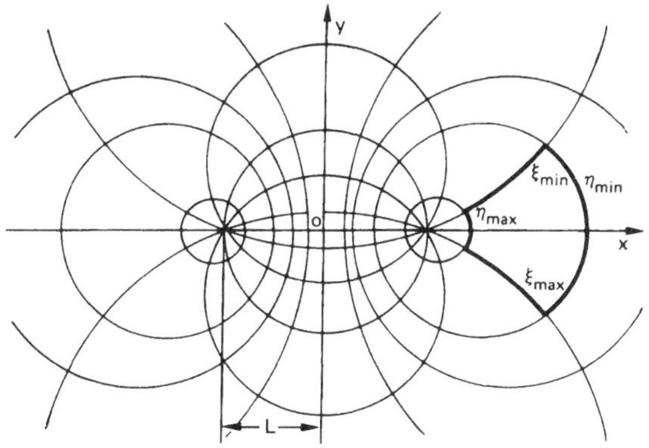

Fig. 2. Scheme of a two-dimensional system (explained in the text)

The boundary conditions correspond to infinite reflecting walls

$$\Psi(x,y)\big|_{x,y\in\sigma} = 0 \tag{33}$$

The chosen form of the billiard with internal convex walls is convenient as it can be transformed into a rectangle with the help of the conformational reflection; in this case, Zhukovsky and Chaplygin transformations (Zhukovsky, 1948)

$$i\zeta = \ln\frac{Z - L}{Z + L} \tag{34}$$

hold good,
where $Z = x + iy$, $\zeta = \xi + i\eta$.
In this case,

$$x = L\cdot\frac{sh\eta}{ch\eta+\cos\xi}, \quad y = L\cdot\frac{\sin\xi}{ch\eta+\cos\xi} \tag{35}$$

where the Jacobian is equal to

$$\left(\begin{vmatrix} \dfrac{\partial x}{\partial\xi} & \dfrac{\partial x}{\partial\eta} \\ \dfrac{\partial y}{\partial\xi} & \dfrac{\partial y}{\partial\eta} \end{vmatrix}\right)^{-1} \equiv w(\xi,\eta) = \frac{(ch\eta+\cos\xi)^2}{L^2}\,. \tag{36}$$

With the variables (ξ,η) the region Ω is transformed into a rectangle

$$\xi_{min} \le \xi \le \xi_{max} \qquad \eta_{min} \le \eta \le \eta_{max}\,. \tag{37}$$

The dimensions of the system are determined by the parameter L .

The initial state, as before, will be set up in the form of a packet of dimension $\ell(\ell \ll L)$, placed inside the system. The eigenfunctions are determined by the equation

$$\hat{H}(\xi,\eta)\Psi_m(\xi,\eta) \equiv w(\xi,\eta)\Delta_{\xi,\eta}\Psi_m(\xi,\eta) = E_m\Psi_m(\xi,\eta) \tag{38}$$

with the boundary conditions

$$\Psi_m(\xi,\eta)\Big|_{\substack{\xi=\xi_{max},\,\xi_{min}\\ \eta=\eta_{max},\,\eta_{min}}} = 0 .$$

The standardization condition has the form

$$\int_{\xi_{min}}^{\xi_{max}}\int_{\eta_{min}}^{\eta_{max}} \Psi_m^*(\xi,\eta)\Psi_m(\xi,\eta)w^{-1}(\xi,\eta)d\xi\, d\eta = 1 . \tag{39}$$

The expression for the energy of an m -th eigenfunction is

$$E_m = \iint \Psi_m^*(\xi,\eta)\hat{H}(\xi,\eta)\Psi_m(\xi,\eta)\frac{1}{w(\xi,\eta)}d\xi\, d\eta = \tag{40}$$

$$= \iint \Psi_m^*(\xi,\eta)\Delta_{\xi,\eta}\Psi_m(\xi,\eta)w(\xi,\eta)\frac{1}{w(\xi,\eta)}d\xi\, d\eta .$$

Equation (38) cannot be solved exactly; it is simplified to some extent in the limiting case, $\eta_{max} \gg 1, \eta_{min} \gg 1, \ \eta_{max} \gg \eta_{min}$. In this case

$$w(\xi,\eta) \simeq e^{2\eta} L^{-2} . \tag{41}$$

The condition $\eta_{max} \gg 1$ means that the radius of the circle $\eta = \eta_{max}$ is sufficiently small, for

$$r(\eta_{max}) = r_{min} = \frac{L}{sh\,\eta_{max}} \simeq L e^{-\eta_{max}} \ll L .$$

At $\eta_{max} \to \infty$ the boundary $\eta = \eta_{max}$ approaches a point; in this case, $w(\xi,\eta)|_{\eta_{max}} \to \infty$ and the transformation becomes irregular. The condition $\eta_{max} \gg 1$, will, further on in the development, play an essential role. The condition $\eta_{min} \gg 1$ means that the radius of the circle $\eta = \eta_{min}$ is not very large; this condition does not play an essential role and can be used only to simplify the calculations. The functions $\Psi_m(\xi,\eta)$ may be chosen as the product

$$\Psi_m(\xi,\eta) = U_k(\xi)\, V_j(\eta) , \tag{42}$$

where the functions $U_k(\xi)$ and $V_j(\eta)$ obey the conditions

$$\frac{U''}{U} = -C_k ; \ U_k(\xi)\Big|_{\substack{\xi=\xi_{min}\\ \xi_{max}}} = 0 ; \quad \int_{\xi_{min}}^{\xi_{max}} L|U_k(\xi)|^2 d\xi = 1 \tag{43 a}$$

$$\frac{V_j''}{V_j} = -E_m L^2 e^{-2\eta} + C_k ; \quad V_j(\eta)\Big|_{\eta=\eta_{max},\,\eta_{min}} = 0 ; \quad \int_{\eta_{min}}^{\eta_{max}} L^2 e^{-2\eta} |V_j(\eta)|^2 d\eta = 1 \qquad (43\ b)$$

(see (49), (38)).
Equation (43 a) has the solution

$$U_k(\xi) = \sin\left(k\pi \frac{\xi - \xi_{min}}{\xi_{max} - \xi_{min}}\right), \quad k = 1, 2, \ldots \qquad (44)$$

From this the meaning of the quantities C_k is clear

$$C_k = \left(\frac{k\pi}{\xi_{max} - \xi_{min}}\right)^2 > 0 \; ; \qquad (45)$$

they represent the squared projections of the wave number on to the axis ξ. The number, k, is equal to the number of extrema of the function $U_k(\xi)$ in the interval (ξ_{min}, ξ_{max}). The minimal value of the quantity C_k (at $k=1$) is equal to

$$C_{(k=1)} = \frac{\pi^2}{(\xi_{max} - \xi_{min})^2} \sim 1 \, . \qquad (46)$$

Each value E_m corresponds to a set of values C_k; this is a consequence of two-dimensions. From (43) it follows that the choice of the values, E_m, and, C_k, completely determines the function V_j. In other words, fixing the indices m and k unabiguously determines the index j. To emphasize this we shall from now on designate the function $V_j(\eta)$ by the indices $V_{m,k}(\eta)$.

Equation (43) cannot be solved analytically, therefore, it is only possible to make some qualitative remarks about the properties of the function $V_{m,k}(\eta)$:

1. the function $V_{m,k}(\eta)$ is small in the region where $C_k > E_m L^2 e^{-2\eta}$; it falls exponentially as η decreases. In the region $C_k < E_m e^{-2\eta} L^2$ the function $V_{m,k}(\eta)$ oscillates. Its absolute value in this region is of the order of $|V| \sim L^{-1/2}$;
2. the characteristic values of $\tilde{E}_m$ and $\tilde{C}_k$ depend on the initial state: the dimensions and the position of the packet. In our case $\tilde{E}_m \sim 1/\ell^2$, and the value $\tilde{C}_k$ depends on the initial position of the packet. If a packet is

situated near a singular point, so that the distance from it to the boundaries ξ_{min} and ξ_{max} is of the order of the length of the packet ℓ, the main contribution to the expansion of the packet is made by functions with a small index, $k \sim 1$; in this case

$$\tilde{C}_k \cong 1.$$

If the packet is far from the singular point, so that the distance to the boundaries ξ_{min} and ξ_{max} is of the order of L, then

$$\tilde{C}_k \cong \left(\frac{L}{\ell}\right)^2.$$

Let us now consider the response of the system to change in the position of the boundary, η_{max}. In this case, it is impossible to calculate directly the eigenvalues and their variations, therefore, we shall make use of perturbation theory.

Let the boundary η_{max} be displaced by the value δ $(\delta \ll 1)$, i. e. $\eta_{max} \to \eta_{max} + \delta$. We introduce a new variable $\eta' = \alpha\eta + \beta$, such that the boundary conditions in the new variables are of similar form to those used earlier

$$\eta'_{min} = \alpha\eta_{min} + \beta = \eta_{min}$$
$$\eta'_{max} = \alpha(\eta_{max} + \delta) + \beta = \eta_{max}. \tag{47}$$

From this

$$\alpha = 1 - \frac{\delta}{\eta_{max} - \eta_{min}}; \qquad \beta = \frac{\delta}{\eta_{max} - \eta_{min}}. \tag{48}$$

Note that the transformation is not conformational (since only one variable is extended). The expression for energy in the new variables has the form

$$E'_m = \int_{\xi_{min}}^{\xi_{max}} \int_{\eta_{min}}^{\eta_{max}} \Psi_m^*(\xi, \eta') \omega\left(\xi, \frac{1}{\alpha}\eta' - \frac{\beta}{\alpha}\right)\left\{\frac{\partial}{\partial \xi^2} + \alpha^2 \frac{\partial^2}{\partial \eta'^2}\right\} \cdot$$
$$\cdot \Psi_m(\xi, \eta') \frac{1}{\omega(\xi, \frac{1}{\alpha}\eta' - \beta/\alpha)} \, d\xi \, d\eta'. \tag{49}$$

We have taken into account here that in first order perturbation theory, the functions, Ψ_m maintain their form.

The energy shifts $\delta E_m = E'_m - E_m$ are equal to

$$\delta E_m = (\alpha^2-1)\iint \Psi_m^*(\xi,\eta)\frac{\partial^2}{\partial\eta^2}\Psi_m(\xi,\eta)\,d\xi\,d\eta = \qquad (50)$$
$$= \frac{2\delta}{\eta_{max}-\eta_{min}}\iint \Psi_m^*(\xi,\eta)\Big\{E_m\frac{1}{w} - \frac{\partial^2}{\partial\xi^2}\Big\}\Psi_m(\xi,\eta)\,d\xi\,d\eta .$$

In the case $\eta_{max} \gg \eta_{min} \gg 1$ the shift can be represented in the form (see 42)

$$\delta E_m = \frac{2\delta}{\eta_{max}-\eta_{min}}\iint |\Psi_m(\xi,\eta)|^2\{E_m L^2 e^{-2\eta} - C_k\}\,d\xi\,d\eta =$$
$$= \frac{2\delta}{\eta_{max}-\eta_{min}}\iint |\Psi_m(\xi,\eta)|^2\frac{1}{L^2}\{E_m L^2 e^{-2\eta} - C_k\}L^2\,d\xi\,d\eta =$$
$$= \frac{2\delta}{\eta_{max}-\eta_{min}}\int |U_k(\xi)|^2 L\,d\xi \int |V_{m,k}(\eta)|^2 L\{E_m L^2 e^{-2\eta} - C_k\}\,d\eta\,\frac{1}{L^2} = \qquad (51)$$
$$= \frac{2\delta}{\eta_{max}-\eta_{min}}\cdot\frac{1}{L^2}\int |V_{m,k}(\eta)|^2 L\{E_m L^2 e^{-2\eta} - C_k\}\,d\eta .$$

Let us investigate this expression for the case when the initial state is set up as a packet near a singular point. Here $\tilde{C}_k \sim 1$, $\tilde{E}_m \sim 1/\ell^2$ and the integrand is positive in the region where $\frac{L}{\ell}e^{-\eta} > 1$.
We choose

$$\eta_{max} = \ln\frac{L}{\ell} . \qquad (52)$$

This means that the boundary, η_{max}, is also close to the initial packet and the dimensions of the arc, η_{max}, in the x, y-plane are of the order of ℓ. Then the integrand is positive over the entire region. The main contribution to the integral is made by the region, where $C_k \ll E_m L^2 e^{-2\eta}$.
Then

$$\delta E_m \cong \frac{2\delta}{\eta_{max}} E_m . \qquad (53)$$

Returning to the coordinates $\{x,y\}$, we see that the shift of the boundary, η_{max}, by an amount δ, corresponds to a change in $\{x,y\}$ of

$$\Delta x = L\cdot\frac{1+\operatorname{ch}\eta_{max}\cos\xi}{(\operatorname{ch}\eta_{max}+\cos\xi)^2}\,\delta \qquad (54)$$
$$\Delta y = L\cdot\frac{\operatorname{sh}\eta_{max}\cdot\sin\xi}{(\operatorname{ch}\eta_{max}+\cos\xi)^2}\,\delta .$$

Introducing the shift measure

$$\lambda \equiv \sqrt{(\Delta x)^2+(\Delta y)^2} = L\delta\frac{1}{(\operatorname{ch}\eta_{max}+\cos\xi)} \simeq L\delta e^{-\eta_{max}}$$

then (see (52))

$$\delta E_m = E_m\frac{2}{\eta_{max}}\cdot\frac{\lambda}{L}e^{\eta_{max}} \simeq \frac{\lambda}{L}E_m e^{\eta_{max}} =$$
$$= \frac{\lambda}{L}\cdot\frac{L}{\ell}\cdot E_m \cong \frac{\lambda}{\ell}E_m . \qquad (55)$$

Comparing (55) with (24) we see that, here, change in self energies as a response to boundary variations increases L/ℓ times.

Let' us now choose $\eta_{max} \gg \ln\frac{L}{\ell}$. This means that the arc, η_{max}, is small compared to ℓ; in the limit it approaches a point. A region of negative contribution appears in the integrand which increases with the increase of η_{max}. The estimate, (55), is not valid in this case; the value δE_m becomes less than $\frac{\lambda}{\ell} E_m$. It is difficult to investigate this case analytically. From physical considerations, it is clear, however, that the variation of the boundary, η_{max}, in this case, must influence the behaviour of a smearing packet (of the initial dimension ℓ) never penetrates into such a narrow angle. However, a small displacement of ξ_{min} or ξ_{max} in this region will lead to the same result as in the previous case.

Let us consider the case when the initial packet is set up far from the boundary, η_{max}. Then $C_k \sim E_m L^2 \sim L^2/\ell^2$. Under this condition the integrand in (51) has a fixed sign, and the estimate, (55) is, in general, not valid. The question of the stability of the system remains open.

Finally, in the case $\eta_{max} \sim \eta_{min} \sim 1$ no additional amplification of the response, compared to the one-dimensional case, is expected. This case corresponds to a curved billiard table which transforms to a rectangle with the aid of regular conformational reflection.

Now we can consider an ensemble of systems within which a packet, at t = 0, is situated near the boundary $\eta_{max} \simeq \ln L/\ell$. Averaging, as before, will be carried out over the values of the shifts λ_ν distributed according to the law (26) $W(\lambda_\nu) = \frac{\Delta}{\Delta^2 + \lambda_\nu^2}$. The averaged density matrix elements are equal to

$$\tilde{\rho}_{mn} = \frac{1}{\pi} \int C_m^{*\nu} C_n^{\nu} e^{i(E_m^\nu - E_n^\nu)t} \cdot \frac{\Delta}{\Delta^2 + \lambda_\nu^2} d\lambda_\nu = \\ = \rho_{mn}^0 \exp\{-|E_m - E_n| \frac{\Delta}{L} t\} \tag{56}$$

for the characteristic energies

$$E_n \sim E_m \sim (E_n - E_m) \sim \frac{1}{\ell^2}: \quad \tilde{\rho}_{mn} = \rho_{mn}^0 e^{-\frac{t}{\tau}\frac{L\Delta}{\ell^2}}. \tag{57}$$

Thus, the time of damping of the nondiagnoal matrix elements, which makes the main contribution in this case, is equal to

$$t_c = \tau \ell^2 / (L\Delta) . \qquad (58)$$

We shall clarify the physical meaning of the result. In the case under consideration, the wave function of the packet reaches the billiard table walls and collides with them before it is smeared about the entire volume. This provides an L/ℓ time decrease of the *stochastization* time. The behaviour of the system in this case may be considered to be highly unstable.

In conclusion we shall consider the basic results:

1. The operation of averaging results in the disappearance of non-diagonal elements in the density matrix and, thus, describes transition from a pure state into a mixed state. The time of transition depends, essentially, on the properties of the system and the position of the initial packet.

2. *Stochastization* occurs if the dimensions of the system are much larger than the particle wave length (or, which is the same, the dimensions of the original packet), and at the same time the quantum can propagate throughout the system. In other words, *stochastization* may occur in the process of quantum delocalization.

3. Boundary conditions in the form of convex walls essentially accelerate the *stochastization* process, which brings correspondence with the classical result.

Note that the interaction of particles (even two) described by a sufficiently steep repulsing potential (more steep than $1/r$, where r is the distance between particles) is equivalent to the presence of a convex wall.

Thus the possibility of *stochastization* of a quantum mechanical system has been demonstrated, and sufficient conditions for this process have been found. In biological systems, these conditions are realized mainly for phonons (since electrons and quanta of normal oscillations are, as a rule, localized in small volumes). Therefore, a phonon subsystem is usually situated near thermodynamic equilibrium, whereas other parts of the system (in particular, the electron) can be far from thermodynamic equilibrium. Energy dissipation and entropy increase are due to the interaction of nonequilibrium subsystems with phonons.

On the Relation Between the H-Theorem and the Principle of Minimum Entropy Production

H. Hasegawa

Introduction

The H-theorem - a statement concerning the natural behaviour of a thermodynamic system approaching its equilibrium position - has a long history of scientific discussion since its introduction by Boltzmann (1896). Surprisingly, however, one finds a relatively small number of reference papers in the literature (Penrose, 1970). Most of them (after the famous controversy between Boltzmann, Ehrenfest, Loschmidt and Zermelo) consider the treatment of Markov chains or continuous Markov processes subject to certain types of equations of time evolution such as the master equation (Lebowitz, Bergmann, 1957); and the Fokker-Planck equation (Green, 1952: demonstrating the monotonic change in time of the properly defined H-function involving the probability or, the probability density in analogy to Boltzmann's H-function $\int f(x) \log f(x)\, dx$. The physical motivation is, of course, the second law of thermodynamics: the entropy of a thermodynamic system never decreases in its approach to equilibrium. Recently, the basic mathematical feature of such monotonic behaviour has been generalized to the framework of quantum Markov processes i. e. the H-theorem for the quantum dynamical semigroup (Lindblad, 1974, 1975, 1976; Spohn, 1978; Hasegawa, Nakagomi, 1980).

The aim of this article is to discuss the possibility that the relation between the H-theorem and the nonequilibrium thermodynamics could be placed in such a position that the two are connected to each other more intimately than in the standard treatment mentioned above. The question is how to place the H-theorem in the thermodynamic version of the principle of minimum entropy production due to Prigogine (1967). Since the thermodynamic

Thermodynamics and Kinetics of Biological Processes

framework does not utilize the ensemble concept and hence neither the probability concept nor the distribution functions, the problem in answering the question may be considered to be a statistical mechanical justification of the proposed thermodynamic principle. This should require a fully developed treatment. One should aim to construct a dynamic framework for a variational principle by which the evolution equation for the distribution can be derived (Hasegawa, 1976 a, b; 1977 a, b).

Briefly, a variational principle can be introduced into classical Markovian processes which characterizes the steady distribution, over the ensemble, towards which every time-dependent distribution approaches under a certain mixing condition (ergodicity). Even the time-dependent distribution itself, once the form of the evolution equation for it is found, conforms to this distribution. The associated H-theorem can then be viewed as a *consequence* of the variational principle.

Thus, one may say that for the H-theorem associated with each Markov process there must exist a variational principle which expresses the principle of minimum entropy production. Or, conversely, one may also note that it is possible to formulate the principle of minimum entropy production statistically within the Markovian framework such that it assures an initial distribution to be either a stationary, or time-dependent solution, of the given evolution equation for that distribution (on which the associated H-theorem is based). This approach can be used to devise a practical method for finding solutions to the equation just as does the variation method in quantum mechanics. In this contribution, we discuss the development of these themes in a rather elementary manner for finite Markov chains with a mixing property. However, the connection between Markovian processes and nonequilibirum thermodynamics can best be seen in systems subject to Fokker-Planck equations, to which a discussion will be added at the end.

The H-theorem for a finite Markov chain

A Markov chain is defined by a set of positive numbers with a pair of discrete indices (i,j) ; $\{P_{ij}\}$, each index specifying a *state* of our (physical) system. A probability shall be assigned to every P_{ij} - the transition probability of the system *from* the state i *to* the state j such that all the transitions from any one state exhaust all the possibilities, i. e. $\sum_j P_{ij} = 1$. The Markovian characteristic of such a probability set can be seen from the requirement that any two successive sets of such transitions of the system under the probability law form another set of admissible transitions. Thus, we have the following complete definition of a Markov chain $\{P_{ij}^{(n)}\}$; $i, j \in Z$, $n = 1, 2, \ldots$:

$$P_{ij}^{(n)} \geqslant 0 \tag{1}$$

$$\sum_j P_{ij}^{(n)} = 1 \tag{2}$$

$$\sum_k P_{ik}^{(n)} P_{kj}^{(1)} = P_{ij}^{(n+1)} . \tag{3}$$

It is easy to see that equation (3) can be replaced by a more general equation

$$\sum_k P_{ik}^{(n)} P_{kj}^{(m)} = P_{ij}^{(n+m)} , \quad P_{ij}^{(0)} \equiv \delta_{ij} \quad n, m = 0, 1, 2, \ldots \tag{4}$$

namely, the Chapman-Kolmogorov equation.

In the following discussion, we shall consider only a finite Markov chain, i. e. the case where the total number of states is finite. Also, without any mathematical justification, we shall make use of the fact that the discrete number, n , of the repeated transitions can be extended to continuous time, t, by replacing $P_{ij}^{(n)}$ by $P_{ij}^{(t)}$ without violation of (1), (2) and (4). The analytic property of $P_{ij}^{(t)}$ as a function of t $(0 \leq t < \infty)$ may be such that, besides continuity

$$\lim_{t \to 0} P_{ij}^{(t)} = \delta_{ij} , \tag{5}$$

a differentiation with respect to t be allowed to give the equation of evolution in the form

$$\frac{d}{dt} P_{ij}^{(t)} = \sum_{h=1}^{d} P_{ih} L_{hj} ,$$

or, more precisely (Feller, 1968) in terms of the transition probability rates w_{ij}

$$\frac{d}{dt} P_{ij}^{(t)} = \sum_{h=1}^{d} (P_{ih}^{(t)} w_{hj} - P_{ij}^{(t)} w_{jh}) \tag{6}$$

where $w_{hj} \geq 0 \quad h, j = 1, 2, \ldots, d.$

The transition probability $P_{ij}^{(t)}$ is the specific solution of the equation

$$\frac{d}{dt} p_j = \sum_{h=1}^{d} (p_h w_{hj} - p_j w_{jh}) \tag{7}$$

with the initial condition $p_i = \delta_{ij}$ at $t = 0$.

Now, a standard statement of the H-theorem associated with any solution $p^{(t)}$ of equation (7) will be presented:

1. Consider two solutions of equation (7), p_j^t and q_j^t. Suppose that the latter solution, q_j^t, is the stationary solution, $q_j^t = p^\infty$, for which $d/dt\, p_j^\infty = 0$. Let $H(x)$ be any real convex function of x on the positive real axis. Then, the quantity $\Phi_H(t)$, defined by

$$\Phi_H(t) = \sum_{j=1}^{d} p_j^\infty H(p_j^t / p_j^\infty) \tag{8}$$

is a non-increasing function of time t, so that $\Phi_H(t) \geq \Phi_H(s)$ $(t<s)$ holds. In the case where the convex function $H(x)$ is smooth and its derivative, $H'(x)$, is finite everywhere $(x > 0)$, then the non-increasing property can be obtained directly from

$$\frac{d}{dt} \Phi_H(t) = \sum_{j=1}^{d} \frac{dp_j^t}{dt} H'(p_j^t/p_j^\infty) =$$

$$= - \sum_{i,j=1}^{d} w_{ij}\, p_j^\infty (H(x_j) - H(x_i) - H'(x_i)(x_j - x_i)) \leq 0$$

with (9)

$x = p^t/p^\infty$

This inequality is due to the fact that for any convex function, $H(x)$, in an interval, I, the following inequality is valid:

$$H(x) - H(x_0) \geq H'(x_0)(x - x_0) \qquad (x, x_0 \in I)$$

(see Penrose, 1970, p. 77).

The equality between the first and the second lines can be obtained from the stationary nature of p^{∞} :

$$\sum_{j=1}^{d} p_j^{\infty} w_{ji} = p_i \sum_{j=1}^{d} w_{ij} .$$

In particular, for Boltzmann's H-function $H(x) = x \log x$ (see Lebowitz, Bergmann, 1957):

$$\frac{d}{dt} \sum_{j=1}^{d} p_j^t \log (p_j^t / p_j^{\infty}) =$$
$$= - \sum_{i,j=1}^{d} w_{ij} \frac{p_j^{\infty} p_i^t}{p_i^{\infty}} \left(\frac{p_j^t p_i^{\infty}}{p_i^t p_j^{\infty}} \log \frac{p_j^t p_i^{\infty}}{p_i^t p_j^{\infty}} - \frac{p_j^t p_i^{\infty}}{p_i^t p_j^{\infty}} + 1 \right) \leq 0. \tag{10}$$

In the above proposition, no additional assumption has been made about the ergodic property of the Markovian process, and therefore the statement in 1. cannot answer the question as to whether or not the given time-dependent solution p_i^t really approaches the supposed stationary distribution p_i^{∞}. Once this question is raised, the problem becomes much complicated depending on the detailed structure of the transition rate matrix (w_{ij}). Let us confine ourselves to the simplest case where the Markovian chain is *irreducible*, that is to say, the case where every transition probability $p_{ij}^{(t)}$ satisfies

$$p_{ij}^{(t)} > 0 \text{ for all pairs of states (i, j).} \tag{11}$$

Suppose that this condition is satisfied at a certain time t. Then, the Chapman-Kolmogorov equation for the continuous time (the integers n, m replaced by the continuous one-parameter values in equation (4)) ensures the satisfaction of the same condition for every later time so that it is plausible to assume them (Lebowitz, Bergmann, 1957) always after a certain time t_0. It is the possible to make the proposition 1. more specified as follows:

2. If the finite Markov chain subject to the evolution equation (7) is irreducible, then there exists only one stationary distribution p_j^{∞} which satisfies (7), with the left-hand side vanishing, and every time dependent solution of (7) approaches this unique stationary solution. Let $H(x)$ be any real *strictly convex* function of X on the positive real axis.* Under the circumstance stated above, the quantity $\Phi_H(t)$ defined in equation (8) is a decreasing function of time, i. e. $\Phi_H(t) > \Phi_H(s)$ for $t < s$ and the equality sign in the inequalities in (9) and (10) holds if and only if

---

* See page 60.

$$p_j^t = p_j^\infty \qquad j = 1, 2, \ldots, d \tag{12}$$

which is realized only in the limit $t \to \infty$.

It is not difficult to prove the above two propositions at least for finite Markov chains, for which the readers are referred to the literature (see Penrose, 1970; Lebowitz, Bergmann, 1957; Moran, 1961; Morimonto, 1963). The literature to date, however, does not seem to provide us with a clear-cut condition of the irreducibility for a finite Markov chain in terms of the transition rate matrix, and so it would be interesting to find the answer provided by recent developments in quantum dynamic semigroup theory (Lindblad, 1974, 1975). This we will discuss in the next section. (We shall use some results of the quantum dynamic semigroup theory for our purpose which, however, is sill within the scope of classical Markov chains).

The irreducibility condition from the dynamic semigroup theory

Let us rewrite the evolution equation (7) by using a vector and matrix notation

$$\frac{d}{dt} p = p L \tag{13}$$

where we have adopted the convention of arranging p_i in a d-dimensional row vector, and

$$L_{ij} = W_{ij} - \left(\sum_{j=1}^{d} W_{ij}\right) \delta_{ij} \ . \tag{14}$$

We investigate effects of the matrix operation by L on the column vectors instead of the row vectors. Let us denote any one of column vectors f, so that

$$(Lf)_i = \sum_{j=1}^{d} L_{ij} f_j = \sum_{j=1}^{d} W_{ij} (f_j - f_i) . \tag{15}$$

---

* A real function $H(x)$ is strictly convex in an interval, if it satisfies $H(\lambda x + (1-\lambda) y) \leq \lambda H(x) + (1-\lambda) H(y)$ for any x and y in the interval and $0 < \lambda < 1$ where the equality sign holds if and only if $x = y$. If $H(x)$ has a finite second derivative there, then $H''(x) > 0$.

The space of such f is a vector space, but multiplication can be allowed within the space by introducing the product operation

$$(fq)_i = f_i q_i \qquad i = 1, 2, \ldots, d \tag{16}$$

so that we may consider f^2, f^3, ... together with f and hence any function $H(f)$ of f on which the matrix operation L can be exerted. It may be remarked that the product fq should be distinguished from the scalar product between a row and a column vector for which the following notation will be used, i. e.

$$\langle p, f \rangle = \sum_{i=1}^{d} p_i f_i \ . \tag{17}$$

Then, the time-dependence of the vector f can be also be defined by requiring

$$\langle p^t, f \rangle = \langle p^{t=0}, f^t \rangle \tag{18}$$

so that, via the evolution equation (13)

$$\frac{d}{dt} f = Lf \qquad \left(\frac{d}{dt} f_i = \sum_{j=1}^{d} w_{ij}(f_j - f_i)\right), \tag{19}$$

which is analogous to the transfer from the Schrödinger to the Heisenberg picture. The transition probability $p_{ij}^{(t)}$ can be expressed by the matrix e^{tL}, i. e.

$$p_{ij}^{(t)} = (e^{tL})_{ij} \ . \tag{20}$$

It has recently been recognized in dynamic semigroup theory that the obvious requirement $p_{ij}^{(t)} \geq 0$, (1), leads to a simple, significant condition to be satisfied by the matrix L (called the generator of the semigroup $\{e^{tL}\}$, for the formation of a Markov chain: The matrix operation must be such that it maps every positive vector into positive (for row vectors and hence for column vectors, too)*, so that for any column vector f and any real λ

$$e^{tL} (\lambda f + 1)^2 \geq 0 \tag{21}$$

---

* The quantum version of the dynamic semigroup theory tells us that the positivity of the mapping e^{tL} is actually the *complete positivity* which is a more limiting condition than that is discussed here. However, for classical Markov chains (dynamic semigroups on commutative algebras) it can be shown that the complete positivity reduces to the present form (Lindblad, 1974, 1976).

must be satisfied. This inequality together with the normalization

$$e^{tL}1 = 1 \tag{22}$$

or, equivalently

$$L1 = 0 \tag{23}$$

which ensures $\sum_{j=1}^{d} p_{ij}^{(t)} = 1$, equation (2), leads to the relation

$$\lambda^2 (e^{tL} f^2) + 2\lambda (e^{tL} f) + 1 \geq 0 .$$

Hence, the condition

$$e^{tL} f^2 \geq (e^{tL} f)^2 \tag{24}$$

must hold as a form of the Schwartz inequality. In order to express the condition independent of the time variable, t, let us subtract f^2 from both sides of (21) and divide by t (>0), so obtaining

$$\frac{e^{tL} - 1}{t} f^2 \geq \left(\frac{e^{tL} - 1}{t} f \right) (e^{tL} + 1) f .$$

The limit $t \to \infty$ provides us with the desired condition

$$L f^2 \geq 2 f (L f) .$$

The inequality is a special case (real, commutative case for f) of that one for the generator of a quantum dynamic semigroup (called a dissipative operator), to be satisfied (Lindblad, 1974, 1976). Physically, one may say that, in contrast to the purely mechanical Liouvillian for which $Lfg = (Lf)g + f(Lg)$ and hence $Lf^2 = 2f(Lf)$, the inequality (25) extracts the dissipative feature of the Markov chain. For this reason Lindblad defines the non-negative quantity $\Phi(f) = Lf^2 - 2f(Lf)$ as the *dissipative function*.

We are now prepared to restate propositions 1. and 2. in a more obvious manner as follows:

1. *A finite Markov chain is a one-parameter family of the matrices* $\{e^{tL} ; 0 \leq t < \infty\}$ *where the exponent matrix* L *is defined in terms of the positive quantities* W_{ij} *by equation (14). It satisfies for any column vector f the inequality (25) and more generally, for any real convex function* $H(f)$ *with finite derivative* $H'(f)$ *defined on any positive column vector f (i. e.* $f_i \geq 0$ *), the inequality*

$$LH(f) \geq H'(f)(Lf). \tag{26}$$

The previous form of the H-theorem, (9), can be deduced from (26) through

(i)
$$\begin{aligned} \frac{d}{dt} \langle p^{\infty}, H(f) \rangle &= \langle p^{\infty}, H'(f)(Lf) \rangle = \\ &= \langle p^{\infty}, H'(f)(Lf) - LH(f) \rangle \leq 0 \end{aligned} \tag{27}$$

by virtue of the adjoint evolution equation (19) for f and the stationariness of p^{∞}

$$p^{\infty} L = 0 \tag{28}$$

together with inequality (26).

(ii) In order to see the time behaviour of $H(p^t/p^{\infty})$, we utilize the following fact: Evolution equation (13) for the row vector p can be transcribed into the adjoint form (19) for f by setting

so that $f = p^t/p^{\infty}$ $\qquad \frac{d}{dt} f = \tilde{L} f,$ (29)

where $\tilde{L} = p^{\infty -1} L' p^{\infty}$ (L' is the transpose of L)

which also satisfies

$$p^{\infty} \tilde{L} = 0. \tag{30}$$

Therefore, applying (ii) to (i) with interpretation (29), we obtain the compact presentation of the H-theorem (9)

$$\begin{aligned} \frac{d}{dt} \langle p^{\infty}, H(p^t/p^{\infty}) \rangle &= \langle p^t, LH'(p^t/p^{\infty}) \rangle = \\ &= \langle p^{\infty}, H'(f)(\tilde{L}f) - \tilde{L}H(f) \rangle_{f = p^t/p^{\infty}} \leq 0 . \end{aligned} \tag{31}$$

The important feature of the dissipation inequality (25) is, however, that it also provides a sufficient condition for a finite Markov chain without recourse to the explicit form (14) and, furthermore, it is capable of predicting the condition of irreducibility, which we now summarize:

2. *A finite dimensional matrix,* L *, generates a one-parameter family of matrices* $\{e^{tL}; 0 \leq t < \infty\}$ *which is an irreducible Markov chain, if, and only if, the following conditions are satisfied by* L *, i. e.*

(i) *the normalization* $L1 = 0$

(ii) *the dissipation inequality* $Lf^2 - 2f(Lf) \geq 0$ *where the equality sign holds only for f = const. x 1 (strict dissipation).*

The proof of this theorem is based on the fact that in finite dimensional matrix algebra the generator, L, of the family $\{e^{tL}\}$ is a bounded operator for which Lindblad's theorem assures the sufficiency of the conditions (i) and (ii) for the family to be a Markov process, which also yields the explicit form of L given by (14). Accordingly, the dissipativity inequality (ii) is expressed as

$$\sum_{j=1}^{d} w_{ij}(f_j - f_i)^2 \geq 0 \qquad i = 1, 2, \ldots, d. \tag{32}$$

Then, it is easy to see that the equality sign leads to $Lf = 0$. The irreducibility implies that such f is equal only to the constant. Conversely, if the equality sign holds only for f = const. x 1, then $w_{ij}(f_j - f_i) = 0$ leads to the structure of L such that it cannot be the direct sum of more than two ergodic components which implies the irreducibility.

The principle of minimum entropy production for finite chain

Recalling the H-theorem in the form of (31), and specifying it by a particular convex function, viz. Boltzmann's H-function, f(x)log f(x):

$$\frac{d}{dt}\langle p^t, \log(p^t/p^\infty)\rangle = \langle p^t, L\log(p^t/p^\infty)\rangle \leq 0. \tag{33}$$

Or, if we use the terminology of relative entropy (Lindblad, 1974, 1976) defined, for the probability p^t with respect to p^∞, by

$$S(p^t|p^\infty) \equiv \langle p^t, -\log p^t + \log p^\infty\rangle = \langle p^t, \log(p^\infty/p^t)\rangle, \tag{34}$$

then, the H-theorem (33) becomes

$$\frac{d}{dt} S(p^t|p^\infty) = \langle p^t, L\log(p^\infty/p^t)\rangle \geq 0. \tag{35}$$

It is therefore natural to call the expression on the right hand side of (35) the rate of entropy production - it might be natural to think that it could be associated with the variational principle in the thermodynamic context (Prigogine, 1967). This is, however, not so simple for the reason we discuss below.

The quantity defined for two arbitrary probabilities p and q (positive row vectors normalized in the present scheme, i. e. $\langle p,1\rangle = \langle q,1\rangle = 1$) by

$$\sigma(p|q) \equiv \langle p, L \log (q/p)\rangle \tag{36}$$

has the following interesting properties (Spohn, 1978):

(i) $\sigma(p|p) = 0$

(ii) $\sigma(p|q)$ is a convex functional of the probability i. e.

$$\sigma(\lambda p_1 + (1-\lambda)p_2|q) \leq \lambda\sigma(p_1|q) + (1-\lambda)\sigma(p_2|q) \qquad 0<\lambda<1 .$$

(iii) If q is the stationary distribution of L , i. e. $qL = 0$, then $\sigma(p|q)$ is lower bounded, the bound being 0 so that $\sigma(p|q) \geq 0$, $\min_p \sigma(p|q) = 0$ with $p = q$, and if L is strictly dissipative (satisfying the irreducibility condition) then this is the unique solution for the minimum value problem of $\sigma(p|q)$ (with respect to the normalized).

(iv) In addition to the stationariness of q, $qL = 0$, if p is the time-dependent solution of $\frac{d}{dt}p = pL$ then the H-theorem (35) holds.

Thus, the minimum property of the quantity $\sigma(p|q)$ in (iii) depends crucially on the premise that the q must be the stationary solution of the evolution equation. This is exactly why we claim that (iii) cannot be the variation principle for a sound statistical basis to the thermodynamic minimum principle: it states the characteristic of the stationary distribution only in terms of the desired characteristic itself. In fact, it is impossible to use the statement for devising a practical method for calculation of the stationary distribution. Also, this type of *variational principle* cannot be extended, by itself, to the time-dependent problem (by adding a term involving $\dot{p}$ to $\sigma(p|q)$), namely the variational principle by which the evolution equation for p is expected to be derived. It is far from this reason that we look for some other formulation without recourse to a stationary initial distribution.

It is possible to make a meaningful variational principle (meaningful in the sense discussed above) in many ways. The non-uniqueness stems from the variation in the convex function $H(f)$. A question often posed is this: What is the special role of Boltzmann's H-function? We believe that a satisfactory answer to this question should be found in the framework of the variational principle. This would require a dynamic variational principle involving two ingredients, both to be formulated so as to obey the equation of time evolution (Hasegawa, 1976 a, b; 1977 a, b). This is, however, beyond the scope of this contribution so that we adopt, here, a more conventional argument in the following.

Let us choose, instead of Boltzmann's $x\log x$, a simpler convex function $H(x) = -\log x / H''(x) = 1/x^2$; strictly convex for $0 < x < \infty$, for which the inequality formula (26) is $-L\log f + \frac{1}{f}(Lf) \geq 0$. Therefore, for any probability p

$$-\langle pL, \log f\rangle + \langle p, \frac{1}{f}(Lf)\rangle \geq 0 \tag{37}$$

holds. Changing the variable from f to $\log f \equiv s$ alters equation (37) to

$$-\langle pL, s\rangle + \langle p, e^{-s}(Le^{s})\rangle \geq 0. \tag{38}$$

We may note that the specific H-theorem (35) is still obtainable from (37) or (38) by setting

$$f = e^{s} = p^{t}/p^{\infty}, \qquad p = p^{t} \tag{39}$$

and

$$p^{\infty} : p^{\infty}L = 0, \quad p^{t} : \frac{d}{dt}p^{t} = p^{t}L. \tag{40}$$

It is, in fact, easy to see that the second term in (37), $\langle p, \frac{1}{f}(Lf)\rangle$, vanishes, reducing to the inequality

$$\langle p^{t}, L\log(p^{\infty}/p^{t})\rangle = \sigma(p^{t}|p^{\infty}) \geq 0. \tag{41}$$

What we observe now is that the essential variational property associated with $-L\log f + \frac{1}{f}(Lf)$ $(= -Ls + e^{-s}(Le^{s}))$ exists in its second term, not in the first term which yields (41). To show this, let us consider the expression defined by

$$\Phi\{s, p\} \equiv \langle p, e^{-s}(Le^{s})\rangle \tag{42}$$

for an arbitrary probability p and a column vector s. Then, we can prove

3. $\Phi\{s, p\}$ *has the following properties:*

(i) It is a convex function of s *for* $-\infty < s < \infty$.

(ii) It is lower bounded, the bound depending on the probability p.

(iii) The lower bound is 0 *so that* $\min_s \Phi\{s,p\} = 0$ *is attained by* $s = 0$ *if, and only, if* $p = p^{\infty}$; $p^{\infty} L = 0$.

Thus, proposition 3. affords a variational principle which, starting from an arbitrary distribution p, establishes how it can be characterized as the stationary distribution. Once this proposition is accepted, it can be extended straight forwardly to a time-dependent distribution p^t by

4. *Consider the variation problem* $-\langle \frac{dp}{dt}, s\rangle + \Phi\{s,p\} = \min$ *with respect to* s *at each fixed time. The minimum is attained by* $s = 0$, *if and only if* p *is a solution of the evolution equation;* $\frac{d}{dt} p = pL$.

A proof of proposition 3. and 4. are now presented by considering the 1st and the 2nd variations of $\Phi\{s, p\}$:

$$\delta^{(1)} \langle p, e^{-s}(Le^{s})\rangle = \langle [-pe^{-s}(Le^{s}) + (L' e^{-s} p) e^{s}]\, \delta s \rangle \tag{43}$$

$$\delta^{(2)} \langle p, e^{-s}\rangle (Le^{s})\rangle = \tfrac{1}{2} \langle p e^{-s} [(Le^{s}) u^2 + L(e^{s} u^2) - 2uL(e^{s} u)]\rangle \tag{44}$$

$$u = \delta s .$$

In particular, we look at these variations at $s = 0$ for which

$$\delta^{(1)} \langle p, e^{-s}(Le^{s})\rangle_{s=0} = \langle pL, \delta s\rangle \tag{45}$$

$$\delta^{(2)} \langle p, e^{-s}(Le^{s})\rangle_{s=0} = \tfrac{1}{2} \langle p, Lu^2 - 2u(Lu)\rangle_{u=\delta s} \geq 0 . \tag{46}$$

<u>Proof of 3. (i):</u> The convexity of $\Phi\{s,p\}$ for $-\infty < s < \infty$ can be ensured from the inequality

$$L(pf^2) - 2f(Lpf) + f^2(Lp) \geq 0 \tag{47}$$

which is valid for any column vector f and any positive column vector p. This is a generalization of the Lindblad inequality, (25), derived from the more general Schwartz inequality:

$$(e^{tL} f^2)(e^{tL} g^2) \geq (e^{tL} fg)^2 .$$

Proof of 3. (ii): To see the lower-bounded property, we use the explicit form of L , writing

$$\langle p, e^{-s}(Le^{s})\rangle = \sum_{ij=1}^{d} p_i w_{ij} (e^{s_j - s_i} - 1) \tag{48}$$

hence $\geq -\sum_{ij} p_i w_{ij}$.

Proof of 3. (iii): The first order variation of $\Phi\{s,p\}$ at s = 0 vanishes for arbitrary δs , if, and only if, $pL = 0$ in (45), which, together with the convexity property (46) yields a value for the minimum property.

Proof of 4.: We consider here the fact that the added term to Φ , for the variation, is linear with respect to s , so that nothing is changed in the second order variation: only the first order variation is changed, so that the minimum condition demanded is

$$\langle -\frac{d}{dt} p + pL, \delta s \rangle = 0 . \tag{49}$$

For practical purposes the above propositions may be formed in an alternative way such that the expression Φ is written in terms of two distributions p and $\hat{p}$ and its variation is carried out over one of them, i. e. a trial distribution $\hat{p} = e^{\hat{s}}$. *Afterwards* it is identified with p , as summarized by the following:

5. $\hat{\Phi}\{s,p\} = \langle p, e^{-s}(L'e^{s})\rangle$ = *min. with respect to* s *or, more generally* $\langle \frac{d}{dt} p, s\rangle + \langle p, e^{-s}(L'e^{s})\rangle$ = *min. with respect to* s *; and the minimum shall be attained at* $s = \log p$. *This is possible if, and only if,* $pL = 0$ *or, more generally,* $\frac{d}{dt} p = pL$.

Concluding remarks

The matter discussed in the preceding section is a clue to the bridge between the H-theorem and the thermodynamic minimum principle, but is certainly not all. For instance we have not yet fully clarified how the results of the variational procedure contained in 3., 4. and 5. are related to the H-theorem as we suggested in the introduction. In fact, if we inspect the time-dependent variational principle in 4, it is easy to see that the minimized value of the expression $-\langle \frac{d\rho}{dt}, s\rangle + \Phi$ always vanishes. It is, however, possible to construct the expression of variation such that it contains two components, one the probability, and the other, the *logarithm* of another probability. It's minimization with respect to the latter component yields a result in agreement with the H-theorem (33) for Boltzmann's f(x)logf(x)*. The general feature of such a framework is that a probability and its logarithm constitute *conjugate* variables to each other: A conditional Markov process (say, in the time interval t_1 and t_2) should have dynamic properties which may be described in terms of such conjugate quantities. The identification of the variable probability with the fixed probability must be considered as the limit $t_2 \to \infty$ of such a conditional process relative to the unconditional one. The variational principle for the unconditional process presented in 5. suggests that the variation must always be performed *before* such a limit is assumed.

Another related problem is that of extending the formulation for finite Markov chains to infinite cases. To be mathematically satisfactory, it is necessary to resort to the measure-theoretical aspect of analysis, which is difficult for non-specialists. However, the algebraic formulation stimulated by quantum dynamic semigroup theory is certainly a good starting point for this. In fact, recent progress in this direction tells us that the condition of strict dissipation (2. (ii)) guarantees the uniqueness of the stationary distribution even in such infinite cases (if it exists at all), although this condition alone is not necessarily sufficient for semigroup formation.**

---

* Recently Spohn and Lebowitz (1978) have discussed a formulation of the principle of minimum entropy production which attempts to characterize the stationary distribution $\rho^{\infty}, \rho^{\infty} L = 0$) for a model of open systems with several thermal reservoirs. Exact characterization is not possible, unless the specific point of the variational procedure discussed here is taken into account. This point has been discussed in detail (Hasegawa, Nakagomi, 1979, 1980 and Hasegawa, Nakagomi, Mabuchi, Kondo, 1980).

** The author is indebted to Dr. A. Kishimoto and Dr. H. Spohn for providing him with the information on this point.

Without extreme rigour, the variational principle stated in 5. can be extended to typical, infinite-dimensional continuous processes, namely, the system subject to a Fokker-Planck equation, a discussion of which would be worth while. Consider the Fokker-Planck equation written in Cartesian coordinate space in terms of the drift vector, $V_\mu(x)$, and the diffusion tensor, $\mathcal{D}_{\mu\nu}(x)$, (symmetric, positive) as follows:

$$\frac{\partial}{\partial t} p = -\frac{\partial}{\partial x_\mu}(v_\mu p) + \frac{\partial}{\partial x_\mu}\left(\mathcal{D}_{\mu\nu}\frac{\partial}{\partial x_\nu}\right) \equiv L' p . \tag{50}$$

Then, the operator adjoint to L' is the adjoint differential operator defined by

$$L = v_\mu \frac{\partial}{\partial x_\mu} + \frac{\partial}{\partial x_\mu}\left(\mathcal{D}_{\mu\nu}\frac{\partial}{\partial x_\nu}\right),$$

and the inequality of the H-theorem (26) for this may be written as

$$LH(f) - H'(f)(Lf) = H''(f)\mathcal{D}_{\mu\nu}\frac{\partial f}{\partial x_\mu}\frac{\partial f}{\partial x_\nu} \geqslant 0 \tag{51}$$

where the inequality is valid because of the positive nature of the diffusion tensor $\mathcal{D}_{\mu\nu}$. We now formally apply the prescription in V to this system:

$$e^{-s}(L'e^{s}) = -\frac{\partial v_\mu}{\partial x_\mu} - v_\mu\frac{\partial s}{\partial x_\mu} + \frac{\partial}{\partial x_\mu}\left(\mathcal{D}_{\mu\nu}\frac{\partial s}{\partial x_\nu}\right) + \mathcal{D}_{\mu\nu}\frac{\partial s}{\partial x_\mu}\frac{\partial s}{\partial x_\nu} .$$

Therefore, writing the first-order variation as

$$\delta\langle p, e^{-s}(L'e^{s})\rangle = \int\left[\frac{\partial}{\partial x_\mu}(v_\mu p) + \frac{\partial}{\partial x_\mu}\left(\mathcal{D}_{\mu\nu}\frac{\partial}{\partial x_\nu} p\right) - 2\frac{\partial}{\partial x_\mu}\left(\mathcal{D}_{\mu\nu}\frac{\partial s}{\partial x_\nu} p\right)\right]\delta s\, dx,$$

and allowing it to vanish for the arbitrary $\delta s(x)$, the result of the minimization becomes

$$\frac{\partial}{\partial x_\mu}(v_\mu p) + \frac{\partial}{\partial x_\mu}\left(\mathcal{D}_{\mu\nu}\frac{\partial p}{\partial x_\nu}\right) - 2\frac{\partial}{\partial x_\mu}\left(\mathcal{D}_{\mu\nu}\frac{\partial s}{\partial x_\nu} p\right) = 0,$$

which is identical to $L'p = 0$ (52)

by invoking the condition $s = \log p .$ (53)

Here, we observe a specific feature of the variational procedure, i. e. the extra condition imposed on the desired probability density, $p(x)$, related to the variation function $s(x)$ through (53), which may be considered as a self-consistency condition. It reminds us of the similar device proposed by Prigogine under the name *local potential* (Glansdorff, Prigogine, 1971).

However, it would be more interesting to investigate how the above variational principle reflects the principle of minimum entropy production in the thermodynamic framework which states (Prigogine, 1967) $\mathcal{J} \cdot \mathcal{X} = min$ where the flux vector $\mathcal{J}$ is related to the force vector $\mathcal{X}$, through J = LX (the Onsager relation), and (some components of) X is varied. An introduction of the concept of *force* and *flux* on the intuitive basis into our system of the Fokker-Planck equation by

$$\mathcal{X}_\mu = \frac{\partial s}{\partial x_\mu}, \qquad \mathcal{J}_\mu = \mathcal{D}_{\mu\nu} \mathcal{X}_\nu, \tag{54}$$

and $\mathcal{X}^\circ_\mu = \partial/\partial x_\mu \log p$

enables us to rewrite the minimization problem as follows:

$$\Phi\{s, p\} = \langle p, e^{-s} (L' e^{s}) \rangle = \langle p, (\mathcal{J} - v)_\mu (\mathcal{X} - \mathcal{X}^\circ)_\mu \rangle = min \tag{55}$$

where the minimization is to be performed with respect to s(x) and afterwards the setting X = X° is made to determine p(x).

A best application of the variational principle for the Fokker-Planck equation to actual problems would be a thermodynamic study of the laser system, a full report of which will be made elsewhere (Hasegawa, Nakagomi, 1979; Hasegawa, Nakagomi, Mabuchi, Kondo, 1980).

Acknowledgements

The author would like to express his gratitued to Professor C. P. Enz of the Department of Theoretical Physics, University of Geneva, Switzerland for a warm hospitality extended to the author during his stay in the institute, 1978 summer, where this article was written. He also wishes to thank Professor N. G. van Kampen and Professor R. S. Ingarden for critical readings of the manuscript and useful comments.

II. The Phase Transition and Nonequilibrium Fluctuations

Nonequilibrium Phase Transitions in Chemical Systems

M. Malek-Mansour, G. Nicolis, I. Prigogine

Introduction

Chemical kinetics is based on the idea of random collisions. Consider, for instance, a gaseous mixture or a dilute solution of chemically active molecules. Each of them undergoes a random thermal motion and, from time to time, it suffers an encounter with one of its partners in the mixture that is, itself, a random event. Depending on the dilution and on the temperature, these encounters belong, in different proportions, to three distinct types of events; elastic ones, conserving kinetic energy; inelastic ones, changing the internal molecular structure; and reactive ones, giving rise to the formation of new chemical species. It is a matter of observation that, despite the enormous complexity associated with motion at the molecular level, the time course of a reaction can be expressed in terms of a limited number of macroscopic observables. The founders of chemical kinetics were quick to realize this point. As early as 1879 Guldberg and Waage state, in their basic paper, that the rate of a chemical reaction is proportional to the average (which was considered to be identical to the most probable) number of collisions between the reacting molecules. The latter is, in turn, proportional to the product of appropriate powers of the concentrations $\rho_1, \rho_2, \cdots, \rho_n$ of the constituents of the mixture. This leads to the well known *law of mass action.* Later advances by Arrhenius, Bodenstein, Polanyi and Eyring among others, have followed this idea. In addition, they have produced workable models for the coefficient of proportionality - the rate constant - and its dependence on the temperature and other physico-chemical parameters.

One of the merits of these ideas was to lead to a simple quantitative formulation. Let w_α stand for the rate of reaction α, k_α for the correspon-

Thermodynamics and Kinetics of Biological Processes

ding rate constant. Then $W_\alpha \equiv W_\alpha(k_\alpha; \rho_1, \ldots, \rho_n)$. Adding up the effects of all reactions one then obtains the time evolution of the composition variables in the form (Prigogine, Defay, 1951):

$$\left(\frac{d}{dt}\rho_i\right)_{reaction} = \sum_\alpha \nu_{i\alpha} W_\alpha \equiv v_i(k_\alpha; \rho_1, \ldots, \rho_n) \qquad (1)$$

The stoichiometric coefficients $\nu_{i\alpha}$ describe the numbers of molecules of i participating in reaction α and can be positive, negative or zero.

In the real world, non-uniformities in the composition are inevitable. Once present, they give rise to the phenomenon of diffusion which tends to re-establish homogeneity. Diffusion is a complex process of molecular origin. Again however, in the absence of heterogeneities like interfaces or membranes, its rate is amenable to a simple expression in terms of the composition gradients. In a sufficiently dilute mixture, this relation takes the form (Prigogine, Defay, 1951; de Groot, Mazur, 1962):

$$\left(\frac{d}{dt}\rho_i\right)_{diffusion} = \mathcal{D}_i \nabla^2 \rho_i \qquad (2)$$

where $\mathcal{D}_i$ is a proportionality coefficient known as the diffusion coefficient. It depends in a similar way to the rate constant k_α on temperature and other state variables.

The assumption involved in the basic laws (1) and (2) are clear. Each event is considered to result from an average over all microscopic characteristics. Everything happens as if, in each subvolume ΔV of the reaction volume V, the molecules feel the average effect of all their partners therein. This effect is itself the result of these very molecules that feel it! We call this the *mean field description*. The term originated in early treatments of equilibrium phase transitions in fluids and in magnetic systems pioneered by Van der Waals and by Weiss and extended since then by the Soviet physicist Landau (Landau, Lifschitz, 1967).

This lumping together of all but the macroscopic degrees of freedom ignores the spontaneous deviations from average behavior, the fluctuations. These deviations are always present in a macroscopic system which, because of the complexity of molecular motions, can be viewed as a spontaneous generator of noise. In addition, macroscopic systems are subject to external noise sources related to the variability of the environment. Our principal goal in this article is to assess the influence of fluctuations and the occurrence of possible deviations from the mean-field description. We first summarize some facts illustrating the necessity to incorporate fluctuations in the traditional description in terms of deterministic rate equations.

Dissipative Structures: Order from Disorder

We have insisted on the chaotic character of the dynamics of macroscopic systems at the molecular level. And yet, experiment and everyday observations show that some spectacular changes leading to coherent behavior at the macroscopic level can take place in such systems. Coherent behavior can manifest itself in the form of regular spatial patterns, or, of rhythmic phenomena emerging abruptly in a hitherto homogeneous and stationary system. Both phenomena imply sharp and reproducible correlations between distant parts of the system. How can this ever happen in a dilute mixture of molecules behaving almost like hard spheres, and hence being incapable of recognising each other over distances of more than a few Angströms?

This problem has been the center of the preoccupation of many authors (see Nicolis, Prigogine, 1977; Haken, 1978). An important step toward its solution was achieved a decade ago by Glansdorff and one of the authors (Glansdorff, Prigogine, 1971). They showed that the physical chemistry of systems involving large deviations from thermodynamic equilibrium may be completely different from traditional physical chemistry. In the latter, one is accustomed to a behavior which is qualitatively similar to the evolution of isolated systems as dictated by the second law of thermodynamics: a tendency to increasing disorganization until a state of maximum entropy known as thermodynamic equilibrium is reached. Nonequilibrium systems are, on the other hand, constrained systems which are traversed by fluxes of matter and energy arising from their interactions with the outside world. Because of this, states displaying spatial or temporal organization can be reached (Glansdorff, Prigogine, 1971). The term *dissipative structure* has been coined for such states to show that they can only exist in conjunction with their environment. If the constraints are relaxed and the system is allowed to approach equilibrium, the entire organization will collapse. Examples of these structures abound in physical, chemical and also biological fields (Nicolis, Prigogine, 1977; Haken, 1978); Rayleigh-Benard instabilities, lasing phenomena, current instabilities in semi conductors (Thomas, 1973), Belousov-Zhabotinski reaction, glycolytic oscillation etc.

Although the concept of dissipative structures encompasses a large variety of problems, we hereafter illustrate it in the case of chemical reactions. Because of the relative simplicity of the formalism involved, the latter will be considered as the prototype of a system capable of switching to coherent behavior.

Of great interest are the restrictions on the kinetics necessary in order to have the formation of dissipative structures. It appears that, in at least one of the steps of the reactions sequence, a feedback must be involved. In other words, a chemical should influence directly or indirectly the rate of its own production or of that of other chemicals in the mixture. The most obvious realization of this condition is through catalytic processes. A particularly striking class is these processes involving cooperative enzymes (Sel'kov, 1968, 1972; Goldbeter, Lefever, 1972).

A simple model fulfilling these conditions is the following trimolecular model, also known as *Brusselator*:

$$\begin{array}{ll} A \xrightarrow{k_1} X & B + X \xrightarrow{k_2} Y + C \\ 2X + Y \xrightarrow{k_3} 3X & X \xrightarrow{k_4} E. \end{array} \qquad (3)$$

A, B, C, E are assumed to be in excess in comparison to X and Y. Ideally, their concentration is controlled throughout the reaction space to a uniform and time-independent value. Note that the simultaneous presence of X and Y enhances the rate of production of X, thus resulting in a positive feedback loop in the system.

In chemical equilibrium, the equality of the chemical potentials of A, X, Y, E and B, C would imply that the inverse reaction rates in (3) cancel the effect of the forward ones. This would rule out any non-trivial dynamic behavior. One may, however, force a nonequilibrium situation by pumping in A and B at constant rate and, by removing C and E as soon as they are produced. We are interested in the behavior of this constrained system. On appropriately scaling the values of concentrations, diffusion coefficients and time variables, one can write the following rate equations:

$$\begin{aligned}\frac{\partial}{\partial t}\rho_X &= \rho_A - (\rho_B + 1)\rho_X + \rho_X^2\rho_Y + \mathcal{D}_X\nabla^2\rho_X \\ \frac{\partial}{\partial t}\rho_Y &= \rho_B\rho_X - \rho_X^2\rho_Y + \mathcal{D}_Y\nabla^2\rho_Y .\end{aligned} \qquad (4)$$

These seemingly simple equations admit of an amazingly large variety of solutions. These cannot be computed exactly (save for a few limiting cases) but only evaluated by approximate methods. One such method is bifurcation theory, associated with perturbative expansions leading to the so-called Poincaré-Lindstedt series. The expansion parameter (s) - also known as bifurcation parameter (s) - is among the parameters appearing in the rate equations. For our purpose it will be sufficient to use ρ_B or one of the characteristic dimensions of the reaction vessel as such a parameter.

The results of this analysis, are as follows (Nicolis, Prigogine, 1977). Under appropriate boundary conditions, there exists a reference solution for all values of the bifurcation parameters which is spatially uniform and time-independent. It plays the role of the state of *maximum disorder* familiar from equilibrium thermodynamics and is given by $\rho_X^o = \rho_A$, $\rho_Y^o = \rho_B/\rho_A$. Other solutions can also exist and depend smoothly on the bifurcation parameters for certain ranges of values. But, for some critical, or bifurcation values of these parameters, various solutions approach equal values and hence exhibit a singular (non-analytic) dependence on the bifurcation parameters (Fig. 3).

Two situations are worth analyzing separately. One refers to the first bifurcation from the reference state. As it turns out, the bifurcating solutions in this case are less symmetrical than the uniform stationary state. For instance, a stationary pattern having a characteristic polarity may emerge as a stable solution, while the uniform solution looses its stability (Fig. 4). We may say that we have compartmentalization of the reaction space resulting from a symmetry-breaking transition. Another possibility is the emergence

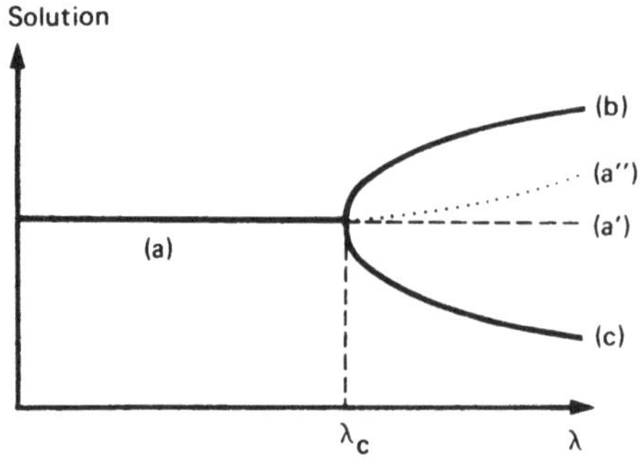

Fig. 3. A typical bifurcation diagram

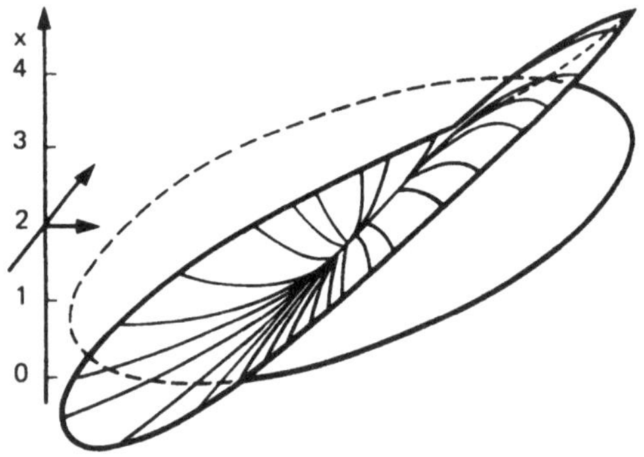

Fig. 4. Polar steady state dissipative structure in a circular domain

of stable time-periodic solutions known as limit cycles (Fig. 5). Contrary to periodic behavior associated with the pendulum and other simple mechanical devices, these solutions are reached for whole families of initial conditions, with an amplitude and period depending on the characteristic parameters appearing in the rate equations. Such chemical clocks exhibit, perhaps, the most ubiquitous difference between the behavior of nonequilibrium systems and that of systems in thermodynamic equilibrium.

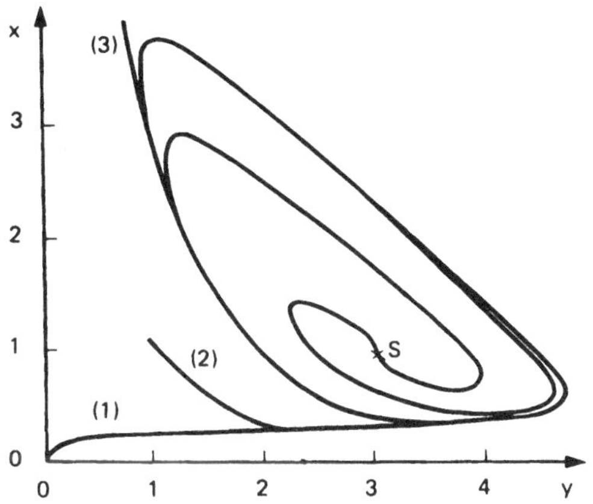

Fig. 5. Approach to a limit cycle for different initial conditions

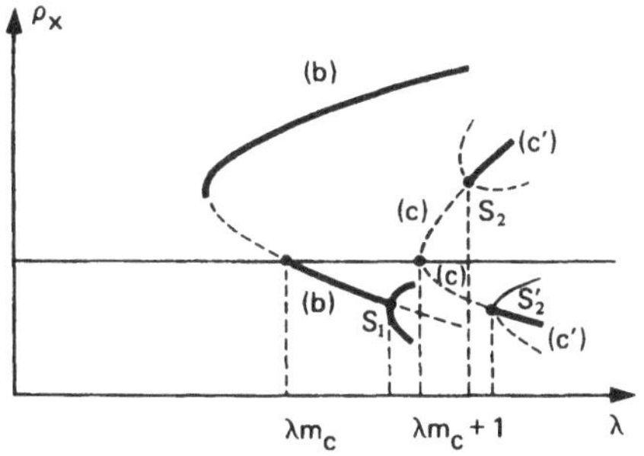

Fig. 6. Illustration of the phenomena of cascading bifurcations

We refer to the situations described by Fig. 4 and 5 as *nonequilibrium phase transitions*. A closer examination of this notion is postponed until the section devoted to fluctuation theory.

The second point of interest refers to the possibility of cascading bifurcations (Fig. 6). In addition to the primary solutions just described (curves (b), (c) ... in Fig. 6) secondary - and as a matter of fact tertiary and higher branches - can bifurcate from an already established pattern. This opens the way to the following fascinating possibility. A primary branch (curve (c) in Fig. 6) emerges as an unstable solution. But, at a point S_2 or S_2' it undergoes a secondary bifurcation which stabilizes it (curve (c') in Fig. 6). We have, therefore, the possibility of progressive complexity of a system by this mechanism of succesive transitions. For instance, it has been shown that such transitions may lead to stable rotating wave solutions for the Brusselator (Erneux, Herschkowitz-Kaufman, 1977), similar to the rotating waves described by Winfree (Winfree, 1974) for the Belouzov-Zhabotinski reaction.

Related to cascading bifurcations is the phenomenon of multiplicity in simultaneously stable solutions. Each of them attracts a different set of initial conditions. The system is, in this way, endowed with the ability to regulate its behavior and, to keep the memory of past events through the characteristics of the pattern that attracted the corresponding initial state.

Fluctuations: The triggers of self-organization

Consider the typical bifurcation diagram depicted in Fig. 3. At $\lambda = \lambda_c$, a stable state, branch (a) looses its stability and is succeeded, for $\lambda > \lambda_c$, by two simultaneously stable branches (b) and (c) which can exist - and even coexist - in macroscopic amounts. A particular system will spend most of its time in the vicinity of either (b) or (c). Yet the statistical average over a large number of replicas of the system will give a value close to the unstable branch (a). This value is completely irrelevant with respect to the observed behavior. One is, therefore, justified in expecting that the spontaneous deviations from the average - the fluctuations - will play a prominent role in fixing the relative weights of the stable states available.

There is a second reason for being interested in fluctuations. Suppose that the system could be prepared initially in the unstable state (a'). As this is a steady state, the system would stay there for ever unless small deviations - the fluctuations - could act to remove it. In other words fluctuations are the triggers of evolution. Eventually, beyond a bifurcation, they even become capable of overwhelming the entire system and drive it to a new macrostate. This is a rather remarkable point, in view of the fact that the bifurcation phenomena, reviewed in section 2, also occur in systems dominated by short range forces or even in ideal systems! Therefore, one would be tempted to consider that any extensive property, X, can be viewed as the sum of partial quantities X_i (referring to a division in small subvolumes):

$$X = \sum_{i=1}^{n} X_i \ .$$

If the dimensions of the latter are chosen to be larger than the range of the forces, these subvolumes will, then, be statistically independent. By virtue of the Central Limit Theorem, one expects X to follow (in the limit $n \to \infty$) a Gaussian distribution which is sharply peaked around the value $\bar{X}$ predicted by the macroscopic analysis;

$$P(X) \sim exp\left\{ -\frac{(X-\bar{X})^2}{2\sigma^2} \right\} \qquad (5)$$

where $\sigma^2 = \langle (\delta X)^2 \rangle \sim O(V)$, and V is the size of the system. The relative magnitude of the fluctuations will therefore behave as:

$$\left(\left\langle \left(\frac{\delta X}{\langle X \rangle} \right)^2 \right\rangle \right)^{1/2} \sim O(V^{-1/2}) \qquad (6)$$

which is negligible in the thermodynamic limit. How can, then, a system evolve spontaneously to a new state showing coherent properties? An answer to this question cannot even be formulated without introducing the concept of localized fluctuations which break spatial homogeneity and introduce in the problem an intrinsic size parameter much smaller than the size, V, of the entire system. Moreover, as shown below, nonequilibrium constraints induce the breakdown of Gaussian behavior, equations (5), and lead, concomitantly to coherent fluctuations, compromizing the picture of a system constituted of statistically independent entities.

All these questions refer to the behavior of the fluctuations generated spontaneously by the system itself. We call them internal fluctuations and we shall study them first. On the other hand, the external environment, in which the system is embedded, is very rarely rigorously constant. Being subject itself to a complex dynamics, the system is also capable of generating its own fluctuations. The latter will be perceived, by the system, as external noise. As we shall see below, under certain conditions this noise can dramatically alter the macroscopic behavior, both by influencing bifurcation phenomena, and, by introducing new transitions not predicted by the deterministic equations of evolution.

Internal fluctuations

The natural approach to the problem of fluctuations is in terms of probability theory. In other words, instead of computing the time evolution of the various quantities of interest - which, because of the extreme complexity of molecular interactions is highly irregular - one deals with probabilities of various types of events. For instance, *reduced* probabilities like that of having at time t a given number of particles of a certain constituent X in a small volume ΔV, $P(X_{\Delta V}, t)$; or, the *multivariate* probability $P(X_{\Delta V_1}, \ldots, X_{\Delta V_n}; t)$, where the partition $\Delta V_1, \ldots, \Delta V_n$ covers the entire reaction space. The construction of the probability function will, then, be achieved by combining the effects of reaction and diffusion and by associating with them an adequate stochastic process.

To fix the ideas, consider an ideal chemical system. Postponing until section C the discussion of diffusion, we focus here on the description of chemical processes alone. The composition of the system can only change through reactive collisions which are, typically, rare events compared to elastic ones because of the existence of activation energies. Thus, one may expect that at each moment the statistical state of the system in a small subvolume will be close to a Maxwellian equilibrium. In irreversible thermodynamics, this is known as the local equilibrium assumption. It suggests that one can lump together all the microscopic (positions-momenta) degrees of freedom and view the evolution as a succession of jumps corresponding to a change of composition through chemical reaction, interrupted by long time intervals. During these intervals, the large number of elastic collisions which will have taken place, will give rise to a randomization and a loss of memory. These remarks strongly suggest that the evolution can be represented by a jump Markov process in an appropriate phase space. Such processes have been investigated extensively by the Soviet School of mathematics, particularly by Kolmogorov (see Feller, 1968; Gikhman and Skorokhod, 1971, for a survey). It can be shown that the probability function descriptive of the process obeys a relation known as the master equation, which derives from Kolmogorov's system of forward equations:

$$\frac{d}{dt} P(X,t) = \sum_{\rho} \{ W_{\rho}(X - r_{\rho} \to X) \times \\ \times P(X - r_{\rho}, t) - W_{\rho}(X \to X + r_{\rho}) P(X,t) \} \tag{7}$$

Here the sum refers to the various reactions and r_{ρ} represents the number of molecules of X produced ($r_{\rho} > 0$) or consumed ($r_{\rho} < 0$) in the reaction ρ . W_{ρ} are transition probabilities per unit time and they are proportional to the frequency of reactive collisions of the constituents involved in reaction ρ , and therefore must be extensive quantities. In other words:

$$W_{\rho}(X) = V w_{\rho}(X/V) \equiv V w_{\rho}(x). \tag{8}$$

Equation (7) constitutes a natural generalization of the macroscopic description as it is based on less restrictive assumptions than the latter. As we see later, in certain situations it will give rise to predictions that cannot be inferred from the macroscopic kinetic laws. Nevertheless, it is of interest to inquire into the possible relationships between the two descriptions under the conditions for which the macroscopic laws are expected to be valid (thermodynamic limit, $V \to \infty$). This problem was investigated extensively by Kurtz (1978). An early result established by this author is as follows: Suppose the initial condition for both the stochastic and the deterministic trajectories X_t / V and $\bar{x}(t)$ to be identical. Then, assuming w_{ρ} to satisfy the Lipschitz condition and the jumps r_o to remain finite:

$$X_t / V = \bar{x}(t) + O(V^{-1/2}) ; \tag{9}$$

true for almost all $t_o \leq t \leq T$, $T < \infty$. In other words, in the thermodynamic limit, a macroscopic deviation between stochastic and deterministic trajectories can only occur for $t \to \infty$. Actually, we shall see explicitly lateron that such deviations may indeed occur in the presence of multiple, simultaneously stable solutions. A similar theorem can be established for the distribution of X_t / V which is shown to be Gaussian, provided it were so initially, for an interval $t_o \leq t \leq T$, $T < \infty$. This result throws new light on the asymptotic method developed some time ago by van Kampen (1961 a, 1976), and by Kubo et al. (1973). A third result by Kurtz relates to the connection between the jump process, X_t/V , and the diffusion process, z(t), which can be derived if the evolution equations were to generate a Markov process with continuous realization. Specifically, one can show that:

$$X_t/V = z(t) + O\left(\frac{\ln V}{V}\right) \qquad (10)$$

which is true for $t_0 \leq t \leq T,\ T < \infty$. These results were recently utilized by Horsthemke and Brenig (1977) to show that the Fokker-Planck equation descriptive of the diffusion process, constitutes an asymptotic representation of the discrete master equation describing a jump process. Note that the Fokker-Planck equation arising in this context features a diffusion coefficient which is weighted by the inverse of the size of the system and which can be a nonlinear function of z. A second type of general result is obtained by constructing the equation for the first moment, $\langle X \rangle$, of the probability functions. Multiplying both sides of equation (7) by X and summing over X from 0 to ∞ one finds:

$$\frac{d}{dt}\langle X\rangle = f\left[\{\langle X\rangle\},\{\langle(\delta X)^2\rangle\},\ldots\right]. \qquad (11)$$

Because of nonlinearity, this relation couples $\langle X \rangle$ to $\langle(\delta X)^2\rangle$ etc. The remarkable thing is that if P were Poissonian, this equation would have the same structure as the deterministic equations of evolution (1) and (2).
We express this fact as follows:

$$\frac{d}{dt}\langle X\rangle = \text{(deterministic rate equation)} \qquad (12)$$
$$+ \text{(terms due to deviation from Poissonian)}.$$

As an example, suppose that we have the chemical step $2X \xrightarrow{k} E$ with rate constant k . The part of equation (12) corresponding to the contribution of this step is:

$$\frac{d}{dt}\langle X\rangle = -k\langle X(X-1)\rangle = k\langle X\rangle^2 - k\left[\langle(\delta X)^2\rangle - \langle X\rangle\right].$$

The first term gives the macroscopic rate law, whereas the second vanishes for a Poisson distribution. Now, the direct solution of the master equation for the model $A \rightarrow X,\ 2X \rightarrow E$ gives a non-Poissonian distribution (Glansdorff, Prigogine, 1971) with $\langle (\delta X)^2 \rangle = 3/4 \langle X \rangle$. Thus, the rate law will differ from the macroscopic one. These deviations from the Poissonian are the triggers that enable the system to choose between many available solutions in the presence of bifurcation phenomena. Normally, in the limit of a large volume, one may neglect them. But in some cases, the deviation from the Poisson behavior becomes comparable to the average value itself. Hence, the first moment equation can no longer be written in a closed form. In other words, the statistical average is no longer representative of the macroscopic behavior of the system.

Note that diffusion, which is always present, tends to re-establish the Poissonian in a scale of the order of the mean free path. However, the competition with chemical kinetics may give rise to a deviation from the Poissonian in a macroscopically large subvolume of the system. Subsequently, this large fluctuations may propagate throughout the system and drive it to a new macroscopic state. We discuss this aspect more fully in section C, devoted to the effect of diffusion.

A. All-or-none transitions in spatially uniform systems: The Schlögl model

The simplest model illustrating the role of fluctuation is a chemical reaction scheme proposed by Schlögl (1972). It involves, just like the trimolecular model, a third order cooperative step, but presents the considerable simplicity of being reducible to a single variable:

$$\begin{aligned} A + 2X &\underset{k_2}{\overset{k_1}{\rightleftharpoons}} 3X \\ X &\underset{k_4}{\overset{k_3}{\rightleftharpoons}} B \;. \end{aligned} \qquad (13)$$

The concentration of A and B are controlled from outside and play the role of *pumping parameters*. Moreover, the system is supposed to remain uniform in space (e. g.: through adequate stirring). This will enable us to neglect the effect of diffusion. According to (13), the rate equation for the concentration ρ_X of X is cubic. Moreover, because of the presence of a single variable, one can always define a potential function $\mathcal{V}$, generating the rate of change of ρ_X through:

$$\frac{d}{dt}\rho_x = f(\rho_x) = -\frac{\partial v}{\partial \rho_x} \, . \tag{14}$$

Obviously v, which plays a role similar to the potential energy of a mechanical system, is quartic in ρ_x . The situation just desribed constitutes a simple case where the catastrophe theory of the French mathematician Thom (1972) applies. This theory, which so far has mostly been developed for systems deriving from a potential function, enables one to determine the number and nature of parameters which suffice to reveal all the qualitatively different types of behavior of a given system. For a cubic rate function, the unfolding - to use Thom's terminology - needs two parameters. Hence, by suitable rescaling of the variables, one may retain only two among the quantities A, B and K_1 to K_4. Letting X = X/N + 1 (N being the average total number of particles) be the redefined (intensive) variable and δ, δ' the scaled parameters, one finds the rate equation

$$\frac{d}{dt}x = -x^3 - \delta x + (\delta' - \delta) \tag{15}$$

with the potential function

$$v = \frac{x^4}{4} + \delta\frac{x^2}{2} - (\delta' - \delta)x \, . \tag{16}$$

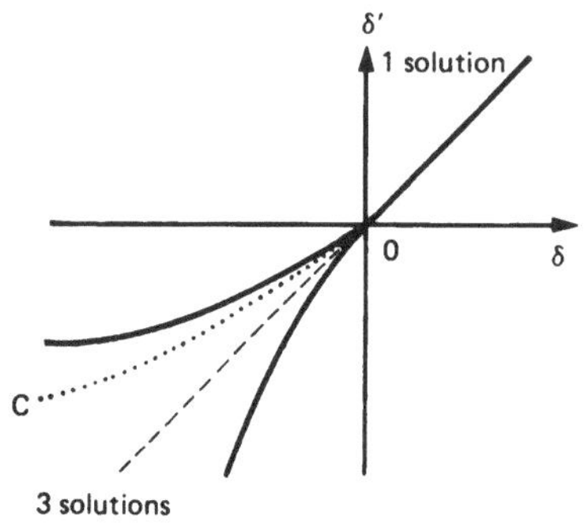

Fig. 7 - Dependence of the deterministic steady state solution on the parameters δ and δ' . C: coexistence line calculated from stochastic theory

Fig. 7 describes the dependence of the steady-state solutions of equation (15) on the parameters δ and δ'. The curve limiting the region of three real roots has a singular point at the origin $\delta = \delta' = 0$, known as a cusp. The system has, at this point, a triple root x = 0. Assuming that one moves in the δ, δ' plane along the line $\delta = \delta'$, one has:

- a root x = 0 for all δ,
- two roots $x_{\pm} = \pm(-\delta)^{1/2}$ for $\delta < 0$.

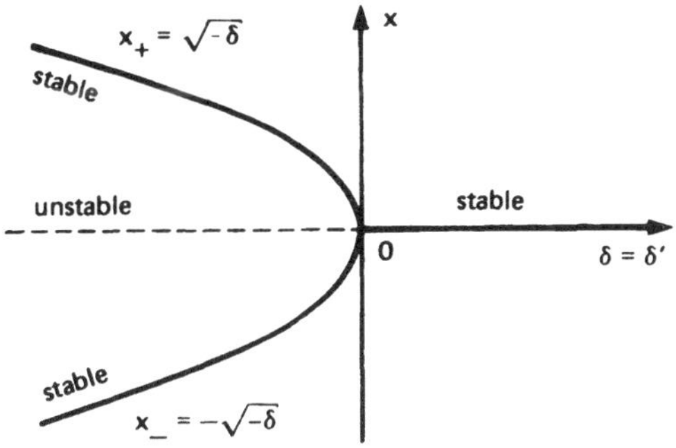

Fig. 8. Stability diagram for the Schlögl model (13). 0 = *Cusp* singularity

This is depicted in Fig. 8 which displays the, by now familiar, phenomenon of bifurcation and loss of stability of the reference solution. We call $|x_{\pm}|$ the order parameter of the problem. If one moves in the multiple steady-state region away from the line $\delta = \delta'$, the *upper* and *lower* branches are no longer symmetrical. Moreover, depending on where the system is initially, different paths will be traced during the time evolution to a final steady state. This is the phenomenon of hysteresis, which is one of the most characteristic features of systems involving multiple steady-state transitions. It can be verified that the entire multiple steady-state region corresponds to a situation which is removed from the state of chemical equilibrium. The behavior outlined so far is strikingly similar to the liquid-vapor transition as described by the classical van der Waals theory. In fact, the steady-state limit of equation (15), has a form identical to the van der Waals equation near the critical point provided δ, δ' and x are taken to be the analogs of pressure, temperature and specific volume, respectively. This analogy prompts us to speak of nonequilibrium transitions in connection with our chemical model (13).

We now come to the behavior of fluctuations, using the theory of Markov process summarized before, and for the purposes of producing exact results we adopt the highly idealized limit of spatially homogeneous fluctuations. It turns out that the master equation can then be solved exactly, in the generating function representation (Nicolis, Turner, 1977). Consider the case where the bifurcation point - the cusp singularity - is approached from the pretransitional region along the line $\delta = \delta'$ (Fig. 7). Then the variance of X can be shown to behave as (Nicolis, Turner, 1977):

$$\lim_{V\to\infty} \langle (\delta X)^2 \rangle / V \propto 1/\delta . \tag{17}$$

For fixed positive δ, this result is in agreement with the central limit theorem. But, as $\delta \to 0$, the variance (reduced in units of volume V) tends to diverge as the inverse power of δ. In equilibrium phase transitions, this is known as a *classical* law of divergence. At the bifurcation point, however, $\delta = \delta' = 0$, the variance behaves in a qualitative different way:

$$\lim_{V\to\infty} \langle (\delta X)^2 \rangle / V^{3/2} = \text{finite} \neq 0 . \tag{18}$$

We see that the conditions of validity of the central limit theorem are not satisfied. Still, in the limit $V \to \infty$ the first moment equation reduces to the deterministic from (15). The situation becomes more spectacular still along a particular line within the multiple steady-state region (line C, Fig. 7). Along this line, which surprisingly turns out not to be the symmetry axis of the cusp curve, we require the probability P to have two equal-height maxima (Nicolis, Malek-Mansour, 1978; Nicolis, Turner, 1979). This corresponds to a situation where the two deterministic stable steady-states x_+ and x_- coexist in macroscopic amounts. If moreover the thermodynamic limit $V \to \infty$ is taken, each of the two humps of the probability function collapses to a delta function. In terms of the intensive variable x one gets, therefore, a stationary probability distribution of the form

$$P(x) = C_+ \delta(x - x_+) + C_- \delta(x - x_-) \tag{19}$$

with $C_+ + C_- = 1$. If now the variance around the mean value - which in the limit $V \to \infty$ coincides with the unstable deterministic solution - is computed, one finds a result fo the same order of magnitude as x_+ or x_- ;

$$\lim_{V \to \infty} \langle (\delta X)^2 \rangle / V^2 = \text{finite} \neq 0 \qquad (20)$$

This introduces macroscopic deviations of the equations for the average from the determinsitic form. Moreover, the equation for the average is no longer closed but is coupled instead to that of the second, third, etc. ... moments. One is therefore forced to describe the system by more and more degrees of freedom, in fact by the entire infinite hierarchy of moments of the probability distribution. Another aspect of this incompleteness of the deterministic description, which was also pointed out in a different context by Landauer (1979), is that the potential function ϑ (equation (16)), does not suffice to determine the form of the probability distribution $P(x)$ (see also Bonifacio et al., 1978).

B. Stochastic theory of limit cyles

We now consider the more complex case of bifurcation of limit cycles. This phenomenon involves at least two coupled variables, and for this reason it is no longer possible to construct exact solutions of the master equation. We hereafter report the results of computer simulations for the trimolecular model (4), whose deterministic behavior was discussed in some length above.

A direct numerical simulation of the stochastic process described by the model can be performed (Gillespie, 1976; Hanusse, 1977; Turner, 1977), using the concept of the minimal process (Karlin, Taylor, 1975). As shown in Fig. 9, one obtains a steady state probability function in the form of a crater-like surface centered on the limit cycle. The portions of the periodic trajectory that correspond to a slow motion are weighted by an absolute maximum in the probability, whereas the portions of fast motion give a much smaller peak. Actually, as shown analytically by Ebeling on a non-chemical limit cycle example, in the three-dimensional representation of the probability crater the fast motion portion corresponds to a saddle rather than to a local maximum (Ebeling, 1978; Feistel, Ebeling, 1978).

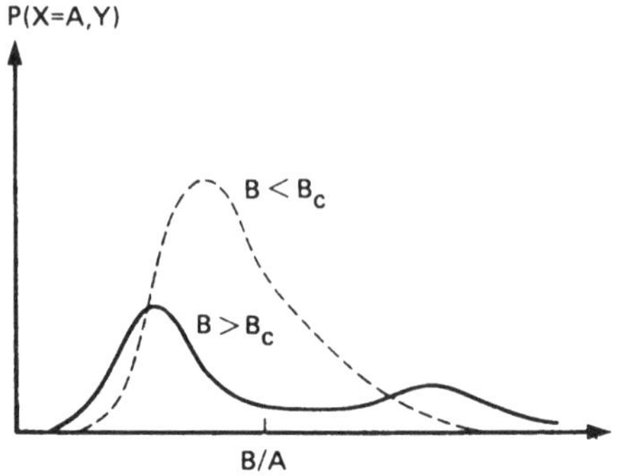

Fig. 9. Probability profile below and above the critical value B_c

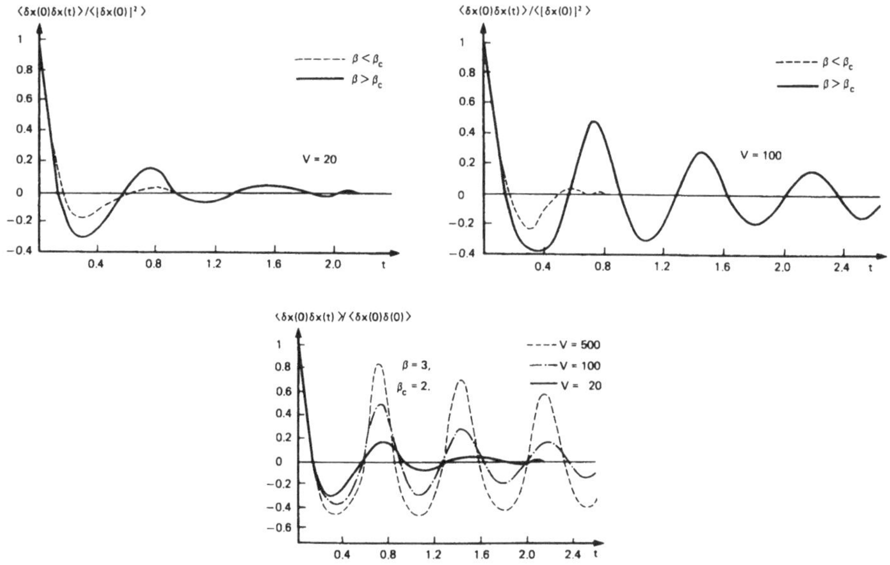

Fig. 10. Time dependence of normalized time-correlation function characterizing the passage to a limit cycle. β = bifurcation parameter; V = volume (in arbitrary unit)

Fig. 10 represents the results of stochastic simulation of the time development of the autocorrelation function of the concentration of X .

We see the appearance of damped oscillations past the bifurcation point. Interestingly, as the size of the system increases a systematic component close to the phenomenological oscillation becomes more and more evident, with a correspondingly diminishing damping. It is tempting to conjecture that in the thermodynamic limit, $N \to \infty$, there would be an undamped oscillation indicating a high phase coherence in the form of a probability peak rotating along the limit cycle. This can be verified directly on the master equation (Nicolis, Malek-Mansour, 1978).

There is of course no discrepancy between these results and the existence of a stationary crater found when the longtime limit $t \to \infty$ is taken before the thermodynamic limit. As above, what is happening is that as N gets larger, the crater gets sharper, so that for $N \to \infty$ and $t \to \infty$ one gets:

$$P(\underset{\sim}{r}) = \sum_{\phi} a(\phi) \delta [\underset{\sim}{r} - \bar{\underset{\sim}{r}}(t; \phi)] . \tag{21}$$

Here $\underset{\sim}{r}$ denotes the couple $(X/N, Y/N)$, $\bar{\underset{\sim}{r}}(t, \phi)$ is the phenomenological trajectory, ϕ denotes the phase along the periodic trajectory and $a(\phi)$ are appropriate weight coefficients. Note the analogy between equation (21) and the situation characterizing ferromagnetic transiton, and indeed, any transition associated with the braking of a continuous symmetry group. A reasonable Ansatz is that $a(\phi)$ is an increasing function of the mean sojourn times along the corresponding part of the limit cycle. This point is currently being investigated.

C. Effect of diffusion

In this subsection we want to allow for spatially inhomogeneous fluctuations. To this end, we subdivide the reaction volume into spatial cells and consider as variables the numbers of particles, $X_{i\alpha}$, of species i within cell α . The random variables, $X_{i\alpha}$, change as a result of two processes. A chemical reaction which will be modeled, as before, by a *birth and death* process and diffusion, whereby, a particle may jump to an adjacent cell; the latter will be assimilated to a random walk. The resulting multivariate probability function obeys now a master equation of the form:

$$\begin{aligned} \frac{d}{dt} P(\{X_{i\alpha}\}; t) = \sum_{\alpha} \Big\{ & \sum_{\{X'_{i\alpha}\}} W(\{X'_{i\alpha}\} | \{X_{i\alpha}\}) P(\{X'_{i\alpha}\}; t) + \\ & + \sum_{i} d_i \big[(X_{i\alpha}+1) P(\ldots, X_{i,\alpha-1}-1, X_{i\alpha}+1, \ldots; t) - X_{i\alpha} P(\{X_{i\alpha}\}; t) + \\ & + (X_{i\alpha}+1) P(\ldots, X_{i\alpha}+1, X_{i,\alpha+1}-1, \ldots; t) - X_{i\alpha} P(\{X_{i\alpha}\}; t) \Big\} , \end{aligned} \tag{22}$$

where d_i are the diffusion rates across cells and W the transition probabilities per unit time for the chemical processes.* We consider the simplest possible bifurcation leading to the all-or-none transitions discussed in Section A. Moreover, we assume that the fluctuating variable $\mathcal{X}$ changes by jumps of ± 1. The transition probabilities have then the more explicit form:

$$\begin{aligned} &W(\mathcal{X}\,|\,\mathcal{X}+1) = \lambda(\mathcal{X}) = &&\text{birth rate} \\ &W(\mathcal{X}\,|\,\mathcal{X}-1) = \mu(\mathcal{X}) = &&\text{death rate} \\ &W = 0 &&\text{otherwise.} \end{aligned} \tag{23}$$

For the Schlögl model (13), they are given by:

$$\begin{aligned} \lambda(\mathcal{X}_\alpha) &= \frac{3}{N}\,\mathcal{X}_\alpha(\mathcal{X}_\alpha - 1) + (1+\delta')N \\ \mu(\mathcal{X}_\alpha) &= N^{-2}\mathcal{X}_\alpha(\mathcal{X}_\alpha-1)(\mathcal{X}_\alpha-2) + (3+\delta)\,\mathcal{X}_\alpha\,. \end{aligned} \tag{24}$$

Equation (22) can be solved exactly only in some particular cases, like the case of linear birth and death transition rates (Gardiner et al., 1976). Otherwise, one must resort to approximations. One such approximation recently used widely (Haken, 1975 b; Gardiner et al., 1976; Lemarchand, Nicolis, 1976), consists in truncation procedures or in Gaussian assumptions. Applying these techniques to the Schlögl model for a one-dimensional array of n cells with periodic boundary conditions, one finds that the space correlation of fluctuations around the steady state and for $\delta = \delta' \geqslant 0$ is

$$\langle \delta\mathcal{X}_\alpha\, \delta\mathcal{X}_\beta \rangle = \langle \mathcal{X} \rangle\, \delta^{k_2}_{\alpha,\beta} + \frac{8R}{d(R^2-1)(R^n-1)}\left[R^{|\alpha-\beta|} + R^{n-|\alpha-\beta|}\right], \tag{25 a}$$

$$|\alpha-\beta| = 0, 1, \ldots, n-1$$

with

$$R = 1 + \frac{\delta}{d} + \left[\left(1+\frac{\delta}{d}\right)^2 - 1\right]^{1/2}. \tag{25 b}$$

As $n \to \infty$, the range of correlations diverges as $|\delta|^{-1/2}$ when the bifurcation point $\delta = \delta' = 0$ is approached. This is a classical law of divergence and leads to equation (17) when integrated over the entire space.

---

*Appropriate boundary conditions must be added.

We want, now, to go beyond this result. To this end, we consider again equation (22) and sum over all values of $X_{\alpha'}$, $\alpha' = \ldots, \alpha-1, \alpha+1, \ldots$. We then find at the stationary state:

$$P(X_\alpha+1) = P(X_\alpha)\,\frac{\lambda(X_\alpha)+\frac{d}{2}\left[E(\alpha+1|X_\alpha)+E(\alpha-1|X_\alpha)\right]}{\mu(X_\alpha+1)+d(X_\alpha+1)} \tag{26}$$

where

$$E(\xi\,|\,X_\alpha) = \sum X_\xi \,\frac{P(X_\xi, X_\alpha)}{P(X_\alpha)} \tag{27}$$

represents the conditional average of X_ξ . A first approximation, developed some time ago by the authors (Malek-Mansour, Nicolis, 1975), consists in adopting a *mean field picture* whereby $E(\xi\,|\,X_\alpha)$ is independent of X_α . This is a reasonable assumption if the size of the cells is comparable to the correlation length. It leads to a nonlinear master equation for the one-cell probability $P(X_\alpha, t)$. We now extend this hypothesis by assuming that $E(\xi\,|\,X_\alpha)$ may depend on X_α and we limit ourselves to a linear dependence (Malek-Mansour, Houard, 1979):

$$E(\xi\,|\,X_\alpha) = \langle X_\xi\rangle + \frac{\langle \delta X_\xi\, \delta X_\alpha\rangle}{\langle(\delta X_\alpha)^2\rangle}\left[X_\alpha - \langle X_\alpha\rangle\right]. \tag{28}$$

It can be shown that this relation is implied by the more stringent assumption that the correlation function

$$G(X_\alpha, r_\alpha; X_\beta, r_\beta) = P(X_\alpha, r_\alpha; X_\beta, r_\beta) - P(X_\alpha, r_\alpha)P(X_\beta, r_\beta)$$

where r denotes the cell position, can be factorized into a part depending on X_α and X_β only and a part depending on r_α and r_β . This is reminiscent of the approximation used in the well-known Enskog equation of the kinetic theory of dense gases. A similar approximation has already been used by Langer in a different context (Langer, 1975).

Using (25) and (28) one may find an expression for the space correlation function $\langle \delta X_\alpha \delta X_\beta\rangle$ at the steady state. For an infinite chain the result is:

$$\langle \delta X_\alpha\, \delta X_\beta\rangle = \langle X\rangle\, \delta^{kr}_{\alpha,\beta} + \frac{ab}{d(a^2-1)}\, a^{-|\alpha-\beta|} \tag{29}$$

with

$$a = 1+\frac{c}{d}+\left[\left(1+\frac{c}{d}\right)^2-1\right]^{1/2} \qquad b = 2\langle X\rangle\left[\frac{\langle\lambda(X)\rangle}{\langle X\rangle} - c\right]$$

$$c = \frac{\langle X\,[\mu(X)-\lambda(X)]\rangle}{\langle(\delta X)^2\rangle} \tag{30}$$

where the averages are taken over the one-cell probability distribution (26).

These expressions can be evaluated explicitly for the Schlögl model (13). Numerical inspection suggests that C cannot vanish, whatever the value of δ, δ' (N being kept finite). Thus, the spatial correlations do not diverge at the bifurcation point $\delta = \delta' = 0$. One cannot rule out divergence in two or three dimensions. Work in this direction is in progress.

Independently of the predictions concerning the critical region around the bifurcation point, equation (29) has been tested away from the bifurcation point by a direct numerical simulation of the stochastic process of reactions and diffusion. The results are shown in Fig. 11 for $\delta = \delta' = 1$, 0.01 and 0. Also plotted is the result obtained from truncations (25). We see that the agreement between the simulations and the correction to the mean-field theory developed in this section is striking.

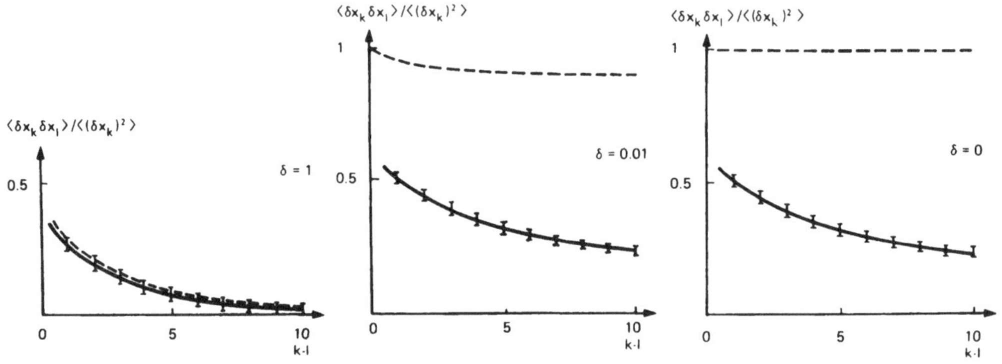

Fig. 11. Spatial correlation function for a periodic chain of 21 cells. Dashed curve represents the result obtained by the truncation of the hierarchy (25); solide curve that of the approximation (28). Vertical bars represent the *computer experiments* result with 99 % confidence intervals.

A different approach to spatially inhomogeneous fluctuations has been advanced recently (Dewel et al., 1977; Walgraef et al., 1980). The starting point is to incorporate the effect of fluctuations by an additive noise term $F(\underset{\sim}{r}, t)$ in the phenomenological rate laws, which then become stochastic differential equations. This term satisfies the usual condition $\langle F(\underset{\sim}{r}, t)\rangle = 0$ and (Guardiner, 1976)

$$\langle F(\underset{\sim}{r},t)F(\underset{\sim}{r}',t')\rangle = 2[\Gamma_1(\underset{\sim}{r},\{X_i\},t + \\ + \mathcal{D}\nabla^2\Gamma_2(\underset{\sim}{r},\{X_i\},t)]\,\delta(\underset{\sim}{r}-\underset{\sim}{r}')\,\delta(t-t'), \tag{31}$$

where $\mathcal{D}$ is the diffusion coefficient matrix.

If Γ_1 and Γ_2 contain a constant term independent of the state variables, then (31) generates a theory which is completely equivalent to a time dependent Ginzburg-Landau theory (Ma, 1976). On converting the stochastic differential equation to a Fokker-Planck equation one then finds a steady-state probability distribution provided a detailed balance condition holds. This is possible in systems involving a single fluctuating order parameter, like systems treated above, or the Brusselator in the vicinity of bifurcation of spatially inhomogeneous steady-state solutions. The result is of the form

$$P \propto \exp \Psi \tag{32}$$

where Ψ is the Landau-Ginzburg functional (Dewel et al., 1977, 1979). Hence, renormalization group methods (Dewel et al., 1977, 1979) and scaling techniques (Mori, 1975) can be applied in the same way as in equilibrium critical phenomena. In systems undergoing all-or-none transitions and described by a cubic rate law (see (15)), the theory is isomorphic with the Ising model (Ma, 1976). Hence, one finds a critical dimensionality of four and divergence of the variance and of the correlation length according to non-classical exponents below this dimensionality. On the other hand, in the Brusselator Ψ turns out to have a structure similar to that encountered in the analysis of anisotropic antiferromagnets, liquid crystal, and Rayleigh-Benard instabilities. As a result, spatial fluctuations may destroy the bifurcation as well as the structures predicted by mean-field theory in infinite one-and-two-dimensional systems.

The connection between this formalism and the approach based on the multivariate master equation has been recently considered by two of the authors (Nicolis, Malek-Mansour, 1980), using a singular perturbation approach. It has been shown that in the vicinity of a bifurcation leading to two simultaneously stable steady states, the steady state probability distribution reduces asymptotically to the exponential of the Landau-Ginzburg functional. The status of symmetry-breaking bifurcation remains, however, open.

Effect of external noise

So far we have been concerned with the dynamics of internal fluctuations in relatively simple systems. We have seen that these fluctuations, which are generated spontaneously by the system itself, tend to be small (in agreement with the central limit theorem), except when the system is in the coexistence region of simultaneously stable states.

In addition to such relatively simple situations, there exist systems which begin to present chaotic behavior once they undergo a certain number of bifurcations far from equilibrium. Turbulence in fluid dynamics is the most typical example. Average behavior and fluctuations become, then, two facets of the same phenomenon. Experimentally, this is reflected by the fact that the variance of fluctuations of intensive variables is of the same order of manitude as the average and is, in particular, independent of the size of the system.

In this section, we examine the coupling between simple chemical systems and external reservoirs undergoing strong fluctuations of the turbulent type. Indeed, in most of the situations involving instabilities, the bifurcation parameter is an externally controlled parameter and is subject, in general, to fluctuations. In the vicinity of the bifurcation point, this environmental noise can influence deeply the macroscopic behavior. A simple experimental set-up giving a random environment was conceived in Hess' laboratory at Dortmund (Boiteux et al., 1975). It concerns an oscillating biochemical reaction sequence, part of the glycolytic pathway, which was subject to a stochastic influx rate of one of the metabolites, in order to simulate the variability of the cellular environment.

Even more spectacular is the possibility of having nonequilibrium transitions generated by environmental fluctuations. This phenomenon was analyzed theoretically, and recently confirmed experimentally on a parametric oscillator circuit driven by a random current additionally supplied to the sinusoidal pumping current.

The traditional approach to environmental fluctuations goes back to Langevin's analysis of the Brownian motion problem. In his view, the rate function (say V(x)) describing the macroscopic evolution of an observable quantity (say x) gives only part of the instantaneous rate of change of x. Because of fluctuations of the surroundings, the system experiences, in addition, a random force F(x,t). Thus, viewing x now as a fluctuating quantity one writes

$$\frac{dx}{dt} = V(x) + F(x,t). \tag{33}$$

If, as in Brownian motion, F reflects the effect of intermolecular interactions, one requires its successive values to be uncorrelated both in time and space (Wax, 1954). Because of this, the conditions of validity of the central limit theorem are not compromised. It has often been imagined that, for appropriate choices of F, the formalism can be applied to the description of internal fluctuations as well (Nitzan et al., 1974; Keizer, 1977). On the other hand, by considering the massive action of a nonequilibrium environment, where the fluctuations are independent of the volume, one can have dramatic modifications of the macroscopic behavior of a system even though the internal fluctuations may still behave in accordance with the central limit theorem.

The appropriate mathematical tool to deal with this situation is given by the theory of Ito-Stochastic-Differential-Equation often referred to as the theory of Diffusion Processes (Stratonovich, 1966; Arnold, 1973). Although from the mathematical point of view, this theory is rather involved, its application to the above problem is nevertheless straightforward. Consider, for instance, this typical example of a phenomenological equation:

$$\frac{d}{dt}\bar{X} = f(\bar{X}) + \lambda G(X) \tag{34}$$

where λ is considered to be an external parameter. According to the procedure described above, λ will be considered as a Gaussian distribution of white noise (this is a necessary and sufficient condition for X_t to be Markov process) with mean value $\bar{\lambda}$ and variance σ^2. This can be written in the form:

$$\lambda_t = \bar{\lambda} + \sigma\xi_t \tag{35}$$

where ξ_t is the standard Gaussian of the white noise with zero mean, and unit variance. This leads to a Langevin type equation:

$$\frac{d}{dt} X_t = [f(X_t) + \bar{\lambda} G(X_t)] + \sigma \xi_t G(X). \tag{36}$$

The association with the Ito-Stochastic-Differential-Equation, is based on the fact that ξ_t can be considered as the time derivative - in the sense of a generalized function - of the Wiener Process W_t which is the usual Brownian motion. Thus we arrive at the following form of the stochastic differential equation:

$$dX_t = [f(X_t) + \bar{\lambda} G(X_t)]\,dt + \sigma G(X_t)\,dW_t. \tag{37}$$

The solution of this equation is defined by the following integral form

$$X_t - X_0 = \int_0^t [f(X_s) + \bar{\lambda} G(X_s)]\,ds + \int_0^t \sigma G(X_s)\,dW_s. \tag{38}$$

Note that in the literature, there exist two different definitions of the stochastic integral, i. e., the second term of the right hand side of equation (38), connected with the names of Ito and Stratonovich respectively. However, in most practical applications, there is no qualitative difference between these two interpretations (Arnold, 1973). In the Ito version, this Stocastic-Differential-Equation is equivalent to the following Fokker-Planck equation:

$$\begin{aligned} \frac{\partial}{\partial t} P(X,t|X_0,t_0) &= -\frac{\partial}{\partial X}[f(X) + \bar{\lambda} G(X)] \times \\ &\times P(X,t|X_0,t_0) + \frac{\sigma^2}{2}\frac{\partial^2}{\partial X^2} G^2(X)\, P(X,t|X_0,t_0) \end{aligned} \tag{39}$$

where $P(X,t|X_0,t_0)$ is the conditional probability for X at time t given that the process had the value X_0 at t_0.

Let us now illustrate the effect of a fluctuating environment by a typical example, given by the following phenomenological equation (May, 1973; Horsthemke, Malek-Mansour, 1976):

$$\frac{d}{dt}\bar{X} = -\bar{X}^2 + \bar{\lambda}\bar{X}. \tag{40}$$

This is known as the Verhulst equation and is frequently used in ecology where $\bar{\lambda}$ is the effective growth rate which strongly depends on the environment.

Using the above assumption, one finds the following Fokker-Planck equation:

$$\frac{\partial}{\partial t} P = -\frac{\partial}{\partial X}[-X^2+\bar\lambda X]P + \frac{\sigma^2}{2}\frac{\partial^2}{\partial X^2} X^2 P, \tag{41}$$

the stationary state solution of which is given by:

$$P_{st} = N X^{(2\bar\lambda/\sigma^2)-2} \exp\{-2X/\sigma^2\} \tag{42}$$

where N is the normalisation constant.

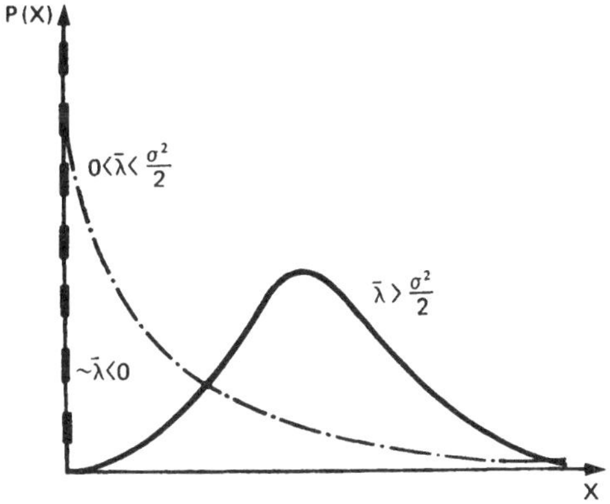

Fig. 12 - Plot of the stationary probability density P(X) (42) versus X.

This leads to the following conclusion (Fig. 12):

i) For $\sigma^2/2 > \bar\lambda$, the above expression is not integrable, meaning that no stationary probability density exists. The stationary probability is concentrated at zero in this case.

ii) The point $\sigma^2/2 = \bar\lambda$, is a transition point which corresponds to the transition $\bar\lambda = 0$ in the deterministic description. At this value, the stationary point zero becomes unstable and a stationary probability density appears which has nevertheless a divergence at zero as long as $\sigma^2 > \bar\lambda > \sigma^2/2$. This means that zero remains the most probable value.

iii) For $\sigma^2/2 < \bar\lambda$, the character of the stationary density changes. The divergence at zero disappears - in fact $P_{st}(0)$ drops to zero - and the density approaches a Gaussian like form around the deterministic value. Thus the point $\sigma^2 = \bar\lambda$ is a second transition point which has no deterministic counterpart. In fact its existence is solely due to external noise. Note that if the Stochastic-Differential-Equation had been interpreted as the Stratonovich equation, the same kind of phenomena is found with $\bar{\lambda}$ replaced by $\bar{\lambda} - \sigma^2/2$.

The existence of an additional transition point is of course a rather surprising result and, since the Stochastic-Differential-Equation was derived by phenomenological arguments and not from a microscopic description, an experimental verification is desirable. Recently, Kawakubo and Kabashima (Kabashima, 1978; Kabashima, Kawakubo, 1979) have devised an electrical circuit which is described by an equation analogue to the Verhulst equation. The existence of a noise-induced transition point was confirmed.

Further work (Horsthemke, Lefever, 1977; Arnold et al., 1978; Bulsara et al., 1978; Lefever, Horsthemke, 1979) has shown the generality of noise-induced transitions and the possibility of extending some of the results to systems subjected to real noise.

Concluding remarks

We are witnessing an extension of the basic concepts of physical chemistry. Coherent behavior becomes possible away from thermodynamic equilibrium. In chemical kinetics, fluctuations in the number of reactive collisions lead to nonequilibrium phase transitions which provide an extension of the ideas familiar from critical phenomena. Nonequilibrium phase transitions illustrate the importance of stochastic elements which begin to manifest themselves at the macroscopic level especially near bifurcation points. They also constitute a useful archetype for the modelling of complex systems of biology, ecology and social interest.

Most of the results discussed in this paper refer to the static behavior of fluctuations. Obviously, the time-dependent characteristics are of considerable interest, as they will determine the time scale of onset of an instability. The initial value problem of instable fluctuations has been considered in a series of paper by Suzuki (1979). An alternative procedure consists in analysing the spectrum of the stochastic evolution operator. Results in this direction are reported by Matsuo (1977) for jump processes (see also Oppenheim et al., 1977) and by van Kampen (1977 b) for diffusion processes. All these results are limited for space-independent fluctuations. The time-dependent behavior of localized fluctuations remains largely open (Chaturvedi et al., 1977; Gardiner, Walls, 1978; Della Donne, Ortoleva, 1978; van den Broek et al., 1980).

Throughout this article, we have assumed that the stochastic master equation provide a satisfactory description of the dynamics of nonlinear systems away from equilibrium. Clearly, this assumption must be modified, since in stochastic theory irreversibility is introduced in a phenomenological way, through the replacement of the exact dynamics by suitable transition probabilities (McQuarrie, 1967).

Recently, the passage from a reversible dynamical description to a probabilistic description has been proved rigorously, on a simple, soluble model (Misra et al., 1979; Prigogine, 1979). It may be hoped that this result can be extended to the analysis of more realistic systems of physico-chemical interest, such as hard sphere fluids or chemical reactions.

Acknowledgement

We would like to thank Drs. W. Horsthemke and J. W. Turner for fruitful discussions. This work has been partially supported by the Belgian Government: Actions de Recherches Concertées, Convention n° 76/81 II.3.

Synergetics. Nonequilibrium Phase Transitions and Selforganization in Biological Systems

H. Haken

From Thermodynamics to Synergetics

The most captivating feature of thermodynamics is its seeming universality. It applies to quite different materials, to different aggregate states and thus applies both to physics and chemistry. Its fundamental laws rest on only very few concepts, the most prominent one being that of entropy. According to the postulates of thermodynamics, in a closed system the entropy tends to a maximum value. For what follows it is useful to think of Boltzmann's definition $S = k \ln W$ where W is the number of microscopic configurations. As is known there are well understood rules, based on the number of degrees of freedom of the system, for counting the number of configurations. When counted properly all degrees of freedom have the same weight or, in other words, no configuration is preferred against another one. As a consequence, maximum entropy implies a maximum degree of disorder. The most elegant way to arrive at these and more general results is most probably by information theory where one seeks the maximum in the information entropy under given constraints, such as given mean energy, mean particle numbers etc. Due to these constraints, different configurations of the system may gain different statistical *weights*, or, more precisely speaking, different probabilities. Such differences are a prerequisite for the possibility of phase transitions. For what follows it is important to note that in the realm of thermodynamics, these differences are of a purely static, i. e. time independent nature. Thermodynamics applies to systems composed of very many sub-systems, which justifies a statistical description. It implies also that one can disregard the individual microscopic motions to a large extent and describe the whole system by few macroscopic observables which again apply to quite different materials. There is, however, a price to be paid for the universality of thermodynamics. It assumes that matter is in, or very close to, thermal equilibrium and, that processes occur infinitely slowly. The extension of thermodynamics to problems where relaxation occurs has been made in irreversible thermodynamics. Here concepts of thermodynamics, such as entropy or temperature, can still be applied to local regions in space, and concepts such as entropy production or entropy flow can be formulated. On this macroscopic or phenomenological level, relaxation and transport processes are treated in a more or less global manner (Lamprecht, Zotin, 1978).

Now let us turn to biology. Here again we have to deal with complex systems, i. e. systems composed of very many sub-systems. But quite obviously these systems are not in the state of greatest disorder even if they are, globally, in thermal equilibrium with their surrounding. On the contrary we observe well defined structures which can perform meaningful functions. One might suspect that there is a pronounced difference in entropy between living and inanimate matter. Such differences have been estimated and, according to Blumenfeld (1974), there is actually no difference between a human being and some piece of rock. How then can we characterize such systems? In the realm of synergetics we have studied numerous systems of physics and chemistry which are driven far from thermal equilibrium and which reveal the formation of spatial, temporal or spatio-temporal structures (Haken, 1973 a, 1974, 1977, 1978). These phenomena are well known, some of them for rather a long time. They include pattern formation in fluids (Benard-Taylor instability); in liquid crystals; coherent oscillations of lasers; macroscopic patterns of chemical reactions etc. The result of this analysis is that in all these systems there occurs a distinction between different degrees of freedom. The new feature which allows us to distinguish between these differing configurations is their be havior over the course of time, i. e. their dynamics. If we cut away couplings between different configurations in a way which we shall discuss below and, if we describe these configurations by well defined, collective variables, the amplitudes of these configurations can either decay or grow. This can be visualized as follows. Imagine a system into which energy continually is fed. Then certain configurations are more capable of using the energy flow than others. Whilst the latter still decay, due to loss processes, the former can compensate the loss processes and thus grow. In this new field of synergetics the Darwinian principle of survival of the fittest, therefore, reappears, but in a much broader sense. It does not only apply to animals or plants but may apply also to their components - down to the constituents of cells. As we will see below the amplitudes of the growing configurations play the role of macroscopic observables. As will transpire later the generalized Darwinian principle does not normally lead to unique developmental criteria. Fluctuations play a vital role in the determination of the direction in which processes eventially go. This does not imply, however, that order is established through fluctuations alone. We will see rather, that it is the interplay between deterministic forces and fluctuations which determines the development.

Nonequilibrium Phase Transitions and Self-Organization

In the realm of synergetics we consider, quite generally, systems composed of many sub-systems. Each sub-system is characterized by certain time dependent variables, $q_j(t)$. When the sub-systems are continuously distributed, the q's may depend on the three-dimensional space coordinate, x , and also on $q_j(x,t)$. Examples of such variables are: the numbers of molecules (especially bio-molecules) in certain spatial regions; firing rates of nerve cells; numbers of cells or animal populations; voltages or currents etc. Because of continuing processes these variables, in general, will change in time. Among these processes are relaxation processes (dealt within irreversible thermodynamics or statistical mechanics) and also, processes leading to growth of certain configurations are included. Explicit examples will be discussed later on. Furthermore, in spatially distributed systems we have to take into account diffusion processes, transport processes or the transmission of waves. Furthermore, we deal with open systems, i. e. we must take into account the coupling with the external world. This introduces so called *control parameters* which can be; the temperature or temperation gradients; power inputs and outputs; input and output of molecules; external signals to the nervous system etc. Finally, the sub-systems can be subject to internal or external fluctuations which we shall take into account by use of fluctuating forces.

The point of view we shall adopt in this article is as follows. We first assume that, for given control parameters a certain state of the system has evolved, i. e. the functions $q_j(x, t)$ are known and well defined, provided we leave aside fluctuations. In the simplest case one has a time-independent and spatially homogeneous state where the q's are constants. But we shall also admit the case in which the q's are space dependent and describe multiperiodic processes. This is actually a rather general assumption. We now change one, or several, control parameters. Under the impact of this change the state of the system described by the q's will, in general, change, too. Then, in principle, two different things may happen. The new state may be qualitatively the same as before. This means, e. g. that the q's are still space and time independent constants but now with somewhat different values which grow, uniquely, out of the old values. More generally a system undergoes multiperiodic motion with somewhat modified frequencies which are still uniquely determined and grow out of the old frequencies by continuous change of the corresponding parameters. The sentence *does not change qualitatively* can be cast into a mathematically rigorous form by invoking the concept of *structural stability*. The other possibility is a qualitative change of the state of the system. This is the situation of most interest in biology. Rather than give a general definition examples now follow. It may happen that the space independent q's become space dependent which entails the evolution of a spatial pattern. It can also mean onset of oscillations or waves etc. Also, the formerly multiperiodic motion can change drastically, for instance an additional period of motion may occur etc.

In the following I shall be concerned with the second type of phenomena, i. e. the occurrence of new patterns. There are, of course, intermediate problems. Such as in observing the growth of organisms it might be difficult sometimes, to tell precisely what is meant by *qualitative change*. However, on the other hand we will encounter below explicit examples where the loss of structural stability is quite obvious. The entity q describes certain macroscopic configurations. Dramatic changes of such configurations are well known in systems at thermal equilibrium, namely phase transitions. It came as a great surprise to many scientists that quite similar transitions occur in nonequilibrium systems. The first example, where this was demonstrated, was the laser (Graham, Haken, 1968 a, b; Haken, 1970).

What mathematical approach may be used to determine the new configurations, i. e. the newly formed spatial, temporal or spatio-temporal patterns or behavior? The method presently used in synergetics (Haken, 1978) can be qualitatively described as follows. We start from a *configuration*, q^{o} , for a certain control parameter, σ , and check the stability of q^{o} by linear stability analysis. The linear equations determine configurations of the system which differ from the old configuration by an infinitesimal amount. There exists a set of such new configurations which we shall call modes. Somewhat above an instability point (in the control parameter space) there are modes whose amplitude grows exponentially. These are the unstable modes. The remaining damped modes are called stable. In the next step of the analysis the actual configuration of the system is expressed as a superposition of unstable and stable modes whose amplitudes, ξ_u , and, ξ_s , respectively, are still unknown, time dependent functions. Starting from the original nonlinear equations it is then possible to derive equations for the set of ξ_u and ξ_s alone. The mathematical analysis shows, that ξ_s , can be uniquely expressed by ξ_u or, in technical terms, the ξ_s are slaves of ξ_u . Thus ξ_s can be eliminated, being replaced by ξ_u . In many nonequilibrium systems close to instability points, the number of ξ_u is much smaller than that of ξ_s . Thus an enormous reduction in the degrees of freedom is achieved. In practical cases, e. g. the laser, a reduction from 10^{14} degrees of freedom to a single degree of freedom has been obtained. However, fluctuations of the slave modes are still present as fluctuating forces acting on ξ_u . Therefore the newly evolving structure can be described by the few *unstable* modes alone. Their amplitude will be called order parameters because they describe macroscopic order. The newly evolving patterns or configurations are determined by several factors:

1. the form of the original equations of the q ;
2. the boundary conditions;
3. the initial conditions;
4. and the fluctuations.

It is, of course, of interest to know which of these factors is most important. While the evolution of new patterns is sensitive to change in the control parameters in the q equations, there are large classes of different equations which eventually lead to the same patterns. This results from the fact that quite different q equations, even those describing entirely different systems, can give rise to the same order parameter equations. Boundary conditions have a most important effect on the shape of the evolving pattern. But under certain symmetric boundary conditions further different patterns are possible for the same parameters of the q equations. Fluctuations will then decide which way the system goes. In general the evolving pattern is determined both by initial conditions and by fluctuations.

The equation (Haken, 1975) of a single real order parameter $\xi_u \equiv \xi$ is (in appropriate units)

$$\dot{\xi} = \lambda\xi - \xi^3 + F(t), \tag{1}$$

F(t) is the stochastic force. Equation (1) can be interpreted as describing the over-damped motion of a particle with coordinate, ξ, under potential V:

$$\dot{\xi} = - \frac{\partial V}{\partial \xi} + F(t).$$

Some features of continuous phase transitions are as follows: for $F \equiv 0$ and $\lambda < 0$, the only stable solution is $\xi = 0$; for $F \equiv 0$ and $\lambda > 0$ the value $\xi = 0$ becomes unstable and is replaced by a new stable value, namely

$$\xi = +\sqrt{\lambda} \qquad \text{or} \qquad \xi = -\sqrt{\lambda}.$$

Either value breaks the symmetry. The change of $\lambda < 0$ to $\lambda > 0$ is associated with a symmetry-breaking instability. As $\lambda \to 0$, ξ relaxes more and more slowly: *critical slowing down*. With $F \neq 0$ and $\lambda \to 0$, the fluctuations become *critical fluctuations*. Provided F(t) is Gaussian and Markovian, the Fokker-Planck equation associated with the *Langevin equation* (1) can be established. For $\langle F(t) \rangle = 0$ and $\langle F(t)F(t') \rangle = Q\delta(t-t')$ the stationary solution of the Fokker-Planck equation is

$$f = N \exp\{Q(\lambda\xi^2 - \xi^4/2)\}.$$

This has exactly the form postulated by Landau (Landau, Lifschitz, 1952) in his theory of phase transitons. When we plot $\xi = \xi_0$ as a function of λ, we obtain Fig. 13, which is typical for bifurcation phenomena. A first order (or discontinuous) phase transition occurs, when the order parameter equation is

$$\dot{\xi} = a + \lambda\xi - \xi^3 + F(t), \quad a \neq 0$$

This equation implies the existence of hysteresis phenomena.

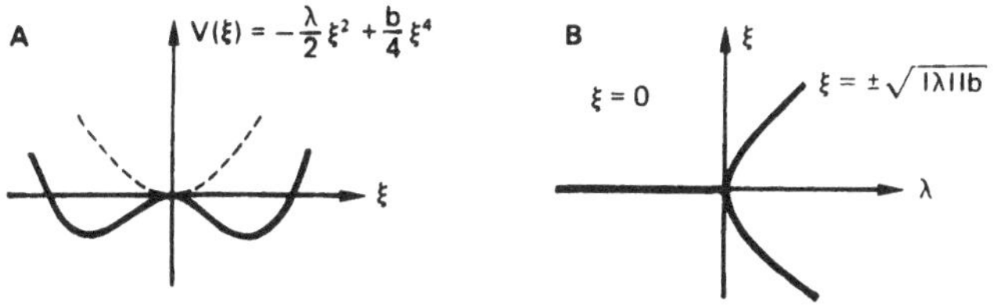

Fig. 13. Scheme of bifurcations. A - Potential as a function of coordinate ξ; dashed line: $\lambda < 0$, solid line: $\lambda > 0$. B - Equilibrium coordinate ξ as function of control parameter λ.
(Further explanations in text.)

Generalized Ginzburg-Landau Equations for Nonequilibrium Phase Transitions in Continuously Extended Media

In this section we deal with equations of motion for continuously extended systems containing fluctuations (Haken, 1978). To explain our procedure we look at the general form of the equations which are used in hydrodynamics, lasers, nonlinear optics, chemical reaction models, morphogenesis and related problems. We denote physical quantities by $U = (U_1, U_2, \ldots)$. U obeys equations of the following type

$$\frac{\partial}{\partial t} U_\mu = G_\mu(\nabla, U) + D_\mu \nabla^2 U_\mu + F_\mu(t) \quad \mu = 1, 2, \ldots, n. \qquad (2)$$

In this equation G_μ are nonlinear functions of U , and, perhaps of a gradient. In most applications G is a linear, or bilinear function of U , though in certain cases (especially in chemical reaction models) a cubic coupling term may equally well occur. The second term in (2) describes diffussion (D : real) or wave type propagation (D : imaginary). In this latter case the second order time derivative of the wave equation has been replaced by the first order derivative by use of the *slowly varying amplitude approximation*. The $F_\mu(t)$ are fluctuating forces. We shall not be concerned with the derivation of equation (2). Rather, our goal will be to derive from (2) equations for the undamped modes which acquire macroscopic size and determine the dynamics of the system in the vicinity of the instability point. These modes form a mode skeleton which grows out from fluctuations above the instability and thus describe the *embryonic* state of the evolving spatio-temporal structure.

We now start with a treatment of (2). We assume that the functions G_μ in (2) depend on control parameters, σ . First we consider values of σ such that $\mathcal{U} = \mathcal{U}_0$ (= independent of space and time) is a stable solution of (2). We decompose $\mathcal{U}$ into

$$\mathcal{U} = \mathcal{U}_0 + q \tag{3}$$

with

$$q = \begin{pmatrix} q_1(x,t) \\ \vdots \\ q_n(x,t) \end{pmatrix} \tag{4}$$

Splitting the right-hand-side of (2) into the linear part, Kq , and the nonlinear part, $g(q)$, we obtain (2) in the form

$$\left(\frac{\partial}{\partial t} - K(\nabla^2)\right) q = g(q) + F(t). \tag{5}$$

The matrix

$$K = (\hat{K}_{\mu\nu}) \tag{6}$$

has the form

$$K_{\mu\nu} = K_{\mu\nu} + \delta_{\mu\nu} D_\mu \nabla^2, \quad \text{where} \quad K_{\mu\nu} = \frac{\partial G_\mu}{\partial \mathcal{U}_\nu}\Big|_{\mathcal{U}_{\nu,0}} \tag{7}$$

Our whole procedure applies, however, to a matrix K which depends, in a general way, on ∇ . g is assumed in the form

$$g_i(q) = \sum_{\mu\nu} q_\mu g^{(2)}_{i\mu\nu}(\nabla) q_\nu + \sum_{\mu\nu\kappa} g^{(3)}_{i\mu\nu\kappa} q_\mu q_\nu q_\kappa . \tag{8}$$

$g^{(2)}$ may, or may not, depend on ∇ , and equally $g^{(3)}$ may, or may not, depend on ∇ (if $g^{(3)}$ depends on ∇ , the sequence of g and q must be properly chosen.

We first deal with K and introduce operators

$$O^{(j)} = O^{(j)}(\nabla) \tag{9}$$

which still depend on ∇ and which are defined as eigenvectors satisfying the equation

$$K(\nabla)O^{(j)} = \lambda_j(\nabla)O^{(j)}. \tag{10}$$

When ∇ is replaced by ik (10) becomes a linear algebraic equation so that (10) can easily be solved without resorting to any operator techniques. We furthermore introduce eigenfunctions of the wave equation

$$\nabla^2 \chi_k(x) = -K^2 \chi_k(x) \tag{11}$$

provided K depends on ∇^2. χ_k can be adequately chosen later on. We shall use a notation which suggests that the χ_k are plane-wave solutions, but it may be advantageous to use other representations as well, e. g. Bessel functions or spherical wave functions, depending on the problem. If K depends on ∇ in odd powers (and in even powers), χ_k are taken as a complex plane wave solutions of (11). We represent $q(x)$ as the superposition

$$q(x,t) = \sum_{kj} O^{(j)} \xi_{kj}\, \chi_k(x). \tag{12}$$

It can be shown by means of explicit examples that there exist small excitation bands for unstable modes. This suggests the building up of wave packets (as in quantum mechanics) so that we take together sums over small regions of K. We thus obtain carrier modes with discrete wave numbers K and slowly varying amplitudes $\xi_{kj}(x)$. Our first goal is to derive a general set of equations for the mode amplitude ξ. To do so we insert (12) into (5), multiply from the left-hand-side by $\chi_k^*(x)\bar{O}^{(j')}$ and integrate over a region which contains many oscillations of χ_k but in which ξ changes very little. After some analysis the basic set of equations has the following structure

$$\dot{\xi}_{kj} - \hat{\lambda}_j(\nabla, K)\xi_{kj} = H_{kj}(\{\xi(x)\}) + \hat{F}_{k,j} \tag{13}$$

where

$$\begin{aligned} H_{kj}(\{\xi(x)\}) \equiv & \sum_{\substack{k'k'' \\ j'j''}} a_{k,k',k'',j,j',j''}\, I_{k,k',k''}\, \xi_{k'j'}\, \xi_{k''j''} + \\ & + \sum b_{k,k',k'',k''',j,j',j'',j'''}\, J_{k,k',k'',k'''}\, \xi_{k'j'}\, \xi_{k''j''}\, \xi_{k'''j'''}. \end{aligned} \tag{14}$$

a, b, I and J are certain constants which are defined explicitly elsewhere (Haken, 1978). λ is essentially an eigenvalue. So far almost no approximations have been made, but to cast (13) into a useful form we have to eliminate the unwanted, or uninteresting, modes. These are the damped modes. Accordingly we put

$$j = u \quad \text{unstable if} \quad \mathrm{Re}\, \hat{\lambda}_{\mu}(0, k) \geq 0 \tag{15}$$

and

$$j = s \quad \text{stable if} \quad \mathrm{Re}\, \hat{\lambda}_{s}(0, k) < 0. \tag{16}$$

It should be noted that, though ξ bears two indices, k and u, these indices are not independent of each other. Indeed the instability usually occrus only in a small region of $k = k_c$ (compare Fig. 14). Thus we must carefully distinguish between the k-values at which (21) is evaluated (see below). Because of the *wave packets* introduced earlier, k runs over a set of discrete values with $|k| = k_c$. The prime at the sums in the following formulae indicates this restriction on the summation over k.

Fig. 14 provides us with an example of a situation where a branch, j, of $\xi_{k,j}$ can become unstable at $k = k_c$. Because undamped modes may grow without limit provided the nonlinear terms are neglected, we expect that the amplitudes of undamped modes are considerably bigger than those of damped modes. Since, on the other hand, close to a *phase transition* point the relaxation time of undamped modes tends to infinity, i. e. the real part of $\hat{\lambda}$ tends to zero, the damped modes must adiabatically follow the undamped modes. Though the amplitudes of the damped modes are small, they must not be neglected completely. Neglect would lead to a catastrophe if in (14) the cubic terms are omitted. It is easy to show that quadratic terms cannot lead to a globally stable situation. Thus cubic terms are necessary for stabilization. Such cubic terms are introduced, even though absent in the original equations, by the elimination of the damped modes. To exhibit the main features of our elimination procedure more clearly, we set

$$\xi_{k,j} \rightarrow (k, j) \tag{17}$$

and drop all coefficients in (13). We assume $|\xi_s| \ll |\xi_u|$ and, in a

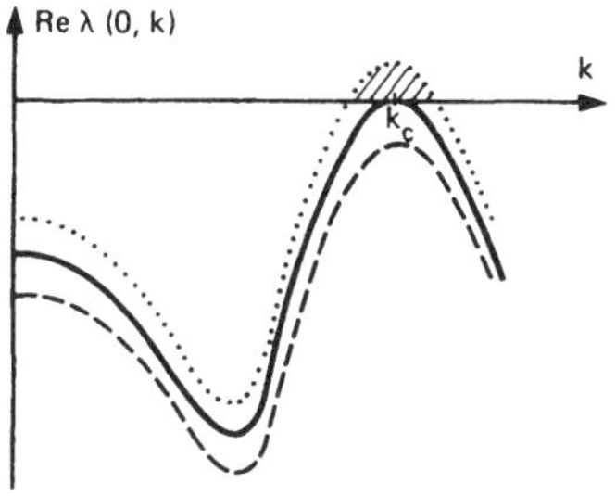

Fig. 14. Real part of the eigenvalue $\hat{\lambda}$ plotted as function of the k-number of modes (example). Dashed line: marginal situation, one mode at $k = k_s$ is about to become unstable. Dotted line: shaded area, a whole range of modes has become unstable giving rise to small band excitations. This situation is taken care of by the formation of wave packets and slowly varying amplitudes.

selfconsistent way, $\xi_s \sim \xi_u^2$. Keeping only terms up to third order in ξ_u in (13), we obtain

$$\left(\frac{d}{dt} - \hat{\lambda}_u\right)(k,u) = \sum_{k'k''u's} (k',u')\cdot(k'',s) + \sum_{\substack{k'k'' \\ u'u''}} (k',u') \times \tag{18}$$

$$\times (k'',u'') + \sum_{\substack{k'k''k''' \\ u'u''u'''}} (k',u')(k'',u'')(k''',u''') + F_{k,u'}$$

Consider now the corresponding equation for j = s. In this equation only terms necessary to obtain an equation for the unstable modes up to third order are retained

$$\left(\frac{d}{dt} - \lambda_s\right)(k,s) = \sum_{\substack{k'k'' \\ u'u''}} (k',u')(k'',u'') + \dots \tag{19}$$

If we adopt an interaction scheme using the inequality $|\xi_s| \ll |\xi_u|$ it is simple to demonstrate that ξ_s is, at least, proportional to ξ_u^2 so that only relevant terms in (19) are those shown explicitly. We now use our second hypothesis, namely, that the stable modes are damped much more quickly than the unstable ones. This is true for the soft mode instability In the case of a hard mode we must be careful to remove the oscillatory part of $(k',u')(k'',u'')$ in (19). This is achieved by retaining the time derivative in (19). We, therefore, write the solution of (19) in the form

(20)

The prescription to evaluate d/dt is this: in the soft mode case it can be neglected, whereas in the case $(k, u) \propto \exp(i\omega_k t)$, d/dt is to be replaced by $i(\omega_{k'} + \omega_{k''})$. Equations of the type of equation (19) can now be readily solved. If time and space derivatives are neglected, the solution of (19) is a purely algebraic problem so that one could retain at least in principle also higher order terms in (19) without difficulty. Inserting this result into (18) we obtain the fundamental set of equations for the order parameters

$$
\begin{aligned}
\left(\frac{d}{dt} - \hat{\lambda}_u(\nabla, k)\right)\xi_{k,u} &= \sum_{\substack{k'k''\\u'u''}} a_{kk'k'';uu'u''} I_{kk'k''} \xi_{k'u'} \xi_{k''u''} + \\
&+ \sum_{\substack{k'k''k'''\\u'u''u'''}} \xi_{k'u'} C_{kk'k''k''';uu'u''u'''} \xi_{k''u''} \xi_{k'''u'''} + \tilde{F}_{k,u} \equiv \\
&\equiv H^{(\eta)}_{k,u}(\{\xi_u(x)\}) + \tilde{F}_{k,u} .
\end{aligned}
\tag{21}
$$

The reason why equations (21) are named generalized Ginzburg-Landau equations becomes clear, with the assumption that only one order parameter exists. In some typical cases of practical importance $\hat{\lambda}_u$ can be written in the form

$$\hat{\lambda}(\nabla, k) = \lambda(0) + \gamma\nabla^2; \quad \lambda(0), \gamma \ \text{real} \tag{22}$$

Furthermore, in a number of systems, a = 0 for symmetry reasons. We then find

$$\frac{d\xi}{dt} = (\lambda(0) + \gamma\nabla^2)\xi - b\xi^3 + F(t) \tag{23}$$

for a real ξ, and

$$\frac{d\xi}{dt} = (\lambda(0) + \gamma\nabla^2)\xi - b|\xi|^2\xi + F(t) \tag{24}$$

for a complex ξ. Equation (24) is the famous Ginzburg-Landau equation for describing the phase transition of a superconductor (Landau, Lifschitz, 1952), but now it applies to nonequilibrium phase transitions (Graham, Haken, 1968 a, b; Haken, 1970, 1978). In most cases of practical interest the original equations (2) cannot be written in the form

$$\dot{q} = -\,\mathrm{grad}_q V(q) \tag{25}$$

where V plays the role of a potential. Quite surprisingly, however, in quite a number of explicit examples the order parameter equations (21) can have the form

$$\dot{\xi}_u = -\frac{\partial V}{\partial \xi_u} \tag{26}$$

or, for complex variables,

$$\dot{\xi}_u = -\frac{\partial V}{\partial \xi_u^*} . \tag{27}$$

In all these cases, however, one can gain an overall picture of the possible configurations. To this end we consider the order parameter equations (26) which includes fluctuating forces. Since systems relax slowly at critical points, we may in many cases assume that the F are δ -correlated in time. Furthermore, very often they can be assumed to be a Gaussian distribution. It is possible to write the Fokker-Planck equation associated with equation (26) when it is interpreted as a Langevin equation. When the fluctuating forces obey the relation

$$< F_\mu(t) F_\nu(t') > = Q \delta_{\mu\nu} \delta(t-t') \tag{28}$$

the time-independent Fokker-Planck equation can be solved explicitly. Its solution, which describes the stationary probability distribution, is

$$f(\{\xi_u\}) = N \exp(-2V(\{\xi_u\})/Q) ; \tag{29}$$

where N is a normalization factor and V is identical with that occurring in (26). By determining the global, or local minima, of $V(\{\xi_u\})$ we find the globally or locally stable configurations.

An Application of Generalized Ginzburg-Landau Equations to Morphogenesis

In the development of organisms one observes that totipotent cells can undergo differentiations giving rise to spatial patterns. To explain this phenomenon, the following mechanism has been proposed: the cells may produce two kinds of substances, an activator and an inhibitor (Gierer, Meinhardt, 1974). Both kinds of chemicals can diffuse in the body. At a high enough concentration of the activator, genes are switched on causing cell differentiation. This process is counterbalanced by the inhibitor molecules. The equations for activator concentration, a, and inhibitor concentration, h, are (Gierer-Meinhardt model):

$$\dot{a} = \rho + \frac{a^2}{h} - \mu a + \Delta a \tag{30}$$

$$\dot{h} = a^2 - h + D\Delta h \tag{31}$$

where a and h are expressed as dimensionless variables. The stationary homogeneous solutions of (30) and (31) are

$$a_o = \frac{1}{\mu}(\rho + 1) \tag{32}$$

$$h_o = a_o^2 \tag{33}$$

Normal linear stability analysis yields the eigenvalue equations

$$\lambda^{\pm}(k) = \frac{\alpha(k)}{2} \pm \sqrt{\frac{\alpha^2(k)}{4} - \beta(k)} \tag{34}$$

$$\alpha(k) = -(D+1)k^2 + \frac{2\mu}{\rho+1} - \mu - 1 \tag{35}$$

$$\beta(k) = (k^2 + \mu)(1 + Dk^2) - \frac{2\mu D k^2}{\rho+1} \tag{36}$$

where k is the wave number of the test mode. The condition for the soft mode instability, i. e. non occurrence of oscillations, is

$$\mathrm{Re}\,\lambda^{+}(k) \geq 0\,, \quad \mathrm{Im}\,\lambda^{\pm}(k) = 0\,. \tag{37}$$

A brief analyis of (34) reveals that (37) is fulfilled if

$$(1)\ \alpha < 0\,, \quad (2)\ \beta \leq 0\,. \tag{38}$$

For a critical ρ_c the instability condition, $\beta = 0$, can first be met for two critical values of k, namely $k = +k_c$ and $k = -k_c$. For $\rho = \rho_{max}$ the condition $\alpha = 0$, which indicates the onset of a hard mode instability, can be fulfilled. In the following analysis we will focus our attention on the soft mode case.

Solution of Nonlinear Equations

We assume a two-dimensional layer with side lengths L_1 and L_2 and $\rho \approx \rho_c$. We first adopt periodic boundary conditions and use the hypothesis (12) such that

$$q = \sum_j O^j(\Delta) \sum_k \xi_k^j(t) e^{ikr} \tag{39}$$

The wave vector k is assumed in the form

$$k = 2\pi \begin{pmatrix} \frac{n}{L_1} \\ \frac{m}{L_2} \end{pmatrix} \qquad n, m = 0, \pm 1, \pm 2, \ldots \tag{40}$$

Since the solution must be real it is required that

$$\xi_k^j = \xi_{-k}^{j*}, \qquad \xi_0^+ = \xi_0^{-*}. \tag{41}$$

We now use the order parameter equations (21) and introduce a new notation by which we replace the vector k by its modulus $|k|$, and, by the angle φ which this vector forms with a fixed axis. We let φ run from 0 to π. The resulting equation is

$$\begin{aligned} \dot{\xi}^+_{k_c\varphi} &= \lambda \xi^+_{k_c\varphi} + c\, \xi^+_{k_c, \varphi+\frac{\pi}{3}} \xi_{k_c, \varphi-\frac{\pi}{3}} + \\ &+ \xi^+_{k_c\varphi} \sum_{\varphi'} d(|\varphi-\varphi'|) |\xi_{k_c\varphi'}|^2, \quad 0 \le \varphi \le \pi, \; 0 \le \varphi' \le \pi \end{aligned} \tag{42}$$

where λ and c are constants. The constants, $d(|\varphi-\varphi'|)$, have been evaluated on a computer. Equation (42) represents a set of coupled equations for the time dependent function $\xi^+_{k_c\varphi}(t)$. These equations can be written in the form of potential equations;

$$\dot{\xi} = -\frac{\partial V}{\partial \xi^*_\varphi} \tag{43}$$

where the potential function is given by

$$\begin{aligned} V = -\sum_{\varphi=0}^{\pi} [\lambda |\xi_\varphi|^2 + \frac{c}{3} (\xi_\varphi \xi_{\varphi+\frac{\pi}{3}} \xi_{\varphi-\frac{\pi}{3}} + c.c.) + \\ + \frac{1}{2} |\xi_\varphi|^2 \sum_{\varphi'=0}^{\pi} d(|\varphi-\varphi'|) |\xi_{\varphi'}|^2. \end{aligned} \tag{44}$$

We assume that the resulting pattern is determined by a configuration of ξ such that the potential acquires a local minimum. Thus we have to find ξ such that

$$\frac{\partial V}{\partial \xi^*_\varphi} = 0, \tag{45}$$

and

$$\sum_{\varphi\varphi'} \frac{\partial^2 V}{\partial \xi_\varphi \partial \xi^*_{\varphi'}} \delta \xi_\varphi \delta \xi^*_{\varphi'} > 0 \tag{46}$$

holds. The system is globally stable if

$$d(|\varphi - \varphi'|) < 0 \tag{47}$$

holds or if the matrix $-d(|\varphi - \varphi'|)$ has only positive eigenvalues. We discuss three typical examples;

1. where the entirely homogeneous state for which all ξ_φ are equal 0 is stable for $\lambda < 0$.

2. where we obtain a *roll* pattern for which

$$\xi_{\varphi_1} = x_1 , \tag{48}$$

$$\xi_\varphi = 0 \quad \text{for } \varphi \neq \varphi_1 \tag{49}$$

where

$$x_1^2 = -\frac{\lambda}{d(\pi)} . \tag{50}$$

φ_1 is a fixed angle which can arbitrarily be chosen. This arbitrary choice is typical of a symmetry breaking instability which is well known from phase transition theory (compare Haken, 1978). This configuration is locally stable for

$$\lambda = 0, \quad d(\vartheta \neq \pi) < d(\pi) < 0 . \tag{51}$$

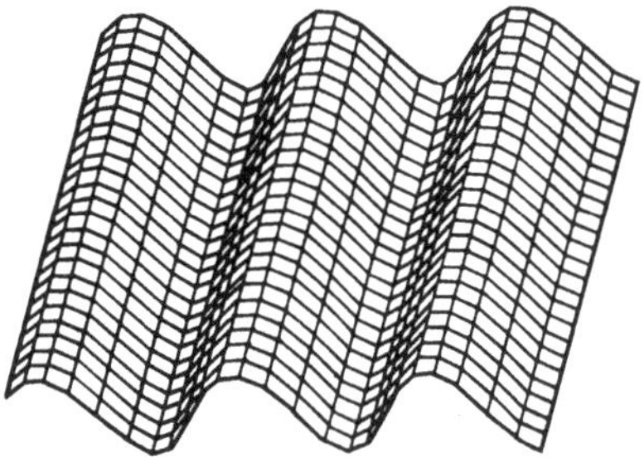

Fig. 15. A roll-type pattern (after Haken and Olbrich - to be published in J. Math. Biol.).

The resulting spatial pattern can be obtained by inserting (48), (49) into (39). The resultant pattern is exhibited in Fig. 15.

3. Another interesting pattern, in which we find hexagons, is realized when the minimum of V is attained for

$$\xi_{\varphi_1} = \xi_{\varphi_1+\frac{\pi}{3}} = \xi_{\varphi-\frac{\pi}{3}} = x_1 \tag{52}$$

or otherwise $\xi_{\varphi} = 0$ (53)

x_1 is given by

$$x_1 = \frac{1}{2F}(-c \pm \sqrt{c^2 - 4F\lambda} \quad \text{and} \quad F = d(\pi) + 2d(\tfrac{\pi}{3}) \tag{54}$$

Again, φ_1 can be arbitrarily chosen. This configuration is locally stable for

$$F_\varphi < F < 0, \quad c x_1 > \frac{c^2}{-F},$$
$$F_\varphi = d(|\varphi-\varphi_1|) + d(|\varphi-\varphi_1+\tfrac{\pi}{3}|) + d(|\varphi-\varphi_1-\tfrac{\pi}{3}|). \tag{55}$$

This approach is also applicable to other boundary conditions. An example illustrating a cylindrical boundary is given in Fig. 16.

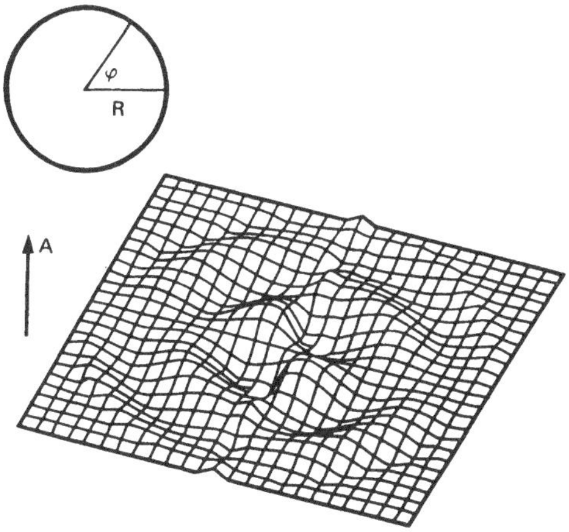

Fig. 16. The activator concentration with cylindrical non-flow boundary conditions described by a rotation symmetric Bessel function with m = 2 (after Haken and Olbrich - to be published in J. Math. Biol.).

Nonequilibrium Phase Transitions and Bifurcation of Limit Cycles

We know from explicit examples (morphogenetic fields, lasers, fluids, chemical reactions etc.) that newly formed structures can become unstable again when external parameters are further changed. This leads to the interesting and important question, can the order parameter notion and the slave principle be extended to situations where the new structure evolves from a different one which is spatially inhomogeneous or is oscillating (or both)? I have shown quite recently that this extension is indeed possible in a quite remarkable way. Since the corresponding mathematics is rather involved, I shall only sketch the basic ideas. Starting from nonlinear differential equations, we assume that for a certain set of control parameters there exists a stable limit cycle solution (i. e. a periodic solution), or a motion on a torus (multiperiodic motion). We then assume that by change of a control parameter that motion becomes unstable. We now perform a stability analysis. Due to the Floquet theorem, or tis generalization to multiperiodic motion, we can again distinguish between stable and unstable modes. But, in addition to our previous development we must include phase shifts along the limit cycle, or, along the torus by phase angles which are introduced as additional variables serving as part of the order parameters. It is then possible to eliminate the stable modes by a procedure which converges rapidly provided the order parameters are still small and the damping constant of the damped modes sufficiently large. The resulting order-parameter equations which include the phases are capable of describing new large classes of higher bifurcation schemes, i. e. new classes of nonequilibrium phase transitions and the evolution of multiperiodic processes.

Concluding remarks

In the foregoing I outlined a method showing to cope with complex systems in which many processes may proceed. The result is that close to situations where the processes change qualitatively the behaviour of a complex system can be treated in terms of a few characteristic variables, the so called *order parameters*. As long as quantum effects can be neglected this approach applies both to processes which lead to dissipative structures, as well as to those which give rise to phenomena closely related to phase transitions in systems at thermal equilibrium (we are here dealing only with finite systems). The methods seem, therefore, to be adequate for treatment of the establishment of functional structures in biology. On the other hand the procedure describes only a first step in a more complex approach. In future some modifications will be required e. g. the slave principle, introduced above, requires in most cases that the order parameters remain sufficiently small. We have studied cases in which order parameters become so large that the slave principle breaks down. Explicit examples show that in such cases so called chaotic motion occurs under certain circumstances in which the system jumps, more or less randomly, between two, or several, states each of which is described by an individual order parameter. In this sense we find an exchange of order parameters. That is the system switches from one order parameter to another. There are other processes in which, after a transition, the first order parameter disappears and is replaced by another. Furthermore we observe that the present approach allows us to treat hierarchical systems in which the order parameters of a lower level are mirrowed by order parameters of a higher level. Systems may also be cross linked, for instance the order parameter of one system may act as the control parameter of another. In general, it appears that the order parameter concept and the slave principle allow us to bring a new ordering principle into the behaviour of complex systems such as are met in biology, where many processes are proceeding simultaneously. This approach decomposes complex systems and complex processes into certain hierarchical levels and allows us to study their interaction.

Brownian Motion in Autooscillation Systems and in Phase Transition Processes

Yu. L. Klimontovich

In recent years in the investigation of a number of physical, chemical and biological problems, in particular the problem of self-organization in non-equilibrium systems, attention has been paid to the analogy of phase transitions and transitions over the generation threshold in autooscillation system-generators (Haken, 1975 a). Natural questions arise here, is such an analogy useful? Does it permit the use of the ideas and methods of phase transition theory for the development of the theory of fluctuations in the development of the theory of fluctuations in autooscillation systems? Is the reverse procedure useful?

The most interesting problems in modern phase transition theory are concerned with the investigation of fluctuations in some critical region. To study fluctuations in autooscillation systems is also of considerable interest. Fluctuations determine, in particular, the oscillation energy distribution within the energy spectrum.

Random oscillations which appear in autooscillation systems may be regarded as a kind of Brownian motion. Random processes, for example, order parameter fluctuations in phase transitions may also be regarded as Brownian motion. This is one of the unifying features of these processes. However, in spite of a certain analogy, the processes of Brownian motion in generators and in phase transitions are essentially different. To visualize this we shall consider three relatively simple examples:

1. Brownian motion in the classical Thomson-type generator;

2. Brownian motion in phase transition in a classical system of nonlinear atoms-oscillators, the interaction of which is achieved by spatial diffusion;

Thermodynamics and Kinetics of Biological Processes

3. Brownian motion in a phase transition in classical (and quantum) systems of atoms interacting through a fluctuating electromagnetic field.

It is clear that in this paper we can present only some of the results. A more detailed description of the problems described above may be found in the literature cited.

Brownian Motion in the Generator

(a) Langevin-type Equations

The fluctuation process in the Thomson-type generator is described by a set of Langevin-type equations for a charge x(t) and a current v(t) (Stratonovich, 1961; Klimontovich, 1980)

$$\frac{dx}{dt} = v; \quad \frac{dv}{dt} + \left(a + \frac{2}{3} b v^2\right) v + \omega_0^2 x = y(t). \tag{1}$$

Here ω_0 is eigenfrequency, a and b are constants. In the generation regime the quantity a is given as a $<$ 0 and on the threshold of generation a = 0. Moments of Langevin's source y(t) are defined by the expressions

$$\langle y(t) \rangle = 0, \quad \langle y(t) y(t') \rangle = 2 \mathcal{D} \delta(t - t'). \tag{2}$$

Thus, the noise is assumed to be of the same intensity, $\mathcal{D}$, over the entire frequency range. In practice, it is sufficient that the noise has a wide band $(\Delta\omega \gg |a|$ near $\omega = \omega_0$.

To derive an exact solution to the set of equations (1), even in the dynamic regime, when y = 0, is a very complicated problem. The equations become much simpler if the relaxation time considerably exceeds the period of self oscillations. In this case the variables x(t), v(t) may be replaced by slowly changing functions defined by the equalities

$$x = \tilde{x} \cos \omega_0 t + \frac{\tilde{v}}{\omega_0} \sin \omega_0 t; \quad v = \tilde{v} \cos \omega_0 t - \omega_0 \tilde{x} \sin \omega_0 t. \tag{3}$$

After averaging over the period $2\pi / \omega_0$ the Langevin equations for the functions $\tilde{x}$, $\tilde{v}$ take the form

$$\frac{d\tilde{x}}{dt} + \frac{1}{2}(a + bE)\tilde{x} = -\frac{1}{\omega_0}\overline{y(t)\sin\omega_0 t}^{T} \tag{4}$$
$$\frac{d\tilde{v}}{dt} + \frac{1}{2}(a + bE)\tilde{v} = \overline{y(t)\cos\omega_0 t}^{T}; \quad E = (\tilde{v}^2 + \omega_0^2\tilde{x}^2)/2 .$$

(b) Fokker-Planck Equation

The equation for the distribution function of the variables $\tilde{x}$, $\tilde{v}$ (the sign " $\sim$ " is omitted later on), which correspond to the set of the Langevin equation (4), has the form

$$\frac{\partial f(x,v,t)}{\partial t} = D\left(\frac{1}{\omega_0^2}\frac{\partial^2 f}{\partial x^2} + \frac{\partial^2 f}{\partial v^2}\right) + \\ + \frac{1}{2}\frac{\partial}{\partial x}[(a + bE)xf] + \frac{1}{2}\frac{\partial}{\partial v}[(a + bE)vf] . \tag{5}$$

This is just the Fokker-Planck equation for the system under consideration.

With the aid of equation (5) one can obtain equations for other distribution functions, for example, for the distribution function of the amplitude and phase values of the values of energy E.

We shall present as an example the expression for the distribution function f(E) in a stationary state

$$f(E) = C\exp\left(-\frac{aE + bE^2/2}{D}\right), \quad \int_0^\infty f(E)dE = 1 . \tag{6}$$

From this there also follows the corresponding distribution of the values of the amplitude f(A).

With the aid of the distributions f(E), f(A) one can find energy and amplitude moments at any values of the parameters, including those on the threshold of generation, when $a = 0$. In particular, for the region of developed generation (here one can use the correlation approximation) the amplitude fluctuation dispersion is defined by the expression

$$\langle (\delta A)^2 \rangle = \frac{D}{2|a|} . \tag{7}$$

From this expression it follows that when approaching the threshold of generation (on the threshold $a = 0$) the fluctuation dispersion increases. At this point there is an analogy with the results of the phase transition theory developed by Landau. Indeed, if transition of the generation threshold is associated with transition of a critical point ($a \leftrightarrow (T-T_c)/T_c$) (see below), then the behaviour of amplitude fluctuations at a $\rightarrow$ 0 will correspond to the behaviour of the dispersion of the order parameter $\langle (\delta\eta)^2 \rangle$ at $T \rightarrow T_c$. However, at low values of a the correlation approximation is broken. The calculation of dispersion with the aid of the distribution function f(A) shows, that, although fluctuations increase when approaching the generation threshold, all the moments are finite at any values of the parameters, including those on the threshold of generation (a = 0).

Brownian Motion under Phase Transition - Interaction Through Diffusion

(a) Initial equations. Landau theory (Landau, Lifshitz, 1976)

Consider a classical system of one-dimensional atoms-oscillators with the eigenfrequency ω_0. These atoms are connected by means of spatial diffusion. The equation for the polarization vector, $P(\vec{R},t) = enx(\vec{R},t)$, of this system with account taken of an effective Lorentz field and allowing anharmonicity of oscillations characteristic of the description of slow processes, may be written in the form (Ma, 1976; Hohenberg, Halperin, 1977; Klimontovich, 1980)

$$\frac{\partial x}{\partial t} + (a + bx^2)x - g\frac{\partial^2 x}{\partial \vec{R}^2} = F(\vec{R},t) + y(\vec{R},t). \qquad (8)$$

The quantity a near the critical point may be represented as

$$a = a_0\tau \equiv a_0\frac{T-T_c}{T_c}, \qquad (9)$$

where a, b, g are constant quantities.

At the phase transition point a = 0. Moments of the Langevin source are determined by the expressions

$$\langle y \rangle = 0; \quad \langle y(\vec{R},t)\,y(\vec{R}',t')\rangle = 2\frac{D}{n}\delta(R-R')\delta(t-t')$$
$$D = \frac{kT}{m\gamma}, \qquad n = \frac{N}{V}. \qquad (10)$$

Expanding the function $x(\vec{R},t)$ in a Fourier series yields

$$x(\vec{R},t)=\sum_{\vec{k}} x_{\vec{k}}(t)\,e^{i\vec{k}\vec{R}};\quad x_{\vec{k}}(t)=\int x(\vec{R},t)\,e^{-i\vec{k}\vec{R}}\frac{d\vec{R}}{V} \tag{11}$$

(V is the volume of the system) and the corresponding equation for the function $x_{\vec{k}}(t)$ is

$$\frac{\partial x_{\vec{k}}(t)}{\partial t}+(a+gk^2)\,x_{\vec{k}}+b(x^3)_{\vec{k}}=y_{\vec{k}}(t). \tag{12}$$

We have used the designation

$$(x^3)_{\vec{k}}=\sum_{\vec{k}_1,\vec{k}_2,\vec{k}_3}\delta_{\vec{k},\vec{k}_1+\vec{k}_2+\vec{k}_3}\,x_{\vec{k}_1}x_{\vec{k}_2}x_{\vec{k}_3}\;. \tag{13}$$

Moments of the random source $y_{\vec{k}}(t)$ are defined by the expressions

$$\langle y_{\vec{k}}\rangle=0;\ \langle y_{\vec{k}}(t)\,y_{\vec{k}}(t')\rangle=2\frac{\mathcal{D}}{N}\,\delta(t-t');\ \mathcal{D}=\frac{kT}{m\gamma}. \tag{14}$$

We shall use, according to Bogolyubov (1971), alongside the usual averages the so-called quasi-averages; designated as $\langle\ \rangle_{\vec{F}=0}$. The designation adopted implies that to obtain a quasi-average value one should first average at $\vec{F}\neq 0$, ($\vec{F}$ is the mean force) and then make the limiting transition.

Designation η for the order parameter requires that

$$\langle x_{\vec{k}}\rangle_{\vec{F}=0}=\eta\,\delta_{\vec{k},0} \tag{15}$$

Thus, the quasi-averages are assumed, here, when $\vec{k}=0$ to be non-zero for the normal mode only. Disregarding fluctuations the order parameter is determined by the expression

$$\eta=\sqrt{\frac{|a|}{b}}\;. \tag{16}$$

Using the equation for the fluctuation

$$\delta\eta_{\vec{k}}=x_{\vec{k}}-\eta\,\delta_{\vec{k},0} \tag{17}$$

which follows from (12) and (16) one can, restricting oneself to the linear approximation, find the basic characteristics: dynamic and static susceptibilities;

$$\chi(\omega,\vec{k})=\frac{1}{-i\omega+a+3b\eta^2+gk^2};\ \chi_{st}(\vec{k})=\chi(0,\vec{k}) \tag{18}$$

the expression for the order parameter dispersion;

$$\langle(\delta\eta)^2\rangle = \frac{\mathcal{D}}{N}\frac{1}{a+3b\eta^2} = \chi(0,0)\frac{\mathcal{D}}{N}\,, \tag{19}$$

and the spatial fluctuation correlation,

$$\langle(\delta\eta)^2\rangle_r = \frac{V\langle(\delta\eta)^2\rangle}{4\pi r_c^2 r}e^{-r/r_c}. \tag{20}$$

The designation for the correlation radius is

$$r_c = \sqrt{\frac{g}{a+3b\eta^2}} = \sqrt{g\chi(0,0)}. \tag{21}$$

The designations exploited here differ from those introduced by Landau and Lifshitz (1976). To construct the formulae corresponding to those of Landau and Lifshitz we need to use the conformity conditions

$$en\eta \leftrightarrow \eta_{\Lambda-\Lambda};\ \frac{e^2n}{m\gamma}\chi \leftrightarrow \chi_{\Lambda-\Lambda}.$$

Then, for example,

$$\langle(\delta\eta)^2\rangle_{\Lambda-\Lambda} = e^2n^2\frac{\mathcal{D}}{N}\chi = \frac{kT}{V}\chi_{\Lambda-\Lambda}.$$

From these it follows that two characteristic dimensionless parameters should be introduced. The first of them, ε_N, characterizes the value of the relative dispersion i. e.

$$\frac{\langle(\delta\eta)^2\rangle}{\eta^2} \sim \frac{\mathcal{D}b}{Na^2} = \varepsilon_N, \quad a = a_0|\tau|. \tag{22}$$

At a finite deviation from the critical point ($|a| \neq 0$) the parameter in the thermodynamic limit vanishes.

The second parameter, G_i, is responsible for the ratio of the correlation function (20) to the order parameter squared η^2 : at the point $r = r_c$. At this point the function $\langle(\delta\eta)^2\rangle_r r^2$ has its maximum value

$$G_i \sim \frac{\langle(\delta\eta)^2\rangle_{r=r_c}}{\eta^2} \sim \frac{V}{r_c^3}\frac{\langle(\delta\eta)^2\rangle}{\eta^2} \sim \frac{V}{r_c^3}\varepsilon_N. \tag{23}$$

The parameter G_i will be called Ginzburg parameter (Patashinsky, Pokrovsky, 1975, 1977). By definition, Ginzburg parameter is finite at $|a \neq 0|$ in the thermodynamic limit.

Formulae (18) and (21) show that as $T \to T_c$ the static susceptibility and the correlation radius tend to infinity. As a consequence of this, according to (19), when approaching the critical point (at a finite N, i. e. before the thermodynamic limiting transition) the order parameter dispersion also tends to infinity. It is natural to expect, however, that at the critical point the quantities $\chi(0,0)$, and r_c, may tend to infinity only after the thermodynamic limiting transition, when $N \to \infty$, $V \to \infty$, but $\frac{N}{V} = const$.
Since in the critical region the fluctuations are not small to obtain appropriate results one needs to overstep the limits of the approximation in question.

(b) Extension of the Results of Landau Theory to the Critical Region

Since in the critical region $\delta\eta \sim \eta$, i. e. the values of the fluctuations are of the same order as the values of the order parameter itself, then for the extension of formulae (18), (19), and (21) to the critical region, it is most natural to substitute in them the mean value of the quantity, η^2, for the order parameter $x^2_{\vec{k}=0}$, i. e.

$$\eta^2 \to \langle x^2_{\vec{k}=0} \rangle \equiv \langle E \rangle . \tag{24}$$

Designating the corresponding new characteristics by $\tilde{\rho}$, $\tilde{r}_c$, $\langle (\delta\eta)^2 \rangle$ then for example,

$$\tilde{\chi}(0,0) = \frac{1}{a + 3b \langle x^2_{\vec{k}=0} \rangle} . \tag{25}$$

To determine $\langle E \rangle$ we make use of the Langevin equation (12), and write the equation for the function $E = x^2_{\vec{k}=0}$. To obtain a qualitative result in the equation for E we take into account the contribution from the natural mode only. As a result we derive the following equation

$$\frac{1}{2}\frac{dE}{dt} + (a + bE)E = y_E ; \quad y_E = x_{\vec{k}=0}\, y_{\vec{k}=0}(t). \tag{26}$$

Whence, taking into account the equality $\langle y_E \rangle = \mathcal{D}/N$, we find the desired equation

$$a \langle E \rangle + b \langle E^2 \rangle = \frac{\mathcal{D}}{N} . \tag{27}$$

Naturally, this equation is not closed since it includes, besides the function $\langle E \rangle$, the higher moment $\langle E^2 \rangle$. A very simple approximation, which at the same time takes into account pecularities of the behaviour in the critical point, is

$$\langle E^2 \rangle = \langle E \rangle \langle E \rangle \tag{28}$$

which leads to the closed equation

$$(a + b \langle E \rangle) \langle E \rangle = \frac{\mathcal{D}}{N} . \tag{29}$$

Here there are two limiting cases: a zero approximation in ε_N ; and for high temperatures when $a > 0$ and $b = 0$ we find that

$$\langle E \rangle = \eta^2 = \frac{|a|}{b} ; \quad \langle E \rangle = \frac{\mathcal{D}}{aN} = \frac{kT}{Nm\omega_0^2} . \tag{30}$$

At the critical point, when $a = 0$

$$\langle E \rangle = \sqrt{\frac{\mathcal{D}}{bN}} \sim N^{-\frac{1}{2}} \sim V^{-\frac{1}{2}} . \tag{31}$$

This result is valid, of course, only for the order of magnitude.

With the aid of formulae (31), (25) and (21) we can find the dependence of the quantities $\tilde{\chi}$, $\tilde{r}_c$ on N on the critical point

$$\tilde{\chi}(0,0) \sim \sqrt{\frac{N}{\mathcal{D}b}} \sim N^{1/2} ; \quad \tilde{r}_c = \sqrt{g\tilde{\chi}} \sim \left(\frac{Ng}{\mathcal{D}b}\right)^{\frac{1}{4}} \sim N^{\frac{1}{4}} . \tag{32}$$

From the latter expression it follows that at the critical point the ratio of the volumes is

$$\frac{\tilde{r}_c^3}{V} \sim N^{-\frac{1}{4}} \sim V^{-\frac{1}{4}} \tag{33}$$

and, consequently, is equal to zero in the thermodynamic limit.

On the contrary, the corresponding values for Ginzburg parameter $\tilde{G}_i$ in the critical point are

$$\tilde{G}_i = \frac{V}{\tilde{r}_c^3} \frac{\langle(\tilde{\delta\eta})^2\rangle}{\langle E\rangle} \sim \frac{V}{r_c^3} \sim N^{\frac{1}{4}} \sim V^{\frac{1}{4}} \tag{34}$$

and, consequently, this tends to infinity in the thermodynamic limit. In deriving this result account has been taken of the result from (19), (31) and (32), for the critical point, that

$$\frac{\langle(\tilde{\delta\eta})^2\rangle}{\langle E\rangle} \sim N^{0}. \tag{35}$$

We see that extension of the results of Landau theory to the critical region leads to unsatisfactory results. It is cureous, however, that this contradiction is eliminated at $d = 4$, i. e. for four-dimensional space. In fact, in this case both the relations (33) and (34) in the thermodynamic limit prove to be finite.

We shall see that we can avoid this contradiction (at d = 3) if we impose the requirement of consistency in two limiting transitions: a limiting transition $T \to T_c$ and a thermodynamic limiting transition. The requirement of consistency of these limiting transtions will lead to re-normalization of the basic characteristics of the system.

(c) Critical Indices. Hypothesis of Scale Invariance

According to the considerations of modern theory of fluctuation in phase transitions the dependence of the quantities η, χ, r_c on $|\tau|$, as in Landau theory, obey a power-law of one, for example

$$\eta \sim |\tau|^{\beta}; \ \chi \sim |\tau|^{-\gamma}; \ r_c \sim |\tau|^{-\nu}. \tag{36}$$

However the values of the exponents - critical indices - differ essentially from the corresponding indices in Landau theory.

Eight indices are altogether introduced: $\alpha, \beta, \gamma, \delta, \nu, \varepsilon, \mu, \xi$ (Landau, Lifshitz, 1976; Balescu, 1975). They are connected by five general relations.

$$2\beta + \gamma = 2 - \alpha; \quad \beta\delta = \beta + \gamma; \quad \varepsilon(\beta + \gamma) = \alpha;$$
$$\mu(\beta + \gamma) = \nu; \quad \nu(2 - \xi) = \gamma. \tag{37}$$

They are, naturally, valid in Landau theory, too. The critical indices in Landau theory have the values

$$\alpha_\Lambda = 0; \ \beta_\Lambda = \frac{1}{2}; \ \gamma_\Lambda = 1; \ \delta_\Lambda = 3; \ \nu_\Lambda = \frac{1}{2}; \ \varepsilon_\Lambda = 0, \ \mu_\Lambda = \frac{1}{3}, \ \xi_\Lambda = 0. \tag{38}$$

A sixth relation for the critical indices may be derived with the aid of the scale invarience hypothesis suggested by Kadanov, Patashinsky and Pokrovsky (Balescu, 1975; Patashinsky, Pokrovsky, 1975; Landau, Lifschitz, 1976). According to this hypothesis, in the critical region the only length parameter characterizing fluctuations is the correlation radius. The corresponding relation among the critical indices has the form

$$\nu d = 2 - \alpha \quad (\text{or } \nu d = \gamma + 2\beta). \tag{39}$$

Exploiting relations (37) and (39) one can express eight critical indices through two independent indices. At present, several methods of approximate calculation of critical indices for the different models have been developed. Wilson and Kogut (1975) and Fisher (1974) were the first to carry out such calculations by the method of the so-called ξ -expansion. The results of the calculation of critical indices for a three-dimensional Ising model by the method of collective variables are presented in the review by Yukhnovsky (1978).

The calculations performed have shown that two of the eight critical indices (namely, α and ξ) are much smaller than the others. Approximately $\alpha = 0$ and $\xi = 0$ in the six relations (37) and (39) then the values of all the other indices may be found. For the three-dimensional space (d = 3) they have the following values

$$\beta = \frac{1}{3}, \gamma = \frac{4}{3}, \nu = \frac{2}{3}, \delta = 5, \varepsilon = 0, \mu = \frac{2}{5} . \qquad (40)$$

The dependence of the parameters η, χ, r_c on $|\tau|$ established by formulae (36) with the critical indices (40) is valid only in the vicinity of the critical point, (at which the correlation radius is much less than the dimension of the system). Only under this condition is the scale invariance hypothesis valid. In other words, the relations described by (36) can be established under the condition that the limiting thermodynamic transition is already over.

To obtain, at least qualitatively, results valid at the critical point itself, both limiting transitions (the transition $T \to T_c$ and the limiting thermodynamic transition) must conform. We shall show that for the case where the conformity conditions are fulfilled, one can obtain the results for any temperature including the critical point itself. When moving away from the critical point towards the region where the parameter ε_N is small, i. e. under the thermodynamic limiting transition we find a region of scale invariance, for which the dependences (36) on the critical indices (40) hold. However, the scale invariance hypothesis is not used in this case.

Proceeding from two conditions (Klimotovich, 1979), namely

1. In the thermodynamic limit at the critical point ($T = T_c$) the correlation volume must increase in proportion to the volume of the system, i. e.

$$\frac{r_c^{*3}}{V} \sim V^0 \sim N^0 . \qquad (41)$$

The sign "*" signifies re-normalized characteristics.

2. In the thermodynamic limit a re-normalized Ginzburg parameter, G_i^*, must remain finite at $T \to T_c$, i. e.

$$G_i^* \sim V^0 \sim N^0 . \qquad (42)$$

We assume first that the relation between the re-normalized parameters r_c^*, χ^* is of the form of (32), i. e.

$$r_c^* \sim \sqrt{g\chi^*}. \tag{43}$$

This relation imposes the condition that $\xi = 0$. Thus, we first consider the approximation when the small indices are equal to zero. More general conditions are considered below.

The first of the conditions presented makes it possible to specify one of the critical indices. In fact, according to (41) the following relation for dimensionless quantities can be written

$$n(r_c^*)^3 \sim (n\tilde{r}_c^3)^{2\nu} \infty V. \tag{44}$$

The convenience of this way of introducing the index ν will be clear later on.

Using relation (32) for $\tilde{r}_c$ and making the exponents of V on the left- and right-hand sides equal, yields

$$\nu = \frac{2}{3} \tag{45}$$

and, consequently,

$$r_c^* \sim n^{\frac{1}{3}} \tilde{r}_c^{4/3}. \tag{46}$$

Employing the second condition we represent the parameter, G_i, as the product of three dimensionless parameters

$$G_i \sim \frac{1}{nr_c^3} \frac{\chi\sqrt{\mathcal{D}b}}{\eta^2\sqrt{b/\mathcal{D}}}. \tag{47}$$

Leaving aside the quantities $\tilde{r}_c$, $\tilde{\chi}$, $\langle E \rangle$ (see (25)) and defining the re-normalized Ginzburg parameter by the expression

$$G_i^* \sim \left(\frac{1}{nr_c^{*3}}\right)^{2\nu} \frac{(\tilde{\chi}\sqrt{\mathcal{D}b})^\gamma}{(\langle E \rangle\sqrt{b/\mathcal{D}})^{2\beta}} \tag{48}$$

requires the introduction of two new critical indices γ, β.

In accordance with the requirement that the parameter G_i^* be finite at the critical point and the thermodynamic limiting transition, we find the following relation among three critical indices

$$3\nu = \gamma + 2\beta . \tag{49}$$

At d = 3 this coincides with equality (39) and, consequently, corresponds to the scale invariance condition.

Finally, from (43) one more relation follows

$$2\nu = \gamma . \tag{50}$$

From the equalities (45), (49) and (50) we obtain values (40) for the critical indices β, γ, ν and, therefore, for the same dimensions

$$\tau^* \sim n^{1/9}\,\tilde{\tau}^{4/3} ; \quad \chi \sim (\mathcal{D}b)^{1/6}\,(\tilde{\chi})^{4/3} ; \quad \langle E \rangle^* \sim \left(\frac{\mathcal{D}}{b}\right)^{1/6} \langle E \rangle^{2/3} . \tag{51}$$

Thus, on the basis of the conditions of the conformity of the two limiting transitions, and making the exponents of V (or N) equal we obtain the values of three critical indices under the approximation $\alpha = 0$, $\xi = 0$. We shall show below that, far from the critical point, (under the expansion in the parameter ε_N) the dependences (36) follow from the general formula.

In the general case, when small critical indices are nonzero, relation (49) remains unchanged, and, in place of (44), (50) and (43) there arise more general relations

$$3\nu = 2 - \alpha , \quad (2 - \xi)\nu = \gamma . \tag{52}$$

To obtain the second of these relations we need to replace the condition of a finite value for the re-normalized Ginzburg parameter at the critical point by a more general condition of finite value, at the critical point, of spatial correlations at any value of τ (so long as $\tau < \tau_c$). Substitution in (20) produces

$$\frac{V\langle(\delta\eta)^2\rangle}{\tau_c^2\,\eta} = \frac{n\mathcal{D}\chi}{\tau_c^2\eta} \longrightarrow \frac{n\mathcal{D}\chi^*}{(\tau_c^*)^{2-\xi}\,\tau^{1+\xi}} \; . \tag{53}$$

Here the new critical index characterizes the deviation from the one half dependence.

Assuming in (52), $\chi^* \sim \tilde{\chi}^{\nu}$, $\tau_c^* \sim (\tilde{\tau}_c)^{2\nu}$ and taking into account the dependence of these quantities on N (see (32)), we obtain, from the condition of finite values for correlations at the critical point at any τ (but at $\tau < \tau_c$), the second of the relations (52). This substitutes for the former condition (50). At $\xi = 0$ both the relations, naturally, coincide.

Since, in the case of $\alpha \neq 0$, $\xi \neq 0$, the values of the indices depend on the details of the model (in particular, on *spin* dimensions) the above qualitative consideration proves insufficient.

(d) Critical Point

The temperature range, within which the relation (51) is valid, may be divided into two sub-ranges. One of them includes the critical point, and the nearest neighbourhood where the parameter, ε_N , is not small.

At the critical point, making use of formulae (31) and (32) we obtain from (51) the following expressions

$$\tau_c^* \sim n^{1/9}\left(\frac{g^2N}{\mathcal{D}b}\right)^{1/3} ; \; \chi^* \sim \frac{N^{2/3}}{\sqrt{\mathcal{D}b}} ; \; \langle E^* \rangle \sim \sqrt{\frac{\mathcal{D}}{b}}\, N^{-1/3} . \tag{54}$$

In accordance with these formulae we obtain, for the re-normalized Ginzburg parameter, the following expression

$$G_i^* \sim \frac{\mathcal{D}b}{g^2 n^{4/3}} \tag{55}$$

Naturally it does not depend on V (or N) but depends only on n .

Expression (55) defines the new dimensionless parameter which characterizes the values of correlations at the critical point.

(e) The Region of Scale Invariance
(Power-Law Dependences of r_c^*, χ^*, η^* on τ)

Consider a second sub-range so remote from the critical point that the parameter

$$\varepsilon_N = \frac{Db}{Na^2} \ll 1\,; \quad i.e.\ \tau^2 \gg \frac{Db}{Na_0^2}\,. \tag{56}$$

According to equation (30), in the zero approximation with respect to the parameter ε_N, the quantity $\langle E \rangle = \eta^2$ and, therefore, the formulae for the quantities $\tilde{\chi}$, $\tilde{r}_c$ coincide with the corresponding expressions from Landau theory, i. e. with formulae (18) and (21). Thus, at $a<0$

$$\tilde{\chi} = \chi = \frac{1}{2a_0|\tau|}\,, \quad \tilde{r}_c = r_c = \sqrt{\frac{g}{2a_0|\tau|}}\,, \quad \langle E\rangle = \eta^2 = \frac{|a|}{b} \tag{57}$$

and, therefore, relations (51) take the form

$$\chi^* \sim \frac{(Db)^{1/6}}{(2a_0)^{4/3}}|\tau|^{-4/3}; \quad \eta^* \sim \left(\frac{a_0}{b}\right)^{1/3}\left(\frac{D}{b}\right)^{1/12}|\tau|^{1/3}; \quad r_c^* \sim n^{1/9}\left(\frac{g}{2a_0}\right)^{2/3}|\tau|^{-2/3}. \tag{58}$$

Thus, we have obtained the desired power-law dependences (36).

Formulae (58) refer to the region below the critical point ($a<0$). For $a>0$ one should use the expressions

$$\tilde{\chi} = \frac{1}{a}\,, \quad \tilde{r}_c = \sqrt{\frac{g}{a}}\,, \quad \langle E\rangle = \frac{D}{Na} \tag{59}$$

instead of (57). These follow from formulae (25) and (29) at $a>0$, $b=0$. As a result the formulae for χ^*, r_c^* take the form

$$\chi^* \sim \frac{(Db)^{1/6}}{a_0^{4/3}}\tau^{-4/3}\,, \quad r_c^* \sim n^{1/9}\left(\frac{g}{a_0}\right)^{2/3}\tau^{-2/3}. \tag{60}$$

It is readily seen that the values of the critical indices, above and below the critical point, coincide. This result is in agreement with the general conclusions of general phase transition theory.

As follows from formulae (48) and (57) the parameter G_i^*, for the sub-range under consideration, is again defined by expression (55).

(f) Transition to the Landau Theory Approximation

As the temperature difference $|T - T_c|$ increases, there appears an intermediate region between the region of scale invariance and that in which Landau theory is valid.

Since Landau theory corresponds to the self-consistent field approximation (when correlations are considered to be ngeligibly small), then as a parameter characterizing the intermediate region it is natural to use the one associated with the Ginzburg parameter. It is designated by ε_B (Wilson parameter) and defined as follows

$$\varepsilon_B = \frac{G_i}{1 + G_i} \; . \tag{61}$$

Since the parameter G_i is small at large $|\tau|$ and tends to infinity, like $|\tau|^{-1/2}$, when approaching the critical point, the parameter ε_B varies in the limits

$$1 \geqslant \varepsilon_B \geqslant 0 \, . \tag{62}$$

In this case the values approaching unity correspond to large values of G_i i. e. the states lie in the scale invariance region. Small values of ε_B correspond to small values of the parameter G_i. So, we are led to the region of applicability of Landau theory.

Thus, the condition of applicability of Landau theory (Landau, Lifshitz, 1976) takes the form

$$G_i \ll 1 \qquad \text{or} \qquad a = a_0|\tau| \gg \frac{\mathcal{D}^2 b^2}{n^2 g^3} \, . \tag{63}$$

(g) Connection of the Critical Indices ν, γ, β with the Corresponding Critical Indices of Landau Theory

One of the major achievements of the modern theory of fluctuation in phase transitions is the method of calculation of critical indices suggested by Wilson.

The calculation of critical indices is based on the values of indices in four dimensions (d = 4) and coincide with the critical indices derived from Landau theory. The unknown indices are calculated in the space with the number of dimensions, $d = 4 - \varepsilon_B$, under the condition $\varepsilon_B \ll 1$ and are represented as series in ε_B. In final analysis the parameter, ε_B, is set equal to unity. By this a formal transition to ordinary three-dimensional space is achieved since $d = 4 - \varepsilon_B = 3$ at $\varepsilon_B = 1$.

On the basis of the preceding we shall try, by using the parameter ε_B (see (61), (62)), to establish (on a qualitative level) the connection of the critical indices ν, γ, β with the corresponding critical indices of Landau theory.

In view of this we should note that, as has already been mentioned, the region of applicability of Landau theory is much wider in four dimensions (d = 4). Indeed, at d = 4 the correlation volume should be determined as r_c^4. Then taking (32) into account, we are led to

$$\frac{\tilde{r}_c^4}{V_{d=4}} \sim \frac{N}{V} \frac{g^2}{D b} \tag{64}$$

and, consequently, this ratio is finite in the thermodynamic limit (cf. (33)). This very fact was known long ago. It cannot, of course, be the ground of Wilson's ε-expansion but it suggests a way of establishing the correlation of the critical indices, ν, γ, β, in real three-dimensional space with the corresponding values of the critical indices, ν_1, γ_1, β_1, of Landau theory (see (38)). This implies the substitution, in formulae (39) (at α = 0), of the quantity, d, for $4 - \varepsilon_B$. This makes it possible to represent the critical indices ν, γ, β in the form:

$$\nu = \frac{2}{4-\varepsilon_B}, \quad \gamma = \frac{4}{4-\varepsilon_B}, \quad \beta = \frac{2-\varepsilon_B}{4-\varepsilon_B}. \tag{65}$$

At $\varepsilon_B = 1$, the values of these indices coincide with (40) and, at $\varepsilon_B = 0$, with the corresponding values of the critical indices in Landau theory. Intermediate values of the indices correspond to the stage intermediate between the region where the scale invariance hypothesis is valid and the one in which the conditions of applicability of Landau theory are met.

This account, naturally, only gives a qualitative account of the character of fluctuations in the critical region. In particular, formulae (65) provide only very slight insight into the connection of the critical indices in the scale invariance region with those of Landau theory. In this connection a more consistent analysis of the initial equation (8) is of undoubted importance.

Now let us turn to the consideration of the last example, mentioned in the introduction.

Brownian Motion under Phase Transition in a Classical System of Atoms-Oscillators Interacting Through Fluctuational Electromagnetic Field

We have seen that for a system of atoms interacting by means of spatial diffusion in the critical region Landau theory is not valid. Several examples are known at present which demonstrate that the region of applicability of Landau theory (or the mean field theory) depends, essentially, on the character of particle interaction in the system. The systems may roughly be divided into two classes. For systems of the first type there exists a comparatively wide temperature range, within which the values of the critical indices differ essentially from those of Landau theory. These are systems described by the Ginzburg-Landau equation for a random source. Our equation (8) serves as an example.

In systems of the other type, Landau theory is valid almost up to the critical region. The word *almost* implies that, in the thermodynamic limit, the width of the region where Landau theory is not valid becomes zero. A system of this kind is the one described by the Hamiltonian of Bardeen-Cooper-Schrieffer (Lifshitz, Pitayevsky, 1978).

Indeed, Bogolyubov, Zubarev and Tserkovnikov (Bogolyubov et al., 1960; Bogolyubov, 1971) have shown that, for a system with such a Hamiltonian, the initial Hamiltonian of Bardeen-Cooper-Schrieffer may be replaced by an approximating Hamiltonian. Such a replacement corresponds to the mean field approximation. It is essential in this case that the solution with an approximating Hamiltonian is asymptotically exact. This means that, after the limting thermodynamic transition, the solution obtained in this way coincides with the exact solution. The results of more recent work on constructing approximating Hamiltonians have been discussed by Bogolyubov (1974).

Of course, there also exist intermediate-type systems. A well-known example is a uni axial ferroelectrium, in which dipole-dipole interaction plays an essential role (Zarkin, Khmelnitsky, 1969).

Consider now the atoms-oscillator system for the case when large-scale correlations, responsible for phase transition, occur not due to the spatial diffusion but due to the interaction of atoms through a fluctuational electromagnetic field (Klimotovich, 1980).

Remember that, up to now, the action of the field on the atoms (F in the right-hand side of equation (8)) has been disregarded and as a consequence, all the correlational characteristics were determined by the diffusion coefficient g. At g = 0 the correlation radius, r_c , becomes equal to zero. In this case the Ginzburg parameter tends to infinity. For this reason phase transition turns out to be impossible.

In equation (8) it is more convenient to use here not the equation for x but directly the polarization vector $P = enx$. At g = 0 this has the form

$$\frac{\partial \vec{P}}{\partial t} + (a + b P^2)\vec{P} = \frac{e^2 n}{m\gamma} \vec{E}(\vec{R},t) + en\vec{y}(\vec{R},t). \tag{66}$$

If one considers, as before, that only one component of the polarization vector is essential, then moments of the random source are determined from expression (10). When the three components in (10) are taken into account, then the substitution $D = kT/m\gamma \rightarrow 3kT/m\gamma$ should be made. The constant b in equation (66) is associated with the former quantity through the relation $b \leftrightarrow b/e^2n^2$.

Equation (66) is not closed. This is why it must be supplemented with field equations.

The role of the order parameter is played by the quantity

$$\vec{\eta} = \langle \vec{P}_{\vec{k}=0}(t) \rangle = \langle \vec{P}(t) \rangle . \tag{67}$$

We have used here Fourier transformation in the form of (11). In the absence of fluctuations, and under the condition that $\langle E \rangle = 0$ then

$$\eta^2 = \frac{|a|}{b} . \tag{68}$$

For the order parameter fluctuations

$$\delta\vec{\eta}_{\vec{k}}(t) = \vec{P}_{\vec{k}}(t) - \vec{\eta}\,\delta_{\vec{k},0} \equiv \delta\vec{P}_{\vec{k}}(t) \tag{69}$$

in a linear approximation, we derive from (66) the following equation

$$\frac{\partial \delta\vec{P}_k}{\partial t} + (a + 3b\langle\eta^2\rangle)\delta\vec{P}_{\vec{k}} = \frac{e^2 n}{m\gamma}\delta\vec{E}_{\vec{k}} + \vec{y}_{\gamma\vec{k}} . \tag{70}$$

From this it follows that the function, $\delta\vec{P}$, may be represented as a sum of two parts

$$\delta\vec{P} = \delta\vec{P}^{ind} + \delta\vec{P}^{sou} . \tag{71}$$

We shall call the second term *a source*. It is defined by the solution of equation (70) at $\delta\vec{E} = 0$ and, consequently, associated with spontaneous fluctuations of the polarization vector. On the contrary, the first, induced, term is defined by equation (70) at $\vec{y} = 0$.

The spectral densities of fluctuations $\delta\vec{P}^{sou}$ are determined by the expressions

$$(\delta\vec{P}\delta\vec{P})^{sou}_{\omega,\vec{k}} = \frac{2e^2 nD}{\omega^2 + (a+3b\eta^2)^2} ; \quad (\delta\vec{P}\delta\vec{P})^{sou}_{\vec{k}} = \frac{e^2 nD}{a+3b\eta^2} . \tag{72}$$

From the latter formula we find the expression for spatial correlation

$$(\delta\vec{P}\delta\vec{P})^{sou}_{R,R'} = \frac{e^2 n}{a+3b\eta^2}\,\delta(\vec{R}-\vec{R}') . \tag{73}$$

It is seen that the intensity of fluctuations, $\delta\vec{P}^{sou}$ increases as τ^{-1} when the critical point is approached. The fluctuations, however, remain δ -correlated and, the correlation radius, therefore, is equal to zero.

The induced fluctuations $\delta \vec{P}^{sou}$ are connected with $\delta \vec{E}$ by the ususal relation $\delta \vec{P}(\omega) = \alpha(\omega)\,\delta E$. Here

$$\alpha(\omega) = \frac{e^2 n}{m\gamma}\,\chi(\omega) = \frac{e^2 n}{m\gamma}\;\frac{1}{-i\omega + a + 3b\eta^2} \tag{74}$$

is the polarizability of the system. When approaching the critical point, the static susceptibility (polarizability) increases according to the Curie law.

To calculate the correlations of field fluctuations, we use the Maxwell equation and since, in phase transitions the oscillations at low frequencies are most important, classical expressions may be used in the calculation.

Fluctuations of a longitudinal (with respect to the wave vector) field in the space are δ -correlated and are, therefore, of no importance in phase transitions. Spatial correlations of transversal field fluctuations are defined by the expression (it follows from formula (90.24) in the book by Landau and Lifshitz (1957) for $\hbar = 0$):

$$(\delta E^{\perp} \delta E^{\perp})_{\omega, \vec{r}} = \frac{2\omega}{ic^2 r}\left(e^{-\frac{\omega}{c}\sqrt{-\varepsilon(\omega)}\,r} - e^{-\frac{\omega}{c}\sqrt{-\varepsilon^*(\omega)}\,r}\right) kT. \tag{75}$$

This expression is finite at $r = 0$. So, under the condition

$$\mathrm{Im}\,\varepsilon \ll \mathrm{Re}\,\varepsilon \tag{76}$$

from (75) there follows the equation:

$$(\delta \vec{E}^{\perp} \delta \vec{E}^{\perp})_{\omega} = \frac{4\omega^2}{c^3}\sqrt{\mathrm{Re}\,\varepsilon(\omega)}\; kT. \tag{77}$$

It differs from the Rayleigh-Jeans formula only by the factor

$$\frac{1}{4\pi^2}\sqrt{\mathrm{Re}\,\varepsilon(\omega)} \tag{78}$$

which, according to (74), increases in a low-frequency region when approaching the critical point.

Under the condition (76) expression (75) can be represented in the form

$$(\delta \vec{E}^{\perp} \delta \vec{E}^{\perp})_{\omega, \vec{r}} = \frac{4\omega}{c^2}\, e^{-\frac{r}{r_c}}\, \frac{\sin\frac{\omega}{c}\sqrt{\mathrm{Re}\,\varepsilon(\omega)}\, r}{r}\, kT. \tag{79}$$

We introduced here a designation for the correlation radius of field fluctuations at a frequency given by

$$r_c(\omega) = \frac{c}{2\sqrt{\mathrm{Re}\,\varepsilon}\cdot\omega}\cdot\frac{\mathrm{Re}\,\varepsilon}{\mathrm{Im}\,\varepsilon}\ . \tag{80}$$

For the normal mode of polarization fluctuations $k_{min} \sim \omega_{min}\sqrt{\mathrm{Re}\,\varepsilon}/c \sim V^{-1/3}$, and therefore the corresponding correlation radius is, using (76)

$$r_c(\omega_{min}) \sim V^{1/3}\left(\frac{\mathrm{Re}\,\varepsilon}{\mathrm{Im}\,\varepsilon}\right)_{\omega=\omega_{min}} \gg V^{1/3}\ . \tag{81}$$

Thus, the correlation radius of the field fluctuations determining induced polarization fluctuations exceeds the dimensions of the system. As a consequence, there exist induced polarization fluctuations spread about the entire volume of the system. However, the phase transition, responsible for the appearance of the state with constant polarization, is possible only if the correlation time depends on V (or N) in such a way that it tends to infinity in the thermodynamic limiting transition.

In accordance with the representation of $\delta\vec{P}$ in the form (71) two corresponding correlation times may be introduced. The correlation time of the fluctuations $\delta\vec{P}^{sou}$ (designated as τ^{sou}) depends on the width of the spectrum. Therefore, from the first formula of (72) it follows that

$$\tau^{sou} \sim \frac{1}{a + 3b\eta^2} = \chi(0)\ . \tag{82}$$

Thus, when approaching the critical point, τ_{sou} increases according to Curie's law.

To obtain the result for the critical point itself we shall operate as in the previous sub-section. Namely, we shall make the substitution (see (24))

$$\eta^2 \to \langle|\vec{P}_{k=0}(t)|^2\rangle = \langle\vec{P}^2\rangle\ ,\ \vec{P}(t) = \int\vec{P}(\vec{R},t)\frac{d\vec{R}}{V} \tag{83}$$

Here $\vec{P}$ is an induced polarization.

Taking into account interactions with the field, we obtain instead of (29) the following equation for the function $\langle P^2 \rangle$

$$(a + b\langle P^2\rangle)\langle P^2\rangle = \frac{e^2 n}{m\gamma}\langle \vec{E}\vec{P}\rangle . \tag{84}$$

This equation is not closed. At zero mean field the correlator $\langle \vec{E}\vec{P}\rangle$ is defined by the expression

$$\langle \delta\vec{E}\delta\vec{P}\rangle = \frac{1}{2\pi}\int \alpha(\omega)(\delta\vec{E}\delta\vec{E})d\omega = \frac{2kT}{V} . \tag{85}$$

Formula (77) is used here, and the normal mode (which alone is left in the sum over $\vec{k}$). As a result, equation (84) takes the form

$$(a+b\langle P^2\rangle)\langle P^2\rangle = \frac{2e^2 n}{3V}\mathcal{D} ; \quad \mathcal{D} = \frac{3kT}{m\gamma_{ef}} . \tag{86}$$

Thus, we are again led to equation (29). From this we find for the critical point, a = 0, that

$$\langle P^2\rangle \sim V^{-1/2} \tag{87}$$

and, consequently, according to (82) and (87)

$$\tau_{sou} \sim V^{1/2} . \tag{88}$$

When moving away from the critical point, the correlation time, τ_{sou} , falls according to Curie's law

$$\tau_{sou} \sim \frac{1}{a} \quad (T > T_c) ; \quad \tau_{sou} \sim \frac{1}{2|a|} \quad (T < T_c) . \tag{89}$$

To find the correlation time of the induced polarization fluctuations τ_{ind}, we use the equations for the function $\langle \vec{P}(t)\vec{P}(t-\tau)\rangle$. With the same approximation as in (86) we have

$$\left(\frac{\partial}{\partial\tau} + (a+b\langle P^2\rangle)\right)\langle \vec{P}(t)\vec{P}(t-\tau)\rangle = 0 . \tag{90}$$

The initial (at $\tau = 0$) condition for this equation is the solution of equation (86).

Thus, from equations (90) and (86) it follows that

$$\tau_{ind} \sim (a+b\langle P^2\rangle)^{-1} = \left(\frac{2e^2 n}{3V}\frac{\mathcal{D}}{\langle P^2\rangle}\right)^{-1} . \tag{91}$$

Whence

$$\tau_{ind} \sim \tau_{sou} \sim \frac{1}{a} \text{ at } T > T_c \,;\; \tau_{ind} \sim \tau_{sou} \sim V^{1/2} \text{ at } T = T_c \tag{92}$$

and finally

$$\tau_{ind} \sim \frac{|a|}{e^2 n \mathcal{D} b} V \sim V. \tag{93}$$

Thus, in the transition of the critical point the time $\tau_{ind} \sim V$ and, therefore, it tends to infinity in the thermodynamic limiting transition. This means that the induced polarization state is long-living below the critical temperature.

From equation (90) it follows that the time, τ_{ind}, characterizes the width of the spectrum of the induced fluctuations, $\delta \vec{P}_{ind}$. Indeed, from equation (90) (with an account taken of equation (86) it follows that

$$(PP)_\omega = \frac{2(a+b\langle P^2\rangle)\langle P^2\rangle}{\omega^2+(a+b\langle P^2\rangle)^2} = \frac{4e^2 n \mathcal{D}}{3V(\omega^2+(1/\tau_{ind})^2)} \tag{94}$$

and, of course,

$$\frac{1}{2\pi}\int_{-\infty}^{+\infty}(PP)_\omega d\omega = \frac{2e^2 n \mathcal{D}/V}{3(a+b\langle P^2\rangle)} = \langle P^2\rangle. \tag{95}$$

Thus, the function $(PP)_\omega$ determines the spectral distribution of the mean energy of induced polarization fluctuations.

The behaviour of the spectral widths of the fluctuations δP_{sou}, δP_{ind} in transition via the critical point is quite different. When approaching the critical point (on the high temperature side) both spectra become narrower ($\Delta\omega_{sou} \sim \Delta\omega_{ind} \sim a$), and at the critical point $\Delta\omega_{sou} \sim \Delta\omega_{ind} \sim V^{-1/2}$. However, in transition via the critical point ($T < T_c$) the fluctuation spectrum $\Delta\omega_{sou}$ becomes even wider whereas the spectrum of induced fluctuations $\Delta\omega_{ind}$ becomes still narrower: $\Delta\omega_{ind} \sim V^{-1}$ and in the thermodynamic limit becomes equal to zero.

Phase Transitions in the System of Two-Level Atoms Interacting through an Electromagnetic Field

Phase transition in the system of two-level atoms and a field has been con-- sidered in a number of papers either on the basis of the Dicke model (Wang, Hioe, 1973; Hepp, Lieb, 1973) or by the model in which the appearance of a soft mode is due to the action of the Lorentz field (Emelyanov, Klimontovich, 1974, 1975; Klimontovich, 1980). For the Dicke model, in the thermodynamic limit, an exact solution is possible, the results of which coincide with those obtained in the mean field approximation.

To estimate the polarization and field fluctuations in the critical region determined by the parameter $\varepsilon_N \sim 1/N$ we shall not represent the calculations made above but use the conditions of correspondence between classical and quantum systems;

$$\gamma \leftrightarrow 2\gamma_{ab}\,;\quad \omega_0 \leftrightarrow \omega_{ab}\,;\quad \frac{e^2 n}{\mu} \leftrightarrow \frac{2}{3}\frac{|d_{ab}|^2 n\,\omega_{ab}|\mathcal{D}|}{\hbar}\,. \tag{96}$$

Here $|\mathcal{D}|$ is the difference between the densities of population of the levels a and b. The rest of the designations are standard. Using these relations, we may write an equation for the function $\langle P^2\rangle$, corresponding to the classical equation (84), in the following form

$$(a+b\langle P^2\rangle)\langle P^2\rangle = \frac{|d_{ab}|^2}{3\hbar\gamma_{ab}}\, n\,\omega_{ab}\,|\mathcal{D}^0|_c\langle EP\rangle\,. \tag{97}$$

The quantities a and b have, here, another meaning. For example, the quantity defined by

$$a = \frac{\omega_{ab}^2}{2\gamma_{ab}}\left(1-\Lambda|\mathcal{D}^0|\right),\quad \Lambda = \frac{8\pi}{3}\frac{|d_{ab}|^2}{3\hbar\omega_{ab}}\, n\,. \tag{98}$$

The critical temperature is now determined by the condition

$$|\mathcal{D}^0|_c = \operatorname{th}\frac{\hbar\omega_{ab}}{2kT_c} = \frac{1}{\Lambda}\,. \tag{99}$$

The quantity b can be found from the Curie-Weys equation by series expansion of the square of the order parameter, η^2. The corresponding expression for polarizability is as follows

$$\alpha_{ind}(\omega) = \frac{|d_{ab}|^2}{3\hbar\gamma_{ab}}\, n\,\omega_{ab}|\mathcal{D}^0|_c\,\frac{1}{-i\omega+a+b\langle P^2\rangle}\,. \tag{100}$$

We see that all the results obtained in this example of a classical system of nonlinear oscillator atoms may be extended to the quantum system of two-level atoms interacting through the field.

Thus, we have considered three quite different examples of Brownian motion in transitions through the generation threshold and the second order phase transition point. From the discussion many common features of these phenomena can be perceived. They become still more evident with a detailed treatment of the phenomena (Klimontovich, 1980). In particular, correlations of different types are also detected in the investigation of fluctuations in the generator.

However, similar as these phenomena seem to be, they still have some essential differences. One of the differences is that the generation is due to a dissipative nonlinearity, whereas the ordinary phase transition is due to a nondissipative nonlinearity. It follows, as a consequence, that the ordinary phase transition may be realized as a transition through a sequence of equilibrium states and, therefore, it is a reversible process. At the same time, the transition through the generation threshold is an irreversible process and is, therefore, possible only in open systems. This is the reason why such transitions are called nonequilibrium phase transitions.

It is natural that the spectrum of nonequilibrium phase transitions is immeasurably richer than that of equilibrium transitions. It is just because of this that they determine, in the main, the possibilities for self-organization, of nonequilibrium processes (Nicolis, Prigogine, 1977; Haken, 1978; Ebeling, 1976), in physical, chemical and biological systems during the course.

Nonequilibrium Fluctuations and Irreversible Processes Far from Thermal Equilibrium

Hiroshi Furukawa

The important, and also interesting, points of fluctuations in an equilibrium state are that fluctuations obey the second law of thermodynamics. The ordinary statistical thermodynamics of irreversible processes is, in fact, based on this property of fluctuations, since the macroscopic characteristic of a nonequilibrium state near to an equilibrium state is closely related to the properties of fluctuations in the equilibrium state. Let us imagine a situation where the thermal equilibrium state is replaced by a reference nonequilibrium state. In such a situation we also expect that the macroscopic characteristic of a nonequilibrium state, near the reference nonequilibrium state, is closely related to the properties of fluctuations in the reference nonequilibrium state. Unfortunately, however, it seems that we have no commonly agreed principle corresponding to the second law of thermodynamics for nonequilibrium fluctuations. The first purpose of this paper is, therefore, to explore a general principle which governs nonequilibrium fluctuations. Our conclusion is that we may admit the existence of a unique law for nonequilibrium fluctuations that corresponds to the second law of thermodynamics for equilibrium fluctuations, and also shows the irreversibility of the temporal evolutions of mean fluctuations. This law has great generality because of its basic premise that a real system is highly complex even in nonequilibrium states. The second purpose is, therefore, to investigate the thermodynamic or variational description of irreversible processes in nonequilibrium states far from the thermal equilibrium.

Thermodynamics and Kinetics of Biological Processes

Nonequilibrium Fluctuations and Their Averages

Suppose a nonequilibrium state, e. g. a transient, unstable or temporally oscillating state, which can be characterized by a set of dynamic variables, $A(t)$, such as coarse-grained particle number, energy and current densities, and initial phase distribution function, ρ_0 . Here A is an n-dimensional column vector whose i-the component is written as A_i , and ρ_0 may not be a thermal equilibrium phase distribution function. In this paper we shall remain within the classical mechanical framework. Then ρ_0 is real and non-negative. The fluctuation of $A(t)$ is defined by

$$\Delta A(t) = A(t) - \langle A(t) \rangle_0 \tag{1}$$

where $\langle A(t) \rangle_0 = \mathrm{Tr}\, A(t)\, \rho_0$ is an average value in the nonequilibrium state. Here Tr denotes a phase space integration.

Onsager (1931 a, b) identified artificial polarization of fluctuations with the spontaneous polarization. Once the spontaneous fluctuation has arisen in the system, this fluctuations will decay or grow according as to whether the nonequilibrium state is stable, or unstable. The average part of the spontaneous fluctuation is evaluated by means of a certain kind of averaging procedure.

Let us assume that a spontaneous fluctuation occurs at time, s . Let the mean value of the spontaneous fluctuation at later time t be $\langle \Delta A(t) ; s \rangle$. Here $\langle \quad ; s \rangle$, which we must determine later, denotes the conditional average; that is the average of $\Delta A(s)$ takes the value $\langle \Delta A(s) ; s \rangle$. If the mean fluctuation is small, then we may assume a linear relation between $\langle \Delta A(t) ; s \rangle$ and $\langle \Delta A(s) ; s \rangle$ as

$$\langle \Delta A(t); s \rangle = U(t;s) \langle \Delta A(s) ; s \rangle, \quad (t \geq s) \tag{2}$$

where U is an n by n matrix. How can we determine the matrix U(t;s)?

The average $\langle \Delta A(t) ; s \rangle$ may be so defined that $\langle \Delta A(t) ; s \rangle$ is identified with the *most probable value*. Let ω denote the phase, i. e. the set of coordinates and moments. Let us consider one of phases of the system at time s and let denote it as ω_s . ΔA for this phase takes the value $\Delta A(s) = \Delta A(\omega_s)$. At later time $t\ (> s)$, starting from $\Delta A(\omega_s)$, ΔA

takes the *true value* $\Delta A(t) = \Delta A(\omega_t)$ determined by the equation of motion, and the initial value $\Delta A(\omega_s)$. Now, the average behavior of ΔA at time t is assumed to be proportional to $\Delta A(\omega_s)$ and is written as $U'(t,s)\,\Delta A(\omega_s)$. It is referred to as the most probable value. Here U'(t, s) is common to all phases at time s. The difference between the true value and the most probable value is reffered to as the *error* and is written as

$$W_\omega(t;s) = \Delta A(\omega_t) - U'(t;s)\Delta A(\omega_s)$$

or, by removing the symbol ω without loss of generality as

$$W(t;s) = \Delta A(t) - U'(t;s)\Delta A(s), \quad (t \geq s). \tag{3}$$

U'(t,s) is determined so that the error W(t;s) is as small as possible. This can be done by the method of least squares. Let 0 be a non-negative definite n by n hermitian matrix. The square of the error is, then, defined by

$$\Psi = \sum_{ij} \langle W_i^*(t;s) O_{ij} W_j(t;s); s \rangle = \\ = \langle W^*(t;s) O W(t;s); s \rangle \geq 0 , \tag{4}$$

where W_i denotes the i-th component of the vector W; O_{ij} denotes the ij-element of the matrix 0; W* denotes a hermitian conjugate row vector of W. Here, as defined, $\langle \;\; ; s \rangle$ denotes the ensemble average; that is the average of $\Delta A(s)$ at time s takes the value $\langle \Delta A(s); s \rangle$. The inequality in (4) can easily be found in the diagonal representation of 0. All the eigenvalues of which are, by definition, real and non-negative. The explicit form of 0 is not essential to the present purpose. Since the mean fluctuation is assumed to be small, we may take the limit $\langle \Delta A(s); s \rangle \rightarrow \langle \Delta A(s) \rangle_0 = 0$ in (4). Thus we may use, instead of Ψ in (4) :

$$\Psi = \langle W^*(t;s) O W(t;s) \rangle_0 \geq 0 . \tag{5}$$

We now require that Ψ be a minimum with respect to the most probable value and therefore with respect to U' and U'*. Here, U'* denotes a hermitian conjugate matrix of U'. We easily find that the minimum condition for Ψ is given by

$$\langle \Delta A_i^*(s) \delta U_{ij}^{\prime *} W_j(t;s) \rangle_0 = \langle W_j(t;s) \Delta A_i^*(s) \rangle_0 \, \delta U_{ij}^{\prime *}(t;s) = 0 \tag{6}$$
$$(t \geq s), \quad (i, j = 1, \ldots, n)$$

or

$$\langle W(t;s)\,\Delta A^{*}(s)\rangle_{0} = 0\,, \qquad (t \geq s)$$

where $\langle W(t;s)\,\Delta A^{*}(s)\rangle_{0} = 0$ is an n by n matrix with $\langle W_{i}(t;s)\,\Delta A_{j}^{*}(s)\rangle_{0}$ as the ij-element and $\delta U'' \equiv O\,\delta U'$.

Once U' is determined from (3) and (6), the most probable value at time t for a given initial value is, by definition, determined by the procedure of putting W(t;s)=0 in (3). The ensemble average $\langle \Delta A(t);s\rangle$ in (2), for a given initial value $\langle \Delta A(s);s\rangle$ is, then, identified with the most probable value. We therefore, find that

$$U(t;s) = U'(t;s)\,. \tag{7}$$

From (5), (3) and (7) we obtain

$$U(t;s) = \langle \Delta A(t)\,\Delta A^{*}(s)\rangle_{0}\,\sigma(s)^{-1}\,. \tag{8}$$

Here $\sigma(s) \equiv \langle \Delta A(s)\,\Delta A^{*}(s)\rangle_{0}$ is the variance matrix at time s in the reference state. With the use of (8) we find that (2) can be rewritten as

$$\langle \Delta A(t);s\rangle = \langle \Delta A(t)\Delta A^{*}(s)\rangle_{0}\,X(s)\,, \qquad (t \geq s) \tag{9}$$

where $X(s) \equiv \sigma(s)^{-1}\langle \Delta A(s);s\rangle$ is referred to as the force conjugate to $\langle \Delta A(s);s\rangle$. It should be noted that (6) can be rewritten as

$$\langle W(t;s);s\rangle \equiv \langle W(t;s)\Delta A^{*}(s)\rangle_{0}\,X(s) = 0, \qquad (t \geq s) \tag{10}$$

which is, however, equivalent to (2) or (9). (10) corresponds to the procedure of putting W(t;s) = 0 to obtain the most probable value.

Before closing this section, we shall derive from (5) an inequality which we shall use later. Let us choose $O \equiv X(t)\,X^{*}(t)$, which can easily be found to be a positive, semi-definite matrix if we use the diagonal representation of 0. By substituting this for 0 in (5) we observe that

$$\begin{aligned} &\langle W^{*}(t;s)\,X(t)\,X^{*}(t)\,W(t;s)\rangle_{0} = \\ &= \sum_{ij}^{n} \langle W_{i}^{*}(t;s)\,X_{i}(t)\,X_{j}^{*}(t)\,W_{j}(t;s)\rangle_{0} = \\ &= X^{*}(t)\langle W(t;s)\,W^{*}(t;s)\rangle_{0}\,X(t) \geq 0\,. \end{aligned} \tag{11}$$

By substituting U' given by (8) and (7) into (11) and then exchanging the arguments t, and s, we find that the last inequality can be rewritten as

$$X^*(s)\sigma(s)X(s) - X^*(s)\langle \Delta A(s)\Delta A^*(t)\rangle_0 \sigma(t)^{-1}\langle \Delta A(t)\Delta A^*(s)\rangle_0 X(s) \geq 0$$

or

$$\Phi(s;s) \geq \Phi(t;s) \qquad (t \geq s) \tag{12}$$

where

$$\Phi(t;s) \equiv \langle \Delta A(t);s\rangle^* \sigma(t)^{-1}\langle \Delta A(t);s\rangle \geq 0. \tag{13}$$

The inequality in (13) is due to the positive semi-definiteness of the variance matrix $\sigma(t)$. (12) plays an important role in later discussions.

Langevin Type Equation for Nonequilibrium Fluctuations

Onsager assumed that the mean fluctuation, corresponding to (9), obeyed a linear equation of motion, if the mean fluctuation was small. He then investigated the mean regression of the fluctuation in a state near equilibrium (Onsager, 1931 a, b). Prior to Onsager's work, Langevin already recognized, in the study of Brownian motion, the importance of a fluctuating force, which is the *rest* part of the equation of motion for fluctuation, $\Delta A(t)$ (Langevin, 1908). The decay, or damping constant, of the mean fluctuation in a state near equilibrium is closely related to the fluctuating force through a simple relation now known as the fluctuation-dissipation theorem (Kubo, 1966).

The progress of modern statistical mechanics of irreversible processes (Kubo, 1957; Zubarev, 1974) necessarily stimulated a new development of Langevin's approach. Mori (1965) proved a general form of the Langevin type equation for the set of dynamic variables, A(t), in equilibrium

$$\begin{aligned} &\frac{d}{dt}A(t) = i\Omega A(t) - \int_0^t \varphi(t-\tau)A(\tau) + f(t) \quad (t \geq 0) \\ &\langle f(t)A^*(0)\rangle_0 = 0, \end{aligned} \tag{14}$$

where $\langle A(t)\rangle_0$, which is independent of t, is set equal to zero. The fluctuating force, f(t), is related to the damping coefficient matrix, φ, by

$$\varphi(t) = \langle f(t) f^*(0)\rangle_0 \langle AA^*\rangle_0^{-1}. \qquad (15)$$

One of the important points of (14) is that $\varphi(t)$ has a memory, which is important, for instance, to the frequency dependence of conductivity. The asymptotic evaluation of the master equation for macrovariables (van Kampen, 1961 a; Kubo et al., 1973) suggests that a linear Langevin equation for macrovariables may be valid in a nonequilibrium state with suitable modification (Keizer, 1975 a, b; Fox, 1978). If memory must be taken into consideration, then one also looks for an equation equivalent to Mori equation (14). The frequency matrix, Ω, can no longer be time-independent and the damping cofficient matrix, φ, can no longer be a function of a single time. However, the modifications should not just be restricted to such cases. The separation of the equation of motion into two parts, i. e. a systematic motion, which is a linear functional of A(t) and a fluctuating force, which is the *rest* part of the equation of motion, contains the uncertainty in selection of an initial time. Thus a fluctuating force has, intrinsically, two time arguments, one of which can be abandoned in equilibrium theory because of the time translation symmetry. In general in a nonequilibrium state we cannot do this.

Let us here present a further generalization of the Langevin-Mori equation (Furukawa, 1976, 1977). The time derivative of $\Delta A(t)$ is formally, and identically, separated into two parts:

$$\frac{d}{dt}\Delta A(t) = F(\Delta A(\tau); t \geq \tau \geq s) + R(t; s) \quad (t \geq s), \qquad (16)$$

where F and R, respectively, denote n-dimensional column vectors. Here, F or R is characterized by the following three conditions:

i) F is an explicit function of two time arguments, t and s, only.

ii) F is a linear function of $\Delta A(\tau)$; $(t \geq \tau \geq s)$.

iii) F is so chosen that R is orthogonal to $\Delta A(s)$:

$$\langle R(t; s)\Delta A^*(s)\rangle_0 = 0 \qquad (t \geq s). \qquad (17)$$

(16) is an identity since R is simply defined by $\frac{d}{dt}\Delta A(t) - F$. Condition i) comes from the reason above. Conditions ii) and iii) give the equation of motion for the mean fluctuation

$$\frac{d}{dt}\langle \Delta A(t); s\rangle = F(\langle \Delta A(\tau); s\rangle ; t \geq \tau \geq s) \tag{18}$$

which corresponds to Onsager's kinetic equation. The Langevin fluctuating force, R(t;s), is related to the systematic motion, (18), through generalized fluctuation-dissipation theorem:

$$\left\langle \frac{dF(\Delta A(\tau); t \geq \tau \geq s)}{ds} \Delta A^*(s)\right\rangle_0 = \langle R(t;s) R^*(s;s)\rangle_0 . \tag{19}$$

There is an infinitely large number of possibilities for an explicit form of F. The form corresponding to the Mori equation is given by

$$F(\Delta A(\tau); t \geq \tau \geq s) \equiv K(t,t)\Delta A(t) - \int_s^t \varphi(t;\tau)\Delta A(\tau)d\tau , \tag{20}$$

where (K(t,t) corresponds to the frequency matrix $i\Omega$ in the Mori equation and is given from (17), by putting t = s as,

$$K(t,t) = \langle \Delta \dot{A}(t)\Delta A^*(t)\rangle_0 \langle \Delta A(t)\Delta A^*(t)\rangle_0^{-1} . \tag{21}$$

Here $\Delta\dot{A}(t)$ stands for time derivative. By differentiating (20) with respect to s, we observe that

$$\frac{dF(\Delta A(\tau); t \geq \tau \geq s)}{ds} = \varphi(t;s)\Delta A(s) . \tag{22}$$

Then by substituting (22) into (19) we obtain

$$\varphi(t;s) = \langle f(t;s) f^*(s,s)\rangle_0 \langle \Delta A(s)\Delta A^*(s)\rangle_0^{-1} \quad (t \geq s) \tag{23}$$

where $R \equiv f$. As is seen from (20) and (23) the fluctuating force, f , necessarily has two time arguments in non-steady states. However, in an equilibrium state, due to the time translation symmetry, (23) reduces to (15), if we put s = 0 and rewrite f(t;0) as f(t).

Irreversibility of Mean Fluctuations

Since Clausius formulated the second law of thermodynamics (that the entropy of an isolated system cannot decrease) this law has been a central guiding principle of all fluctuation theories near thermal equilibrium states. Let $\langle A \rangle_o$ denote an equilibrium average. Then the entropy S , near the equilibrium state, may be expanded, for simplicity, for a single component, A , as

$$S(t) = S_o + \frac{\partial S_o}{\partial \langle A \rangle_o} \langle \Delta A(t) \rangle + \frac{1}{2} \frac{\partial^2 S_o}{\partial \langle A \rangle_o^2} \langle \Delta A(t) \rangle^2 + \ldots \tag{24}$$

where $\langle \Delta A(t) \rangle$ denotes an average deviation from the equilibrium value. In equilibrium states, the second law of thermodynamics demands that

$$\frac{\partial S_o}{\partial \langle A \rangle_o} = 0, \quad \frac{\partial^2 S_o}{\partial \langle A \rangle_o^2} < 0, \tag{25}$$

and

$$\frac{d}{dt} \left\{ \frac{\partial^2 S_o}{\partial \langle A \rangle_o^2} \langle \Delta A(t) \rangle^2 \right\} \geq 0. \tag{26}$$

Inequality (26) states that the average deviation $\langle \Delta A(t) \rangle$ should always decay. Therefore, the second law of thermodynamics guarantees the stability of a thermal equilibrium state. However, we note that this is a peculiarity of the thermal equilibrium state.

Let us here replace in (24):

$$\frac{\partial^2 S_o}{\partial \langle A \rangle_o^2} \rightarrow -\sigma(t)^{-1} \tag{27}$$

and also

$$\langle \Delta A(t) \rangle \rightarrow \langle \Delta A(t); s \rangle. \tag{28}$$

We then obtain the positive semi-definite quadratic form $\Phi(t; s)$, defined by (13), which corresponds to the third term on the right hand side of (24). Although the replacement (27) is, in general, valid only for macrovariables, A , the quantity $\Phi(t; s)$ thus defined should be recognized as a starting point for later discussions of non-macrovariables A. In what way does $\Phi(t; s)$ evolve in nonequilibrium states? Nicolis and Prigogine (1977) classified, using an inequality similar to (26), nonequilibrium steady states into three cases, i. e., stable, marginal and unstable states, according as the quantity corresponding to the left hand side of inequality (26) is positive, zero or negative. All possibilities thus appear and therefore it seems that there is no unique principle for nonequilibrium fluctuations corresponding to the second law of thermodynamics. Nicolis and Prigogine (1977) considered only the case where the second derivative of S is a constant, independent of time. What however, is an unstable nonequilibrium state? We can write the coefficient of $\partial^2 S_0/\partial\langle A\rangle_0^2$ as $\partial^2 S/\partial\langle A(t)\rangle^2$. It may no longer be a constant, independent of time, if $\langle A(t)\rangle$ is not a constant independent of time as in an unstable nonequilibrium state. Another fact which contradicts the Nicolis and Prigogine classification has been discussed by many authors on the basis of H-theorem and stochastic models. This is, that the theoretical information entropy, relative to that of a nonequilibrium reference state, cannot decrease; even in nonequilibrium states (Schlögl, 1974; Pfaffelhuber, 1977). Such a result suggests that $\Phi(t;s)$ would have a definite direction of asymptotic temporal evolution.

Recall that $\langle\Delta A(t);s\rangle$ denotes the most probable value which can be obtained by neglecting the fluctuating force f(t;s). A real system interacts with an infinitely large number of degrees of freedom. Then, as usual, in a real system, we assume that the fluctuating force, f(t;s), has a finite correlation time:

$$\varphi(t;s) = 0 \quad \text{for} \quad t-s \geq \tau . \tag{29}$$

We find that $\Phi(t;s)$ asymptotically decreases with time t

$$\Phi(t;s) \leq \Phi(t';s) \qquad t-t' \gg \tau_1 , \quad t,t' > s . \tag{30}$$

Indeed, with the aid of (20) and (17) we can find that U(t;s) defined by (8) obeys the equation of motion

$$\frac{d}{dt} U(t;s) = K(t,t)U(t;s) - \int_s^t \varphi(t;\tau)U(\tau;s)d\tau \qquad (t \geq s) \tag{31}$$

For $t-s \geq \tau_1$, $U(t;s)$ is of the form

$$U(t;s) = V(t)C(s),$$

since the lower limit of the integration (s) in (31) can be, due to (29), put to be $-\infty$. Thus, by putting $t = \tau + s$, we obtain

$$C(s) = V^{-1}(s+\tau)U(s+\tau,s) \qquad \tau > \tau_1$$

and therefore

$$\mathcal{U}(t;s) = V(t)V(s+\tau)^{-1}\,\mathcal{U}(s+\tau,s) \qquad t-s>\tau_1\,,\ \tau>\tau_1\,.$$

This leads to

$$\mathcal{U}(t;s) = \mathcal{U}(t;t')\,\Theta(t'+\tau,t')\,\mathcal{U}(t';s) \qquad \tau,\,t-t',\,t'-s>\tau_1 \tag{32}$$

where

$$\Theta(t'+\tau,t') = \mathcal{U}(t'+\tau,t')^{-1}\,V(t'+\tau)\,V(t')^{-1}\,.$$

If we neglect Θ, then we obtain

$$\mathcal{U}(t;s) = \mathcal{U}(t,t')\,\mathcal{U}(t',s) \tag{33}$$

from (32). Notice that (12) can be rewritten as

$$\begin{aligned}\Phi(t;s) &\equiv [\sigma(s)X(s)]^{*}\,\mathcal{U}^{*}(t,s)\,\sigma(t)^{-1}\,\mathcal{U}(t,s)[\sigma(s)X(s)] \leq \\ &\leq [\sigma(s)X(s)]^{*}\sigma(s)^{-1}[\sigma(s)X(s)] \equiv \Phi(s;s).\end{aligned} \tag{34}$$

We then find, in the same way as (34), with the use of (33) that

$$\begin{aligned}\Phi(t;s) &= [\mathcal{U}(t,t')\,\mathcal{U}(t',s)\sigma(s)X(s)]^{*}\sigma(t)^{-1}\,[\mathcal{U}(t,t')\,\mathcal{U}(t',s)\sigma(s)X(s)] = \\ &= [\mathcal{U}(t',s)\sigma(s)X(s)]^{*}\,\mathcal{U}^{*}(t,t')\sigma(t)^{-1}\,\mathcal{U}(t,t')[\mathcal{U}(t',s)\sigma(s)X(s)] \leq \\ &\leq [\mathcal{U}(t',s)\sigma(s)X(s)]^{*}\sigma(t')^{-1}\,[\mathcal{U}(t',s)\sigma(s)X(s)] = \Phi(t';s)\,.\end{aligned} \tag{35}$$

It can easily be seen that $\Theta(t'+\tau,t')$ is independent of τ for $\tau>\tau_1$. Thus, $\Theta(t'+\tau,t')$ is a microscopic quantity. Therefore, in (32) we may neglect $\Theta(t'+\tau,t')$ compared with U(t;t') for a macroscopic time interval, $t-t'\gg\tau_1$. Then, (30) may be derived from (35). It is obvious that (30) plays the same role as the second law of thermodynamics for nonequilibrium fluctuations. The fact that $\Phi(t;s)$ always decreases with time, t, even in an unstable state means that the increment of the variance, $\sigma(t)=\langle\Delta A(t)\,\Delta A^{*}(t)\rangle_0$, in the reference state always exceeds that of the mean fluctuation, $\langle\Delta A(t);s\rangle\langle\Delta A(t);s\rangle^{*}$.

It is not possible to determine the stability or instability of a system by $\Phi(t;s)$ since this quantity always decreases. What, then, is the statistical mechanical signification of $\Phi(t;s)$ in nonequilibrium state? Inequality (30) shows a kind of irreversibility. This becomes clearer if we identify the logarithm of the probability distribution function for fluctuation, ΔA, with $-\frac{1}{2}\Phi(t;s)$. This, in general, is only possible for a macrovariable A. Boltzmann's interpretation of the second law of thermodynamics is then that the system evolves to states with higher probabilities. Another significance of (30) is the mixing property. If $\Phi(t;s)$ really tends to zero, then this can be equally expressed as

$$\lim_{|t-s|\to\infty} \langle \Delta\widetilde{A}_i(t)\Delta\widetilde{A}_j^*(s)\rangle_0 = 0 \qquad (i,j=1,2,\ldots,n). \tag{36}$$

Here $\Delta\widetilde{A}(t)$ is defined as

$$\Delta\widetilde{A}(t) = \xi(t)^{-1}\Delta A(t), \tag{37}$$

where $\xi(t)$ is an n by n matrix satisfying the relation

$$\xi(t)\xi(t)^* = \sigma(t) \tag{38}$$

It is to show that

$$\langle \Delta\widetilde{A}_i(t)\Delta\widetilde{A}_j^*(t)\rangle_0 = \delta_{ij}\,. \tag{39}$$

That is, $\langle \Delta\widetilde{A}(t)\Delta\widetilde{A}^*(t)\rangle_0$ is a unit matrix. Then, (36) can easily be found since $\Phi(t;s)$ can be rewritten with the use of $\Delta\widetilde{A}(t)$, as

$$\Phi(t;s) = [\langle \Delta\widetilde{A}(t)\Delta\widetilde{A}^*(s)\rangle_0 \widetilde{X}(s)]^*[\langle \Delta\widetilde{A}(t)\Delta\widetilde{A}^*(s)\rangle_0 \widetilde{X}(s)] \geq 0 \tag{40}$$

where

$$\widetilde{X}(s) = \xi^*(s)X(s). \tag{41}$$

(36) is the mixing condition. It should also be noted that (36) can be reduced to

$$\langle \widetilde{A}_{i,u}\rangle_0 = \widetilde{A}_{i,u}\,, \tag{42}$$

where

$$\widetilde{A}_{i,u} \equiv \lim_{T\to\infty}\frac{1}{2T}\int_{-T}^{T}\widetilde{A}_i(t)U_i(t)dt \tag{43}$$

and where $U_i(t)$ $(i=1,\ldots,n)$ are arbitrary functions of t. Namely, the time average of $\Delta A(t)$ with a weighted function U(t) is equivalent to its ensemble average, if the mixing property (36) is satisfied. It is obvious that in equilibrium the ergodic condition is a special case of (42), since $\langle A(t)\rangle_0 = \langle A(0)\rangle_0$ at thermal equilibrium.

Thermodynamic Description of Irreversible Processes Far from Equilibrium

The first step in achieving a thermodynamic description of irreversible processes from the above statistical mechanical arguments is to assume that the fluctuating force, f(t;s), has a short time correlation viz.

$$\langle f(t;s) f^*(s;s)\rangle_0 = 2\,D(t)\,\delta(t-s). \tag{44}$$

This assumption is justified for a suitable set of macrovariables, A, since macrovariables are functions of a very large number of microscopic degrees of freedom. Also the origin of the fluctuating force are the individual motions of microscopic variables whose time scale can be well separated from that of macrovariables. Then, it is possible to derive, from (18), (20) and (23), that

$$\frac{d}{dt}\langle \Delta A(t);s\rangle = [K(t,t)\sigma(t) - D(t)]\sigma(t)^{-1}\langle \Delta A(t);s\rangle \tag{45}$$

and, also from (21) that

$$\frac{d}{dt}\sigma(t) = K(t,t)\sigma(t) + \sigma(t)K^*(t,t). \tag{46}$$

Since $\tau_1 = 0$, we find from (30) that $\Phi(t,s)$ is a monotonically decreasing function of t. With the aid of (45) and (46) we find that (30) can be written as

$$\frac{d}{dt}\Phi(t;s) = -\,2X^*(t)\,\mathcal{D}(t)\,X(t) \leq 0, \tag{47}$$

where

$$\mathcal{D}(t) \equiv D(t) + D^*(t)\,.$$

Here $X(t) = \sigma(t)^{-1}\langle \Delta A(t);s\rangle$. Notice that $X(t)$ does not depend explicitly on s for (44). From (47) we can find that D(t) is a non-negative matrix.

We define the kinetic coefficient, $L(t)$, by

$$L(t) \equiv -K(t,t)\sigma(t) + \mathcal{D}(t) = -\langle \Delta\dot{A}(t)\Delta A^*(t)\rangle_0 + D(t). \tag{48}$$

Onsager (1931 a, b) proved, for equilibrium situations, that L is symmetric if A_i $(i=1,\ldots,n)$ are even with respect to momenta. He also showed that relation (45) between the set of flows defined by $-\langle\Delta\dot{A}(t);s\rangle$ *, and the set of forces, $X(t)$, can be reproduced through the variational principle. The variational function of this is written in terms of entropy production. Such a thermodynamic argument of irreversible processes has been recognized as a peculiarity of a thermal equilibrium state (together with the reciprocity relation of the kinetic coefficient). In fact, in a nonequilibrium state, $L(t)$ cannot be symmetric, even if A_i $(i=1,..,n)$ are even with respect to momenta. However, suitable extension of the variational principle can exist even in a nonequilibrium state. Let $\mathfrak{J}$ denote a set of flows corresponding to $-\langle\Delta\dot{A}(t);s\rangle$ *. Here, however, $\mathfrak{J}$ is not related to the force X by relation (45). Let us introduce a variational function Ψ' such that,

$$\Psi' \equiv [\mathfrak{J} - L(t)X(t)]^* O' [\mathfrak{J} - L(t)X(t)] \geq 0, \tag{49}$$

where O' is a non-negative n by n matrix. Then, by minimizing Ψ' with respect to $\mathfrak{J}$, we obtain the same relation as (45) between the flow $-\langle\Delta\dot{A}(t);s\rangle$ and the force X:

$$\mathfrak{J}(t) = L(t)X(t). \tag{50}$$

If we choose O' to be $L^*(t)^{-1}\mathcal{D}(t)L(t)^{-1}$, a non-negative matrix for a non-negative matrix D(t), then the minimum condition of Ψ' for a given $X(t)$ is written as

$$\mathfrak{J}^*\{L^*(t)^{-1}\mathcal{D}(t)L(t)^{-1}\}\mathfrak{J} - \{X^*(t)\mathcal{D}(t)L(t)^{-1}\mathfrak{J} + \mathfrak{J}^*L^*(t)^{-1}\mathcal{D}(t)X(t)\} = \tag{51}$$

= minimum with respect to $\mathfrak{J}$ (and $\mathfrak{J}^*$) . The first term corresponds to what is called the Rayleigh dissipation function, and the second term on the left hand side of (51), if (50) is substituted, denotes the excess en-

---

*Strictly speaking, $-\langle\Delta\dot{A}(t);s\rangle$ is a set of excess flows and L is the differential kinetic coefficient.

tropy production and twice the excess entropy production defined by $-\frac{1}{2}\frac{d}{dt}\Phi(t;s)$. Thus (51) is a generalization of Onsager's variational principle (Onsager, 1931 a, b; Nakano, 1973). Onsager's variational principle can be obtained by putting $L = D$, which is, in general, valid only in a thermal equilibrium state. On the other hand, if we choose O' to be $\mathcal{D}(t)^{-1}$, which is also non-negative, the minimum condition of Ψ' is written as

$$\mathcal{J}^*\mathcal{D}(t)^{-1}\mathcal{J} - \{X^*(t)L^*(t)\mathcal{D}(t)^{-1}\mathcal{J} + \mathcal{J}^*\mathcal{D}(t)^{-1}L(t)X(t)\} = \tag{52}$$

= minimum with respect to $\mathcal{J}$ (and $\mathcal{J}^*$).

This also reduces to Onsager's variational principle for a thermal equilibrium state. In this case, however, $-\frac{1}{4}\Psi'$ gives the logarithm of the two gate probability distribution functions for a short time interval Δt (Hashitsume, 1952; Onsager, Machlup, 1953; Machlup, Onsager, 1953):

$$P(\Delta A(t+\Delta t), t+\Delta t; \Delta A(t), t) \quad ,$$

$$\mathcal{J}\Delta t = \Delta A(t+\Delta t) - \Delta A(t) .$$

Each term of (52) is not, in general, equivalent to the excess entropy production or its multiple.

The development of this section can be also obtained from a more primitive basis (as in section 2) by defining the variational function as

$$\Psi' = \langle h^*(t;s) O' h(t;s); s \rangle \geq 0, \tag{53}$$

where

$$h(t;s) = \Delta\dot{A}(t) - \Lambda(t;s)\Delta A(s). \tag{54}$$

Concluding Remarks

The main purpose of this contribution was to examine a statistical-themodynamical view of nonequilibrium states far from equilibrium. This has been done by studying fluctuations of dynamic variables around a reference nonequilibrium state. The macroscopic characteristics of nonequilibrium states near the reference nonequilibrium state are found to be represented by the properties of fluctuations in the reference state. This is the same result as in treatments of the ordinary statistical thermodynamics of irreversible processes. For ordinary statistical thermodynamic treatments of irreversible processes, however, the reference state is chosen as the thermal equilibrium state and then the irreversible processes around the thermal equilibrium state are considered.

Two questions have thus arisen. Firstly, how can we derive a general law governing fluctuations in a state far from equilibrium? Secondly, what is the practical role of a reference, nonequilibrium state? The first question has been discussed in the text by treating the deviation from the most probable value as the *error*, and, assuming that the fluctuating force, which is the deviation from the systematic or the most probable motion, has a finite correlation time. Thus an extremum law (30), which corresponds to the second law of thermodynamics, has derived. Since, in a real system, a large number of degrees of freedom exist, such an assumption for the fluctuating force would be plausible even in nonequilibrium states.

A brief introduction to the answer to the second question is as follows. Consider an open system which is supplied with energy, matter etc. from it's surroundings. The state parameter of the surroundings are the nonequilibrium average $\langle A(t) \rangle_0$, and these also characterize a reference state. A change of state of the system, i. e., the open system in this case, is described by the fluctuation $\Delta A(t)$. For instance, let us consider a steady nonequilibrium state. Then $-\langle \dot{A}(t) \rangle_0$ denotes a set of constant flows, which we regard as the external conditions, and $-\langle \Delta\dot{A}(t);\mathcal{A}\rangle$ denotes the set of excess flows which is artificially produced in the system. The dynamic aspect of $\Delta A(t)$ under the external condition, $\langle A(t) \rangle_0$, is described by the Langevin type equation (16) and the absolute evolution of mean fluctuation, $\langle \Delta A(t) ;\mathcal{A} \rangle$, is determined by the extremum principle (30).

III. Kinetic Equations

Generalized Fokker-Planck Equation in the Theory of Irreversible Processes

D. N. Zubarev

The problem of the derivation of the Fokker-Planck equation for a nonequilibrium system from *first principles* occupies a special place in the theory of nonequilibrium processes because many important theoretical results were first obtained for just this problem.

In the derivation of the Fokker-Planck equation, in the theory of stochastic processes, the existence of some probabilities for the dynamic system to pass from one state to another is assumed (Chandrasekar, 1943; Prokhorov, Rozanov, 1973). In 1939 Bogolyubov and Krylov (see also Bogolyubov, 1970) derived the Fokker-Planck equation for the distribution function of a small sub-system interacting with a thermal bath from *first principles*, i. e. from the Liouville equation by means of perturbation theory with respect to small interaction with the environment. A more complete solution of this problem is given in a recent paper by Bogolyubov (1978).

The generalized Fokker-Planck equation for distribution functions of rough hydrodynamic variables is widely used in modern phase transition theory (Kawasaki, 1970, 1971). These equations are usually derived by Zwanzig's method (Zwanzig, 1960, 1961) which is based on the introduction of projection operators into the space of rough variables, and, in the choice of projection operators one usually follows the method suggested by Mori (1965). For the case when the nonequilibrium state is determined by the average values of some rough variables, $\hat{a}_i$ (functions of the coordinates and momenta of all the particles) and by all their possible fluctuations, Kawasaki (1970, 1971) has used Mori's projection operator (Mori, 1965) for all possible products of $\hat{a}_i$ by introducing appropriate infinite matrices.

Thermodynamics and Kinetics of Biological Processes

Following the treatment in the paper by Zubarev and Khazanov (1978), it is possible to show

i) how the generalized Fokker-Planck equation can be derived with the aid of a nonequilibrium statistical operator (Zubarev, Kalashnikov, 1970; Zubarev, 1974) and,

ii) how this method leads to a certain form of projection operator. In this example, we shall give a presentation somewhat different of the use of a nonequilibrium statistical operator, which proceeds from the formal solution of the initial problem for the Liouville equation and averaging it over the initial state. We shall present this problem as a simple example to illustrate the derivation of the equation that describes an irreversible process which proceeds from the mechanics of properties of a many-particle system in the form of a time-reversible Liouville equation for the particle distribution function. In this example, we wish to show how one can express, exactly, the idea of irreversible loss of the information about the initial state of the system. This is the fundamental problem of nonequilibrium statistical mechanisms. This discussion may be used as a general scheme for the consideration of many other problems of nonequilibrium statistical mechanics.

Let the nonequilibrium state be characterized by a distribution of some set of rough variables $\hat{a}_1, \ldots, \hat{a}_n$ designated by $\hat{a}$. These may be dynamic variables - Fourier components of the density of energy, momentum, and the number of particles.

In classical statistical mechanics, we derive the Fokker-Planck equation for the rough variables distribution function as

$$f(a,t) = \langle \delta(\hat{a}-a) \rangle^t = Sp(\rho(t)\delta(\hat{a}-a)), \tag{1}$$

where $\delta(\hat{a}-a) \equiv \prod_i \delta(\hat{a}_i - a_i)$ may be regarded as the main dynamic variable whose average value characterizes a nonequilibrium state of the system. Sp indicates integration over the coordinates and momenta of all the particles with the normal factor $1/(\hbar)^{3N} N!$. $\rho(t)$ is a nonequilibrium distribution function corresponding to our choice of dynamic variables which itself has to be defined as a particular solution of the Liouville equation that depends only on time through f(a, t) in the form of the functional

$$\rho(t) = \rho(\ldots f(a,t)\ldots). \tag{2}$$

Consequently, the definition of the function of the rough variables (1) is, in fact, an equation for it.

Distribution function (2) may be far from equilibrium. It is, however, convenient to construct an auxiliary distribution, which corresponds to the extremum of the information entropy, $S_u = -Sp(\rho \ln \rho)$, under the supplimentary conditions that the mean value, $\langle \delta(\hat{a} - a) \rangle$, is knwon and the normalization condition, $Sp\,\rho = 1$, is fixed. This distribution will be referred to as quasi-equilibrium and

$$\begin{aligned} \rho_q(t) &= exp\{-\Phi(t) - \int \mathcal{F}(a,t)\delta(\hat{a}-a)da\} = \\ &= exp\{-\Phi(t) - \mathcal{F}(\hat{a},t)\}, \end{aligned} \tag{3}$$

where $\Phi(t)$ is found from the normalization condition

$$\Phi(t) = \ln Sp\, exp(-\mathcal{F}(\hat{a},t)). \tag{4}$$

The function $\mathcal{F}(a,t)$ is found from the condition of coincidence of quasi-equilibrium averages of $\delta(\hat{a}-a)$ with their true averages $\langle \delta(\hat{a}-a)\rangle_q^t$, which are obtained with the aid of $\rho(t)$ as

$$\langle \delta(\hat{a}-a)\rangle_q^t = \langle \delta(\hat{a}-a)\rangle^t = Sp(\rho(t)\delta(\hat{a}-a)) = f(a,t). \tag{5}$$

Equation (5), which is the self-consistency condition of the solution of the Liouville equation, may be written in the form

$$f(a,t) = e^{-\Phi(t) - \mathcal{F}(a,t)} Sp(\delta(\hat{a}-a)). \tag{6}$$

With the aid of (6), one can exclude the normalization functions, $\Phi(t)$ and $\mathcal{F}(a,t)$, and write the quasi-equilibrium distribution in the form

$$\rho_q(t) = \left.\frac{f(a,t)}{Sp\,\delta(\hat{a}-a)}\right|_{a=\hat{a}} = \int \frac{f(a,t)}{W(a)}\delta(\hat{a}-a)da, \tag{7}$$

where

$$W(a) = Sp\,\delta(\hat{a}-a) \tag{8}$$

is the structure function of the surface. The distribution function of rough variables f(a, t) has been normalized to unity

$$\int f(a,t)\,da = 1, \tag{9}$$

which follows from normalization of the quasi-equilibrium distribution $Sp\,\rho_q(t) = 1$.

It is obvious that $\rho_q(t)$ does not implie that it is close to the equilibrium distribution. The name quasi-equilibrium distribution function has been selected since, by choosing the densities of energy, momentum and the number of particles as the main system of dynamic variables, we arrive at a local equilibrium distribution.

Knowledge of the quasi-equilibrium distribution function makes it possible to define the entropy of a nonequilibrium state as the mean logarithm (with a negative sign) of the quasi-equilibrium distribution function

$$S = -\langle \ln \rho_q(t) \rangle^t = \Phi(t) + \int F(a,t) f(a,t)\,da. \tag{10}$$

From which, making use of (6) and (9), we derive

$$S = \int f(a,t) \ln\left(\frac{f(a,t)}{W(a)}\right) da. \tag{11}$$

Now we shall need the nonequilibrium distribution function derived from the Liouville equation,

$$\frac{\partial \rho(t)}{\partial t} + iL\rho(t) = 0, \tag{12}$$

where L is the Liouville operator

$$iL\rho = \{H, \rho\} = \sum_k \left(\frac{\partial H}{\partial p_k} \cdot \frac{\partial \rho}{\partial q_k} - \frac{\partial H}{\partial q_k} \cdot \frac{\partial \rho}{\partial p_k}\right); \tag{13}$$

H is the Hamiltonian function of the system; p_k, q_k are particle momenta and coordinates. The desired solution is that which depends on time through f(a, t).

The solutions of the Liouville equation satisfying the initial condition of its coincidence with the quasi-equilibrium equation at $t = t_0$ which is

$$\rho(t)\big|_{t=t_0} = \rho_q(t_0) \tag{14}$$

has the form

$$\rho(t,t_0) = e^{-iL(t-t_0)}\rho_q(t_0) = \int \frac{f(a,t_0)}{W(a)}\delta(\hat{a}(t_0-t)-a)\,da \tag{15}$$

where

$$\hat{a}(t_0-t) = e^{-iL(t-t_0)}\hat{a}(\ldots p_k, q_k \ldots) = \hat{a}(\ldots p_k(t_0-t), q_k(t_0-t)\ldots).$$

This contains a non-physical dependence on the initial moment, t_0 , which must be eliminated. To this end we average the solution of (15) over the initial moments within the interval from t_0 to t ;

$$\rho(t) = \frac{1}{t-t_0}\int_{t_0}^{t} e^{-iL(t-t_0)}\rho_q(t_0)\,dt_0 . \tag{16}$$

It is assumed that $t-t_0$ is sufficiently large to damp the initial non-physical states. Formally this requires that $t-t_0 \to \infty$ after the thermodynamic limit has been found in calculating the averages with the aid of $\rho(t)$.

Solution (16) satisfies the Liou ville equation with the right-hand side - an infinitely small source that loses symmetry under time reversal i. e.

$$\frac{\partial \rho(t)}{\partial t} + iL\rho(t) = -\frac{\rho(t)-\rho_q(t)}{t-t_0} , \tag{17}$$

where $t_0 - t \to -\infty$ after the thermodynamic limiting process in the calculation of the averages. This break-down of symmetry is essential since the majority of paradoxes in the theory of nonequilibrium processes and statistical mechanics are connected with the symmetry of the Liou ville equation under time reversal. As is known from the theory of quasi-averages in statistical mechanics (Wagner, 1966; Bogolyubov, 1971, 1973) infinitesimal break-down of Hamiltonian symmetry may lead to a finite result for equilibrium averages. This is so since the operations of the perturbation that destroys the symmetry and declines to zero the calculation of the thermodynamic limit are not permutable. It is natural to expect that such quasi-averages also exist in the nonequilibrium case (Zubarev, 1973).

The limiting theorem below follows: from Laplacian integral theory (Ditkin, Kuznetsov, 1951): if the function f(t) is limited from below and there exists one of the limits

$$\lim_{T\to\infty}\frac{1}{T}\int_0^T f(t)\,dt \quad \text{or} \quad \lim_{\varepsilon\to+0}\int_0^\infty e^{-\varepsilon t} f(t)\,dt ,$$

the other exists too, and both are equal. Hence, the limiting transition, $t_0 - t \to -\infty$, may be replaced by $\varepsilon \to +0$ and one obtains

$$\rho(t) = \varepsilon \int_{-\infty}^{0} e^{\varepsilon t_1} e^{i t_1 L} \rho_q(t + t_1)\, dt_1 , \tag{18}$$

where $\varepsilon \to +0$ after the thermodynamically limiting process. Formula (18) admits a simple probability interpretation. Let us rewrite it in the form

$$\rho(t) = \int_{-\infty}^{t} \frac{1}{T} e^{-\frac{t-t_0}{T}} e^{-iL(t-t_0)} \rho_q(t_0)\, dt_0 , \tag{19}$$

where $T \to \infty$.

Thus, the system may be considered to evolve from the state $\rho_q(t_0)$, beginning at time t_0 as an isolated state with the Liouville operator, L, making random transitions from one phase trajectory to another with the exponential probability $W(t-t_0) = \frac{1}{T} e^{-\frac{t-t_0}{T}}$. These random transitions may be interpreted as the influence of the thermal bath. The mean interval, T, between temperature jumps in the thermal bath tends to infinity in the averages calculated with the aid of $\rho(t)$.

Solution (18) satisfies the Liouville equation with the infinitesimal source in the right-hand side of the equation

$$\frac{\partial \rho(t)}{\partial t} + iL\rho(t) = -\varepsilon(\rho(t) - \rho_q(t)), \tag{20}$$

which breaks down its invariance under time reversal. Note that the source (20) has the form of the relaxation term that describes the relaxation of $\rho(t)$ to a quasi-equilibrium with the mean time, $T = \varepsilon^{-1}$, after the thermodynamically limiting process.

Equation (20) may be written in the form

$$\left(\frac{\partial}{\partial t} + iL + \varepsilon\right)\Delta\rho(t) = -\left(\frac{\partial}{\partial t} + iL\right)\rho_q(t), \tag{21}$$

where $\Delta\rho(t) = \rho(t) - \rho_q(t)$.

Differentiating the quasi-equilibrium distribution with respect to time, we are lead to

$$\frac{\partial}{\partial t}\rho_q(t) = -\int W^{-1}(a)\,Sp(\delta(\hat{a}-a)iL\Delta\rho(t))\,\delta(\hat{a}-a)\,da - \\ -\int W^{-1}(a)\,Sp(\delta(\hat{a}-a)iL\rho_q(t))\,\delta(\hat{a}-a)\,da\,, \tag{22}$$

which may be written in the form

$$\frac{\partial}{\partial t}\rho_q(t) = -\mathcal{P}_q iL\Delta\rho(t) - \mathcal{P}_q iL\rho_q(t) = \mathcal{P}_q\,\dot{\rho}(t)\,, \tag{23}$$

where

$$\mathcal{P}_q A = \int W^{-1}(a)\,Sp(A\delta(\hat{a}-a))\,\delta(\hat{a}-a)\,da \tag{24}$$

which is the projection operator that realizes the transition to tough variables. Indeed, $\mathcal{P}_q^2$ is equal to $\mathcal{P}_q$ by virtue of the relation

$$Sp(\delta(\hat{a}-a)\delta(\hat{a}-a)) = W(a)\,\delta(a-a'). \tag{25}$$

The operator $\mathcal{P}_q$ does not change the quasi-equilibrium distribuion function, $\mathcal{P}_q\rho_q(t) = \rho_q(t)$ and transforms the nonequilibrium distribution function to the quasi-equilibrium function $\mathcal{P}_q\rho(t) = \rho_q(t)$. This keeps the set of rough variables, $\hat{a}_i$, and their products and functions unchanged. In particular, $\mathcal{P}_q\,\delta(\hat{a}-a)$ equals $\delta(\hat{a}-a)$. Thus, $\mathcal{P}_q$ realizes projection to the space of rough variable functions. This is necessary since the functions of rough variables are rough variables themselves, and when the fluctuations of rough variables are not small, we cannot restrict ourselves to projection into $\hat{a}_i$ only.

Substituting (23) into (21), we have

$$\left(\frac{\partial}{\partial t} + (1-\mathcal{P}_q)iL + \varepsilon\right)\Delta\rho(t) = -\int da\,\mathcal{X}(a)\frac{\partial}{\partial a}\left(\frac{f(a,t)}{W(a)}\right), \tag{26}$$

where

$$\mathcal{X}(a) = (1-\mathcal{P}_q)\,\delta(\hat{a}-a)\,iL\hat{a} \tag{27}$$

is the density of the generalized forces corresponding to the random forces in Mori's theory (Mori, 1965). In our case, they are not random forces but dynamic variables depending on the phase variables of all the particles. The action of the operator $1-\mathcal{P}_q$ means that the slowly changing part is subtracted from the total force. This justifies the term *random force*. Projections of generalized force densities into the space of rough variables are equal to zero:

$$\mathcal{P}_q X(a) = 0 . \quad (28)$$

This property corresponds to orthogonality of random forces and rough variables in Mori's theory (Mori, 2965).

By integrating equation (26), we find the nonequilibrium distribution function

$$\rho(t) = \rho_q(t) - \int da \int_{-\infty}^{t} dt' e^{(t-t')[(1-\mathcal{P}_q)iL+\varepsilon]} X(a) \frac{\partial}{\partial a}\left(\frac{f(a,t')}{W(a)}\right). \quad (29)$$

The role of a generalized flow in the space of rough variables is played by the quantity

$$\mathcal{J}(a) = (1-\mathcal{P}_q) iL\delta(\hat{a}-a) = -\frac{\partial}{\partial a} X(a) , \quad (30)$$

since $L\delta(\hat{a}-a) = -\frac{\partial\delta(\hat{a}-a)}{\partial a} L\hat{a}$.

In a quasi-equilibrium state, the mean generalized flow is zero

$$\begin{aligned} \langle \mathcal{J}(a)\rangle_q &= Sp(\rho_q(1-\mathcal{P}_q) iL\delta(\hat{a}-a) = \\ &= \langle iL\delta(\hat{a}-a)\rangle_q - Sp(\rho_q \mathcal{P}_q iL\delta(\hat{a}-a)) = 0 \end{aligned} \quad (31)$$

and its true mean value is associated with $\frac{df(a,t)}{dt}$ by the relation

$$\begin{aligned} \langle \mathcal{J}(a)\rangle^t &= \langle \dot{\delta}(\hat{a}-a)\rangle^t - \langle \dot{\delta}(\hat{a}-a)\rangle_q = \\ &= \frac{df(a,t)}{dt} - \langle iL\delta(\hat{a}-a)\rangle_q . \end{aligned} \quad (32)$$

The mean flow (30) is calculated with the aid of (29) as

$$\langle \mathcal{J}(a)\rangle^t = \frac{\partial}{\partial a} \int da' \int_{-\infty}^{t} dt' e^{\varepsilon(t'-t)} K(a,a',t-t') \frac{\partial}{\partial a'}\left(\frac{f(a',t')}{W(a')}\right) , \quad (33)$$

where the function

$$K(a,a',t-t') = Sp(X(a) e^{(t-t')(1-\mathcal{P}_q)iL} X(a')) \quad (34)$$

determines the probability of transition from the state a to the state a' for the time $t-t'$ under the influence of the random forces (27). The mean value of $\dot{\delta}(\hat{a}-\alpha)$ in the quasi-equlibrium state is equal to

$$\langle iL\delta(\hat{a}-a)\rangle_q = -\frac{\partial}{\partial a}\int \frac{f(a',t)}{W(a')}\,Sp\,(\delta(\hat{a}-a)\delta(\hat{a}-a')\,iL\hat{a})\,da' = -\frac{\partial}{\partial a}\left(\frac{f(a,t)}{W(a)}\right)Sp\,(\delta(\hat{a}-a)\,iL\hat{a}), \tag{35}$$

since

$$Sp\,(\delta(\hat{a}-a)\delta(\hat{a}-a')\,iL\hat{a}) = Sp\,(\delta(a'-a)\,iL\hat{a})\,\delta(a-a'). \tag{36}$$

Consequently,

$$\langle\dot{\delta}(\hat{a}-a)\rangle_q = -\frac{\partial}{\partial a}\,v(a)\,f(a,t), \tag{37}$$

where

$$v(a) = \frac{Sp\,(\delta(\hat{a}-a)\,iL\hat{a})}{W(a)} \tag{38}$$

are the rates of change of the rough variables.

Taking (32), (33), (37) into account, we finally obtain the generalized Fokker-Planck equation in the form

$$\frac{\partial f(a,t)}{\partial t} + \frac{\partial}{\partial a}\,v(a)f(a,t) = \frac{\partial}{\partial a}\int da'\int_{-\infty}^{t} e^{\varepsilon(t'-t)}K(a,a',t-t')\frac{\partial}{\partial a'}\left(\frac{f(a',t')}{W(a')}\right). \tag{39}$$

Equation (39) is a retardation-type integrational-differential equation, since integration proceeds over past, i. e. negative times. As distinct from the corresponding Zwanzig equation, the integration is performed from $-\infty$ to t, but not from zero to t. This is associated with the choice of boundary conditions in a remote past and not of the initial time, t = 0. In the latter case, in order that the equation be valied we should assume that t is sufficiently large when compared to the random force correlation time. The projection operator, (24), is the same as Zwanzig's operator. With the aid of the method of nonequilibrium statistical operators, the generalized Fokker-Planck equation has been obtained (Bashkirov, Zubarev, 1970; Mishnev, 1977; Zubarev, Khazanov, 1978). For other conclusions see Hashitsume (1952); van Kampen (1954 a, b); and Sewel (1965).

Equation (39) is exact, therefore. It seems surprising that equation describing a nonequilibrium process has been obtained from a time-reversible Liouville equation. This is due to the fact that we have smoothed the solution by averaging it over the initial moments of time. This is the same procedure as setting a boundary condition of coincidence between $\rho(t)$ and

$\rho_q(t)$ in a remote past. Such a solution has causal character. If we set up the condition of coincidence between $\rho(t)$ and $\rho_q(t)$ in a remote future, we would obtain, not a retarded, but an advanced Fokker-Planck equation with the opposite sign for the differential operator and with integration limits from ∞ to t. It is rather interesting that such anti-causal solutions lead, not to an increase, but to a decrease of entropy. This property of transport equations was first noticed by Cohen and Berlin (1960) in a derivation of the Boltzmann kinetic equation for a gas using Bogolyubov's condition of correlation weakening, not in a remote past as is commonly admitted, but in a remote future. This property was also demonstrated by us in an example of the derivation of the master equation by the method of nonequilibrium statistical operators (Zubarev, Kalashnikov, 1971). Thus, the derivation of an irreversiblel Fokker-Planck equation from the reversible Liouville equation is a consequence of the choice of retarded solutions - a simple example of reduced description of the nonequilibrium state.

If we assume that the correlation of the random forces (27) decreases rather rapidly with time and for different states, i. e. the kernel, (34), of the integral equation has a sharp peak at $t \sim t'$, $a \sim a'$, then equation (39) may be simplified by taking the slowly changing quantities from under the integral sign, and representing them in the form of the usual Fokker-Planck equation

$$\frac{\partial f(a,t)}{\partial t} + \frac{\partial}{\partial a} v(a) f(a,t) = \frac{\partial}{\partial a} \mathcal{L}(a) W(a) \frac{\partial}{\partial a} \left(\frac{f(a,t)}{W(a)} \right), \tag{40}$$

where the equation coefficient

$$\mathcal{L}(a) = [W(a)]^{-1} \int da' \int_{-\infty}^{t} e^{\varepsilon(t'-t)} K(a,a',t-t')dt \tag{41}$$

is expressed through the time correlation function of the generalized force densities (27). Equation (40) is a differential equation and does not contain a retardation term. In the case when the rough variables, $\hat{a}_i$, are Fourier components of the densities of energy, momentum and the number of particles, equation (40) is functional in the space of variables. This corresponds to the real and imaginary parts of complex collective variables, i. e. Fourier components. By definition, the δ-function of the complex variable is considered to be equal to the product of δ-functions of the real and imaginary parts of the argument.

Using the scheme presented one can obtain practically all the known types of equations of nonequilibrium statistical mechanics.

For example, in a hydrodynamic regime of a nonequilibrium system, when the state is described by the mean values of the densities of energy, $H(x)$, momentum, $p(x)$, and the number of particles, $n(x)$, their Fourier components H_k, p_k, n_k are conveniently chosen as $\hat{a}_i$. This choice implies that we neglect fluctuations in these quantities. This is admissible far from the phase transition point. Then

$$\rho_q(t) = \exp\left\{-\Phi(t) - \sum_i \mathcal{F}_i(t)\hat{a}_i\right\}, \tag{42}$$

$\mathcal{F}_i(t)$ being found from the condition $\langle \hat{a}_i \rangle_q = \langle \hat{a}_i \rangle$. Following the above scheme, we can obtain generalized hydrodynamic equations, i. e. heat conductivity and Navier-Stokes equations with non-locality and memory (Sergeev, 1974; Tishchenko, 1976), and, for $\mathcal{P}_q$ the Kanasaki-Gunton projection operator (Kawasaki, Gunton, 1973).

For the case of two interacting sub-systems, which correspond to the statistical operators

$$\rho_1(t) = Sp_{(2)}\rho(t), \quad \rho_2(t) = Sp_{(1)}\rho(t) \tag{43}$$

the product

$$\rho_q(t) = \rho_1(t)\,\rho_2(t) \tag{44}$$

may be chosen as a quasi-equilibrium distribution.

Then, for ρ_1 and ρ_2 we obtain a set of bound master equations with the only difference that integration over time is performed, not from zero to t, but from $-\infty$ to t. For a system of weakly interacting particles obeying the correlation weakening condition, then we need

$$\rho_q(t) = \prod_{i=1}^{N} \rho_1(i), \tag{45}$$

where $\rho_1(i)$ is a one-particle distribution function. We obtain, in the lowest approximation with respect to density, a kinetic Boltzmann equation (Zubarev, Novikov, 1972).

The method suggested is particularly convenient for constructing transport equations for very complicated systems, when we are interested only in the general form of equations and not in the numerical values of the constituent constants.

The Diffusion Approximation for Markov Processes

N. G. van Kampen

"Warum willst du dich von uns allen
Und unsrer Meinung entfernen?"
Ich schreibe nicht, euch zu gefallen,
Ihr sollt was lernen!

Goethe

Markov processes in physics, chemistry, and biology are often regarded as generalized diffusion and accordingly described by a Kolmogorov equation or a Langevin equation. This procedure is exact if the Lindeberg condition is obeyed. It is here argued, however, that it is at best only approximately obeyed and that therefore the mathematical theory derived from it should be replaced with an approximation procedure. Such a procedure, in the form of a power series expansion in a parameter Ω is described. The conclusion is that the diffusion approximation is inconsistent, except for a special subclass of Markov processes. This distinction has nothing to do with the distinction between discrete and continuous variables.

Markov Processes and the Diffusion Approximation

Let $Y(t)$ be a continuous time Markov process whose range consists of real numbers. The range may be a *discrete* countable set, or it may be *continuous*, covering a finite or infinite interval on the real line. A combination of both may also occur but will not be considered here.

Thermodynamics and Kinetics of Biological Processes

For $t > t_0$ let $P(y,t\,|\,y_0,t_0)$ be the transition probability, i. e., the probability (or probability density) of $Y(t)$ conditional on $Y(t_0) = y_0$. As a consequence of the Markov property it obeys the Chapman-Kolmogorov or master equation

$$\frac{\partial P(y,t\,|\,y_0,t_0)}{\partial t} = W\,P(y,t\,|\,y_0,t_0), \tag{1}$$

where W is a linear operator acting on the y-dependence. Its matrix element $W(y\,|\,y')$ for $y \neq y'$ is the transition probability per unit time from y' to y and for simplicity is here supposed to be independent of time.

The continuous range Markov processes contain a subclass of those for which W has the form of a differential operator

$$\frac{\partial P}{\partial t} = -\frac{\partial}{\partial y}\,a(y)\,P + \frac{1}{2}\frac{\partial^2}{\partial y^2}\,b(y)\,P, \tag{2}$$

with two functions $a(y)$, $b(y)$ with $b(y) > 0$. It has been proved that higher order differential operators cannot occur (Pawula, 1967 a). The conditions that characterize this subclass are

$$\lim \frac{\langle \Delta Y \rangle}{\Delta t} = a(y)\,;\ \lim \frac{\langle (\Delta Y)^2 \rangle}{\Delta t} = b(y)\,,\ \lim \frac{\langle (\Delta Y)^\nu \rangle}{\Delta t} = 0$$

where y is the value of Y at any time t, $\Delta y = Y(t+\Delta t) - Y(t)$, the averages are taken with fixed y, the limits refer to $\Delta t \to 0$, and $\nu = 3, 4, \ldots$ Under these conditions Kolmogorov (Kolmogorov, 1931) derived (2), which is called the (second) Kolmogorov equation*, the (nonlinear) Fokker-Planck equation, or the (generalized) diffusion equation. The third condition was later replaced with the Lindeberg condition

$$\text{Prob.}\{|Y(t+\Delta t) - Y(t)| > \delta\} = O(\Delta t) \tag{3}$$

for any $\delta > 0$ (Feller, 1971). Conversely, when (2) holds these conditions are satisfied.

Unfortunately the impeccable mathematics of Kolmogorov's proof have obscured the question of the validity of the Lindeberg condition for physical processes. Some physicists are under the impression that (2) must hold

---

*(but this name is also used for the general master equation (1))

whenever the range of Y is continuous (Horsthemke, Brenig, 1977). This misconception is abetted by the fact that the term *continuous process* has been used as a name for precisely that subclass of *continuous range processes* that obey the condition (3) (Gnedenko, 1962).

We shall argue that *for physical processes the Lindeberg condition is at best only approximately oebeyed*. Hence replacing W with a second order differential operator is not just a mathematical identity but an approximation. To justify it one must perform a systematic expansion of the master equation. It then turns out that one is *not* led to (2), except in a special case treated at the end of this article. In other cases (2) is inconsistent, inasmuch as it contains terms of the same order in the expansion as those that are neglected by omitting higher derivatives.

Incidentally, it may be added that (2) is mathematically equivalent with the (nonlinear) Langevin equation

$$\dot{y} = a(y) + \sqrt{b(y)}\, l(t)$$

where $l(t)$ is Gaussian white noise. This equation may alternatively be written in the Itô form

$$dy = a(y)dt + \sqrt{b(y)}\, dw(t)$$

where $w(t)$ is the Wiener process. As a corollary to our result we therefore also find that this equation is at best only approximately fulfilled and only in a special case. It is merely a step in the systematic approximation scheme, rather than a fundamental starting point for fluctuation theory (Graham, 1973, 1978). Thus the discussion about the correct mathematical interpretation of the Itô equation (Mortensen, 1969) is moot, because the expansion of the master equation automatically leads to well-defined equations (van Kampen, to be published).

To avoid confusion we add the obvious remark that for any given stochastic process $Y(t)$ one may write

$$\dot{y} = a(y) + f(t)$$

thereby *defining* the stochastic term f. But not only will f in general not be Gaussian white noise, also *its stochastic properties depend on those of* Y (Akcasu, 1977; Bedeaux, 1977; Agarwal et al., 1978; Onuki, 1978). This device is therefore of no use for finding the fluctuation of Y.

Critique of the Lindeberg Condition

For a continuous range process one must have

$$P(y,t|y_0,t_0) \to \delta(y-y_0) \quad \text{as} \quad t \to t_0 .$$

The question is: how is this "limit" approached, i. e., how does $P(y,t|y_0,t_0)$ behave when $t-t_0 = \Delta t$ is small? There are two fundamentally different classes.

Class (i). Jumps in Y have some average size, independent of Δt, but the probability of a jump to occur becomes small as $\Delta t \to 0$. (The only difference with discrete range processes is that here the jump size is not restricted to discrete values). For very small Δt the probability that a jump from y_0 to y occurs is $W(y|y_0)\Delta t$, and the probability that two or more jumps occur is of higher order in Δt. The transition probability per unit time $W(y|y_0)$ is the probability distribution for jumps of various sizes. For any $\delta > 0$ and small Δt one has therefore

$$\text{Prob}\{|Y(t+\Delta t)-Y(t)|>\delta\} = \text{proportional to } \Delta t,$$

and therefore the condition (2) does not hold.

Class (ii). Even for small Δt there are still very many, very small jumps in Y. The central limit theorem leads one to expect a Gaussian:

$$P(y,t_0+\Delta t|y_0,t_0) = \frac{1}{\sqrt{4\pi D\Delta t}} \exp\left[-\frac{(y-y_0)^2}{4D\Delta t}\right]$$

with some constant D. In this case the Lindeberg condition is obeyed and of course this P does obey a differential equation

$$\frac{\partial P}{\partial t} = D\frac{\partial^2 P}{\partial y^2} . \tag{4}$$

However, this is the ordinary diffusion equation, which is homogeneous in y. The essential assumption implied by the Lindeberg condition is that one can choose Δt so small that: (a) many jumps occur during Δt ; and (b) the width $\sigma^2 \Delta t$ covered by them is small that the dependence of the coefficients on y, is not yet felt.

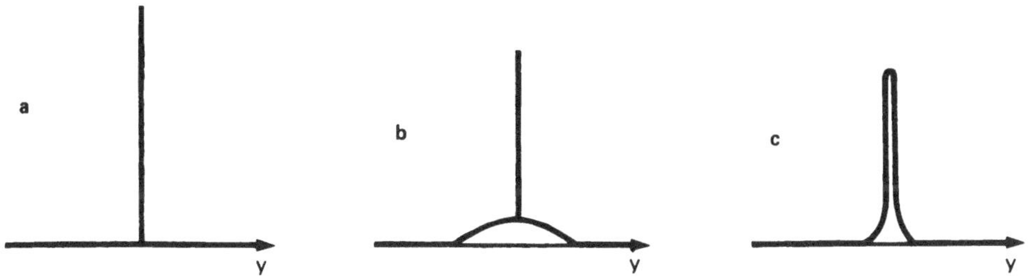

Fig. 17 a) Initial delta distribution at $t = 0$.
b) Distribution at $\Delta t > 0$ for class (i) obeying (5).
c) Distribution at $\Delta t > 0$ for class (ii) obeying (2).

The difference between both classes is clarified in Fig. 17. In class (i) after a small Δt the initial delta function is still visible since there is a chance that no jump took place. However, it has developed a foothill whose shape is $W(y|y_0)$ and accordingly the height of the delta function has decreased. In fact, to order one has

$$P(y, t+\Delta t | y_0, t_0) = W(y|y_0)\Delta t + \delta(y - y_0)\left[1 - \Delta t \int W(y'|y_0)\, dy'\right]. \tag{5}$$

In class (ii), however, the initial delta function starts to broaden immediately, and after any Δt , however small, it has changed into a narrow but smooth Gaussian-like peak. In other words, $Y(t)$ varies through infinitely many infinitely small jumps. Which picture applies to physics?

Here is an example of a physical Markov process with continuous range that belongs to class (i) and does not obey the Lindeberg condition (Fig. 18). Let $Y(t)$ be the velocity V of a Brownian particle, in this connection often called a Rayleigh particle (Rayleigh, 1891; Hoare, Rahman, 1973). Let M be its mass, m the mass of the surrounding fluid molecules, T their temperature, and $v = \sqrt{kT/m}$ their thermal velocity. Obviously the number of collisions in any finite time interval will be finite (with probability one) and the jumps in V are of the order

$$\frac{m}{M} v = \frac{1}{M} \sqrt{mkT} . \tag{6}$$

For small intervals Δt the probability for a jump to occur is proportional to Δt, but the size of the jumps that do occur is determined by (6).

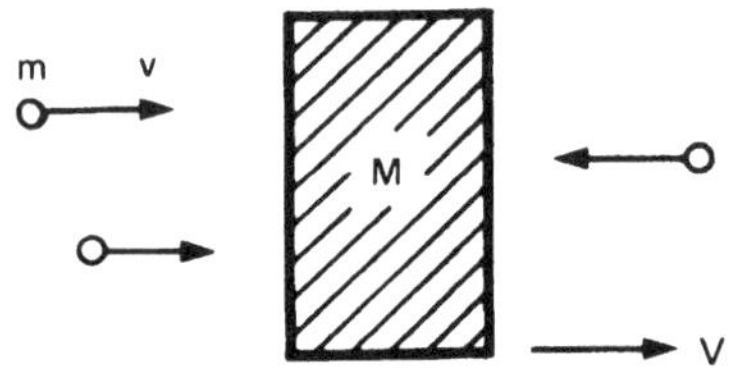

Fig. 18 The one-dimensional Rayleigh particle.

By similar considerations one can convince oneself that the Lindeberg condition is not satisfied either by the *position* of a Brownian particle, the current fluctuations in a resistor, the fluctuations of the electromagnetic field in a medium, etc. The basic reason is that fluctuations in physical processes are due to the particulate nature of matter, and the magnitude of the individual jumps is determined by the mass, charge, or spin of the particles involved and cannot therefore be infinitely small. It is obviously not satisfied by such processes as radio-active decay, emission and absorption of photons, and chemical reactions, because they have a discrete range. For the same reason problems in population dynamics and epidemiology do not satisfy Kolmogorov's requirement.

Linear and Nonlinear Problems

Write the transition probability $W(y|y')$ as a function $W(y';\Delta y)$ of the starting point y' and the length $\Delta y = y - y'$ of the jump. As a function of Δy it is a more or less narrow peak, whose width is a measure for the size of the jumps. As to the dependence on y' we distinguish various cases.

1. *Homogeneous case:* $W(y';\Delta y)$ for fixed Δy does not depend on y'. The only quantity with which the jump size can be compared has to be a property of the solution, viz., the distance over which P(y,t) varies appreciably. For homogeneous problems any solution becomes flat in the course of time, and hence (2) holds in the limit of long time. This is the reason why Einstein and Smoluchowski could use this equation for the position of a Brownian particle.

2. *Linear case:* $W(y';\Delta y)$ depends linearly on y'. In this case the solutions tend typically to an equilibrium solution $P^{eq}(y)$ with a finite width. The validity of the diffusion approximation is now determined by the ratio of the jump size to the variation distance of P^{eq}. This ratio is an inherent property of W; i. e., it is determined by the parameters in W. In this case, therefore, a systematic approximation must be based on the expansion in powers of such a parameter. For instance, for the Rayleigh particle the equilibrium distribution is

$$P^{eq}(V) = \text{const} \exp[-MV^2/2kT],$$

and has therefore a variation distance $\sqrt{kT/M}$. In order to ensure that the jump size (6) is small one is forced to consider values of M for which $\sqrt{m/M} \ll 1$. This is the well known condition for the validity of a Fokker-Planck equation for $P(V,t)$.

3. *Nonlinear case:* $W(y';\Delta y)$ is a nonlinear function of y'. This gives rise to the additonal requirement that the jump size be small compared to the variation distance of W itself as a function of y'. Unless the jump size is much smaller one cannot hope to construct a general approximation method. Hence one has to find an expansion parameter by which one can reduce the jump size with respect to the scale inherent in the nonlinearity.

Evidently that will lead in lowest approximation to a description of the fluctuations as if the system were linear. That is the reason for the success of the linear noise approximation in many nonlinear electronic devices. However, in order to include the effect of the nonlinearity on the fluctuation one has to go to the next approximation. Inevitably this also brings corrections due to the fact that the jumps are not infinitely small. As a result it turns out to be inconsistent to use (2) with nonlinear functions $a(y), b(y)$ without at the same time adding higher order derivatives. These inconsistencies were responsible for the confusion in the early attempts to treat noise in nonlinear resistors (van Kampen, 1965).

For example, the velocity fluctuations of a Rayleigh particle in equilibrium are described in linear noise approximation by the familiar linear Fokker-Planck equation

$$\frac{\partial P(V,t|V_0,t_0)}{\partial t} = \gamma\left\{\frac{\partial}{\partial V} VP + \frac{kT}{M}\frac{\partial^2 P}{\partial V^2}\right\} . \tag{7}$$

This corresponds to the linear damping law $\dot{V} = -\gamma V$. In the case of an ideal gas one can compute from kinetic theory

$$\gamma = \frac{4\rho A}{M}\left[\frac{2mkT}{\pi}\right]^{\frac{1}{2}},$$

where A is the surface area of the particle and ρ the number density of the gas molecules. One can also find the first nonlinear correction

$$\dot{V} = -\gamma V\left[1 + \frac{m}{6kT}V^2 + \ldots\right] . \tag{8}$$

This correction is of relative order m/M which is the same parameter that measures the size of the jumps. In fact, we had to suppose $m/M \ll 1$ in order to approximate the master equation by a second order differential equation. *It is therefore inconsistent to modify (7) so as to take the nonlinear correction (8) into account without adding at the same time higher derivatives to account for the fact that the jumps are no longer infinitely small.* In fact, m/M is the only parameter in the problem, and one should therefore start from the master equation, expand systematically in powers of m/M, and expect to obtain successive nonlinear correlations and simultaneous higher derivative terms (van Kampen, 1961 a).

The Expansion of the Master Equation

This section describes the expansion of the master equation in general terms; for details the reader is referred to the original literature (van Kampen, 1961 a, 1976; McNeil, 1972; Kubo et al., 1973; Kurtz, 1976; Fox, Kac, 1977). Our starting point is the master equation (1). It is not necessary to differentiate between discrete and continuous range. For most actual cases in physics, chemistry, and biology it turns out that W involves a parameter Ω in such a way that for $\Omega \to \infty$ the magnitude of the individual jumps goes to zero relative to the distance in the Y -scale over which the nonlinearity becomes appreciable. In many cases, for instance the kinetics of chemical reactions, Ω is simply the size of the system, but in the example of the Rayleigh particle $\Omega \equiv M$. In electric circuits Ω is often the capacity of a condenser (van Kampen, 1960, 1961 b).

The precise condition on how W must depend on Ω is specified in the literature. For chemical reactions it roughly says that the kernel $W(y|y') \equiv W(y';\Delta y)$ must depend on the jump size Δy through an extensive variable, and on the starting point y' through an intensive variable. When no such parameter can be found the following expansion method cannot be used. Examples where no such parameter is available: the random velocity of one of the molecules of an ideal gas; the random walk of a diatomic molecule along its vibrational levels due to collisions (Montroll, Shuler, 1958).

When a parameter with the required properties is available it is possible to give a systematic expansion of the master equation in powers of $\Omega^{-1/2}$. In order to describe the result we define the jump moments for $\nu = 1, 2, \dots$ in the ususal way by

$$\alpha_\nu(y) = \int (\Delta y)^\nu W(y;\Delta y) d(\Delta y) = \alpha_{\nu,0}(y) + \Omega^{-1} \alpha_{\nu,1}(y) + \dots$$

The expansion then leads to the following consequences.

1. The largest terms (lowest order in $\Omega^{-1/2}$) yield

$$\frac{\partial P(y,t)}{\partial t} = -\frac{\partial}{\partial y} \alpha_{1,0}(y) P. \qquad (9)$$

This is a Liouville equation corresponding to the deterministic equation

$$\partial_t y = \alpha_{1,0}(y). \tag{10}$$

This equation contains no fluctuations because it is obtained in the thermodynamical limit $\Omega \to \infty$. It is therefore the macroscopic equation, which in the case of a chemical reaction is the familiar rate equation. Thus the deterministic, macroscopic, or phenomenological equation is a consequence of the master equation. Evidently it may well be nonlinear even though the master equation itself is always linear in its unknown variable P .

2. The next order in $\Omega^{-1/2}$ describes to lowest order the fluctuations of Y about the macroscopic value determined by (10). More precisely, let $\varphi(t)$ be a solution of the macroscopic equation (10). If one sets

$$y = \varphi(t) + \Omega^{-1/2}\xi \tag{11}$$

the Ω -expansion gives for the probability distribution $\Pi(\xi,t)$ of ξ

$$\frac{\partial \Pi(\xi,t)}{\partial t} = -\alpha'_{1,0}(\varphi(t))\frac{\partial}{\partial \xi}\xi\Pi + \frac{1}{2}\alpha_{2,0}(\varphi(t))\frac{\partial^2 \Pi}{\partial \xi^2}. \tag{12}$$

(The prime denotes the derivative of $\alpha_{1,0}$ with respect to its argument φ). This is a Fokker-Planck equation whose coefficients are linear in ξ but depend on time. The solution is a Gaussian and describes the fluctuations in linear noise approximation. In the special case that φ is a stationary solution of (10), that is, $\alpha_{1,0}(\varphi^{st}) = 0$, the coefficients of (12) are independent of time and it reduces to the familiar equation for equilibrium fluctuations, of which (7) is an example.

3. Higher orders in $\Omega^{-1/2}$ add higher order terms to (12). For example, to order

$$\begin{aligned}\frac{\partial \Pi}{\partial t} = &-\frac{\partial}{\partial \xi}\left[\alpha'_{1,0}\xi + \frac{1}{2}\Omega^{-1/2}\alpha''_{1,0}\xi^2 + \frac{1}{3!}\Omega^{-1}\alpha'''_{1,0}\xi^3\right]\Pi + \\ &+ \frac{1}{2}\frac{\partial}{\partial \xi^2}\left[\alpha_{2,0} + \Omega^{-1/2}\alpha'_{2,0}\xi + \frac{1}{2}\Omega^{-1}\alpha''_{2,0}\xi^2\right]\Pi - \\ &- \frac{1}{3!}\frac{\partial^3}{\partial \xi^3}\left[\Omega^{-1/2}\alpha_{3,0} + \Omega^{-1}\alpha'_{3,0}\xi\right]\Pi + \frac{1}{4!}\frac{\partial^4}{\partial \xi^4}\Omega^{-1}\alpha_{4,0}\Pi - \\ &- \Omega^{-1}\left[\alpha'_{1,1}\frac{\partial}{\partial \xi}\xi\Pi + \frac{1}{2}\alpha_{2,1}\frac{\partial^2 \Pi}{\partial \xi^2}\right].\end{aligned} \tag{13}$$

All α^j 's are functions of $\varphi(t)$. This equation leads to the following comments.

a) The higher order terms do not only add nonlinear corrections to the coefficients of (12), but *at the same time higher derivatives appear.* At no stage of the expansion does the nonlinear Fokker-Planck equation (2) emerge: it is inconsistent to take the nonlinearity of $a(y)$ and $b(y)$ into account while neglecting higher derivatives.

b) The fluctuations are no longer Gaussian, and the equilibrium auto-correlation function is no longer a simple exponential. Rather, each order of Ω^{-1} adds one more exponential, and therfore one more Debye relaxation term to the spectral density of the fluctuations.

c) To order Ω^{-1} it is no longer true that $\langle \xi \rangle = 0$, and hence $\langle Y(t) \rangle$ no longer coincides with the solution $\varphi(t)$ of the macroscopic equation. To put it differently, $\langle Y(t) \rangle$ is not a solution of the macroscopic equation (10). Nor can one find some modified differential equation for $\langle Y(t) \rangle$ that includes the corrections of order Ω^{-1}. Instead one has to solve two coupled equations for the average and the variance of $Y(t)$ (Kubo et al., 1973).

d) As mentioned before, the fourth order equation (13) cannot serve as an exact master equation of any Markov process. It is therefore not the exact equation for a Markov process that in some way approximates the original process; rather it is an approximate equation for the exact P. Only those solutions of (13) are meaningful for which the higher derivative terms, not included in (13), are actually small. Taken at face value, the equation has also other solutions, which may become negative or infinite, but they are not approximations to the actual solutions of (1). The safest way to avoid them is to treat the higher order terms in (13) by means of perturbation theory (Siegel, 1960).

The Diffusion Type

The Ansatz (11) is based on the anticipation that the fluctuations will be of relative order $\Omega^{-1/2}$. Its a posteriori justification provided by the fact that according to (12) the values of ξ are in fact of order unity. Nonetheless it can happen that they grow in time and become so large that the formal expansion in powers of $\Omega^{-1/2}$ becomes illusory after a certain initial period. It is easily seen from (12) that this will not be so provided that $\alpha'_{1,0}(\varphi(t)) < 0$. This inequality is also the condition that the solution $\varphi(t)$ of (10) is stable (asymptotically stable in the sense of Lyapunov). When the macroscopic solution is stable the fluctuations do not grow and the Ω-expansion can be used for all times. The physical reason for this is clear. For the precise statement of this criterion one has to distinguish between local and global stability, but these complications are not relevant here. Nor do we want to discuss the various cases in which the criterion is violated, such as critical points (Kubo et al., 1973) and bistable systems (Sizuki, 1977, 1978; van Kampen, 1978), but only one special case.

If it happens that $\alpha_{1,0}(y) \equiv 0$, the macroscopic solutions are: $\varphi(t)$ = constant. They are not asymptotically stable, not even locally. According to the linear noise approximation (12) the fluctuations grow proportionally with $\sqrt{t}$, just as in Brownian motion. The above expansion can therefore be used only for a limited time, roughly for $t < \Omega$. For master equations of this type, however, the Ω-expansion takes a different form.

The point is that it is no longer true that (9) is the leading term in the expansion, because this term vanishes. Consequently it is no longer possible to split off a macroscopic part as in (11), and this affects the orders of the remaining terms. Rearranging them accordingly one now obtains as the leading term in the Ω-expansion of the master equation (van Kampen, 1977)

$$\frac{\partial P}{\partial t} = \Omega^{-1}\left[-\frac{\partial}{\partial y}\alpha_{1,1}(y)P + \frac{1}{2}\frac{\partial^2}{\partial y^2}\alpha_{2,0}(y)P\right], \qquad (14)$$

which is precisely the generalized diffusion equation (2). The variation occurs on a slower time scale than in (9), because it is now purely a net effect of fluctuations, rather than due to a concerted drift.

To show that the occurrence of $\alpha_{1,0} \equiv 0$ is not just a freak, consider the following model for diffusion in a solid (Fig. 19). A particle moves in a periodic potential V(x) by jumping from each equilibrium site n to either n + 1 or n - 1 with probabilities per unit time a_n and b_n. Thus it performs a continuous time random walk with jump probability per unit time

$$W(n|n') = a_{n'}\delta_{n,n'-1} + b_{n'}\delta_{n,n'+}$$

$$a_n = b_n = A\exp(-W/kT).$$

Now let the particles be charged and subject to an external field U. The fact that U is supposed to vary on a macroscopic scale is expressed by writing it as a function of $y = n/\Omega$, where Ω is the number of sites n per cm. The field U(y) modifies the height of the potential peaks, so that the new jump probabilities are

$$a_n = A\exp\frac{-1}{kT}\left[W + U\left(\frac{n-\frac{1}{2}}{\Omega}\right) - U\left(\frac{n}{\Omega}\right)\right]$$

$$b_n = A\exp\frac{-1}{kT}\left[W + U\left(\frac{n+\frac{1}{2}}{\Omega}\right) - U\left(\frac{n}{\Omega}\right)\right].$$

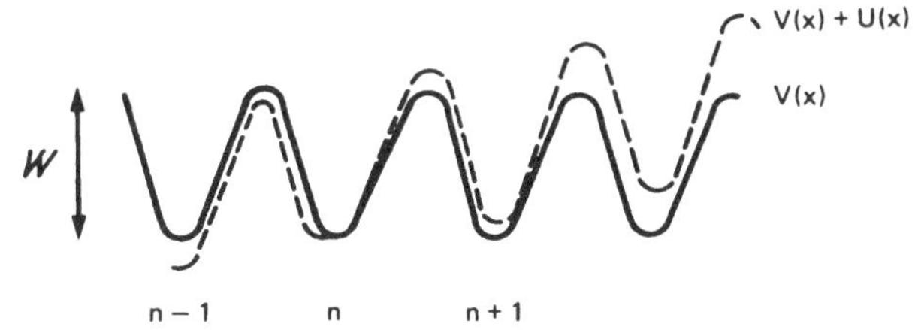

Fig. 19 The potential for a particle diffusing through a crystal

Hence the first jump moment is

$$\alpha_1(y) = -a_n + b_n = A\exp\left[-\frac{W}{kT}\right]\times$$

$$\times\left\{\exp\left[\frac{1}{2kT\Omega}U'(y) + O(\Omega^{-2})\right] - \exp\left[\frac{-1}{2kT\Omega}U'(y) + O(\Omega^{-2})\right]\right\}.$$

On expanding the exponential one sees that the lowest order $\alpha_{1,0}(y)$ cancels so that the diffusion approximation applies.

To find the explicit formula we need the next term

$$\Omega^{-1}\alpha_{1,1}(y) = A\exp\left[-\frac{W}{kT}\right]\times\frac{U'(y)}{kT\Omega}.$$

Also one has

$$\alpha_2(y) = a_n + b_n = 2A\exp\left[-\frac{W}{kT}\right] + O(\Omega^{-2}).$$

Hence the diffusion equation (14) has the form

$$\frac{\partial P}{\partial \tau} = -\frac{\partial}{\partial y}U'(y)P + kT\frac{\partial^2 P}{\partial y^2}, \tag{15}$$

where the constant factor $(A/\Omega kT)\exp(-W/kT)$ has been absorbed in the time scale. The conclusion is that (15) is a consistent first approximation for diffusion in an external field U(x) when the jump size is small compared to the variation distance of U(x). In the limit it is of course exact (Feller, 1968).

The example indicates the basic reason for the applicability of the diffusion approximation. In contrast with e. g. the Rayleigh particles here the nonlinearity is due to some external agency, physically distinguished from the noise source. That makes it possible to choose Ω in such a way that for large Ω the jumps become arbitrarily small compared to the variation length of the nonlinear field U. Another example is a nonlinear pendulum with damping and fluctuations caused by the surrounding air. It has not yet been investigated whether or not the nonlinearities of hydrodynamics can be disentangled in this way from the fluctuations associated with the transport coefficients.

Summary

1. In physical situations that can be described by a continuous range Markov process the Lindeberg condition is not satisfied. One can therefore not conclude as a mathematical theorem that the diffusion equation (2) must hold, nor that the corresponding Langevin equation by means of a systematic approximation method, and the Ω-expansion appears to be rather universally applicable.

2. For the Ω-expansion no distinction need be made between continuous and discrete ranges. The first term yields the macroscopic equation (10), and the next one the linear Fokker-Planck equation (12) for the fluctuations. Higher orders add nonlinear terms to the coefficients of this equation but simultaneously also higher derivatives. The result is valid for all times provided that $\alpha'_{1,0} < 0$.

3. At no stage does (2) appear: it is not a consistent approximation in this expansion. Of course (2) may lead to correct results, namely if one merely uses it in linear approximation, such as to make use only of those features of it which coincide with the linear noise approximation. But it is incorrect to take it seriously beyond that, and for instance to conclude from it

$$P^{eq}(y) = \frac{\text{const}}{b(y)} \exp\left[-2\int^{y} \frac{a(y')}{b(y')}\, dy'\right]. \tag{16}$$

4. In the special case $\alpha'_{1,0} \equiv 0$ however, the lowest term in the Ω -expansion does have the form (2), or more precisely (14). Typically this happens when the nonlinearity stems from a different physical cause than the fluctuations, so that both can be disentangled by a suitable choice of Ω. In this case (16) is valid apart from corrections of order Ω^{-1}.

Extremal Principles and Catastrophe Theory for Stochastic Models of Nonlinear Irreversible Processes

W. Ebeling, H. Engel-Herbert

This paper is devoted to a discussion of the evolution principles governing the approach of nonequilibrium stochastic systems to the stationary state and to the topological investigation of the stationary probability distribution itself. Stochastic evolution principles are the counterparts of the well-known thermodynamic evolution principles developed e. g. by Onsager and Machlup (1953), Vojta (1967), Gyarmati (1970), Glansdorff and Prigogine (1971), Lamprecht and Zotin (1978); Presnov and Zotina (1978) and many other workers. Stochastic evolution principles were - to our knowledge - first developed by Lebowitz and Bergmann (1957), and by Felderhof (1961), van Kampen (1970), Graham (1973), Uhlmann (1977), Crell et al. (1978), Haken (1970), etc. The relation between stochastic and thermodynamic evolution principles has been extensively discussed by Schlögl (1971 a, b; 1972, 1973) and Lurie (1978). In this paper, which is based on earlier investigation (Ebeling, 1976, 1978; Ebeling, Engel-Herbert, 1977) we consider only stochastic processes with continuous variables and concentrate on the study of the stationary distribution which appear as the target of evolution. The application of catastrophe theory to stochastic models of nonlinear dynamic systems is proposed. We believe that catastrophe theory could be very useful in classifying different instabilities occurring in dynamic systems under the influence of fluctuations since the instabilities must be reflected in topological changes of the probability distribution. The general conditions under which instabilities occur in nonlinear systems were investigated by Glansdorff and Prigogine (1971) and many other workers.

It is well known that, in the deterministic description when all fluctuations are neglected, a nonlinear system may be in a variety of stable stationary states. Of great interest are the transtions between different branches of

Thermodynamics and Kinetics of Biological Processes

stationary states due to instabilities of the system. If the dynamic system is of the gradient type the transitions between the stationary states lead to qualitative changes in the local form of the potential function, $V(x, u)$.

If fluctuations are included, the description by a unique and global stable stationary probability distribution, $P^{o}(x, u)$, over all states exists, and instabilities manifest themselves in the local form of $P^{o}(x, u)$.

We consider both $V(x, u)$ and $P^{o}(x, u)$ as real hyperfaces spanned over the product space $R = X \times C$ (where X is the state space of the system under consideration and C is the space of external parameters; control space). The qualitative changes in the local behavior of these surfaces are closely connected with the transitions between the stable stationary states, the most probable states, of the dynamic system which are caused by instabilities when the external parameters cross some critical values.

Extremal Principles for the Probability Distribution

Throughout this contribution we restrict ourselves to homogeneous, continuous, nonlinear systems. Non-homogeneous problems are considered elsewhere (Ebeling, Malchow, 1979). The random process is assumed to be Markovian and is characterized in the usual way by the probability, $P(x,u;t)dx$, of finding the system at time, t, in the dx neighbourhood of the point, $x = \{x_1, x_2, \ldots, x_f\}$, in the state space, X, and by the transition probabilities, $G(x|x_0; u)$, which reflect the properties of the system under consideration. The system changes with the probability, $G(x|x_0; u)dt$, within dt from $x_0 = x(t_0)$ to $x = x(t)$. $u = \{u_1, u_2, \ldots, u_\lambda\}$ are external parameters which, in general, describe the influence of the environment of the dynamic system; in catastrophe theory u and C are often called control parameters and control space respectively.

The time evolution of the probability density is determined by the Master equation (Stratonovich, 1963; van Kampen, 1970; Ebeling, 1976; Tikhonov, Mironov, 1977).

$$\frac{\partial P(x,u;t)}{\partial t} = \int dx_0 \left[G(x|x_0;u)P(x_0,u;t) - G(x_0|x;u)P(x,u;t)\right]. \quad (1)$$

It is well known that for small stochastic transitions, $\Delta x = x(t+\tau)-x(t)$, an expansion of G in Δx transforms the right-hand side of (1) into an infinite series of differential operators (Kramers-Moyal-expansion).

$$\frac{\partial P(x,u;t)}{\partial t} = \sum_{n=1}^{\infty} \frac{(-1)^n}{n!} \left[\frac{\partial^n [M_{i_1,i_2,\ldots,i_n}(x;u)P(x,u;t)]}{\partial x_{i_1}\partial x_{i_2}\ldots\partial x_{i_n}} \right] \quad (2)$$

where

$$M_{i_1,i_2,\ldots,i_n}(x;u) = \lim_{\tau\to 0} \frac{m_{i_1,i_2,\ldots,i_n}(x;u)}{\tau} \quad (3)$$

and the

$$m_{i_1,i_2,\ldots,i_n}(x;u) = \langle (x_{i_1}(t+\tau)-x_{i_1}(t))(x_{i_2}(t+\tau)-x_{i_2}(t))\ldots(x_{i_n}(t+\tau)-x_{i_n}(t))\rangle \quad (4)$$

are the moments of the increments. The angular brackets define the mean values of the enclosed quantities and summation over repeated indices is always implied.

For a continuous stochastic Markov process two alternative cases are possible (Pawula, 1967 a, b; Horsthemke, Brenig, 1977). If all limits in (3) are finite

$$M_{i_1,i_2,\ldots,i_n}(x;u) < \infty$$

and

$$M_{i_1,i_2,\ldots,i_n}(x;u) = 0$$

for some even n, then

$$M_{i_1,i_2,\ldots,i_n}(x;u) = 0$$

for all $n \geq 3$ (Pawula's theorem (1967 a, b)).

Therefore, we have to consider two essentially different cases: (1) two coefficients in the Kramers-Moyal-expansion are different from zero, (2) the coefficients of all orders are different from zero.

In the first case, equation (2) takes the form of the well-known Fokker-Planck equation

$$\frac{\partial P(x,u;t)}{\partial t} = -\frac{\partial}{\partial x_i}\left[M_i(x,u)P(x,u;t)\right] + \frac{1}{2}\frac{\partial^2}{\partial x_i \partial x_k}\left[M_{ik}(x,u)P(x,u;t)\right], \quad i,k=1,\dots,f \tag{5}$$

$M_i(x,u)$ and $M_{ik}(x,u)$ are the so called drift and diffusion coefficients respectively. The Fokker-Planck equation is closely connected with another well-developed concept for inclusion of fluctuations (Stratonovich, 1963; Haken, 1975 a). In the deterministic model of the nonlinear dynamic system given by the equations

$$\frac{dx_i}{dt} = f_i(x,u) \tag{6}$$

fluctuations may be included by adding noise terms in the following way;

$$\frac{dx_i}{dt} = f_i(x,u) + g_{ik}(x,u)\xi_k(t). \tag{7}$$

The $\xi_i(t)$ are independent Gaussian, δ-correlated fluctuating quantities with

$$\begin{aligned} \langle \xi_i(t) \rangle &= 0 \\ \langle \xi_i(t+\tau)\xi_k(t) \rangle &= \delta_{ik}\,\delta(\tau) \end{aligned} \tag{8}$$

and the $g_{ik}(x,u)$ known functions of x and u.

Stratonovich (1963) has proved that the stochastic differential equation (7) is equivalent to the Fokker-Planck equation (5) if the drift and diffusion coefficients are chosen to be

$$\begin{aligned} M_i(x,u) &= f_i(x,u) + \frac{1}{2}\frac{\partial g_{ik}(x,u)}{\partial x_l} g_{lk}(x,u) \\ M_{ik}(x,u) &= g_{il}(x,u)\,g_{kl}(x,u). \end{aligned} \tag{9}$$

The probability distribution P(x, u;t) has to satisfy the normalization condition that

$$\int P(x, u; t)\, dx = 1 . \tag{10}$$

If we assume an arbitrary initial probability density such as

$$P(x, u; t_0) = P_0(x, u) \tag{11}$$

then at some initial time, t_0 , we can determine P(x,u;t) for $t > t_0$ by solving the Fokker-Planck equation provided it is supplemented with appropriate conditions. In the following we shall always assume that the probability flow vector

$$\Pi_i(x, u; t) = - M_i(x; u) P(x, u; t) + \frac{1}{2} \frac{\partial}{\partial x_k} [M_{ik}(x; u) P(x, u; t)]. \tag{12}$$

disappears at the boundaries of the state space or is periodic in the case of angle variables. Later we shall assume that under these conditions the stationary boundary value problem

$\frac{\partial \Pi_i(x, u)}{\partial x_i} = 0$ if $x \in X$, and $\Pi_i(x_b, u) = 0$ on the boundary where $x = x_b$. (13)

and has a well defined normalizable solution $P^0(x, u)$.

In the following we shall prove that under given conditions from the existence of $P^0(x, u)$ follows the uniqueness of $P^0(x, u)$ and the convergence of any solution to the unique, stationary solution (Green, 1952).

In order to prove this we consider an evolution criterion for the Fokker-Planck equation. Let us consider any two solutions, $P_1(x, u; t)$ and $P_2(x, u; t)$, of the Fokker-Planck equation under the given boundary conditions and the positive semi-definite function (K-function)

$$K(t) = \int P_1(x, u; t) \ln \frac{P_1(x, u; t)}{P_2(x, u; t)}\, dx . \tag{14}$$

It can easily be shown that $K(t) > 0$ for $P_1 \neq P_2$ and, K(t) = 0 if, and only if, $P_1 \equiv P_2$. (In the following the x, u, t dependence of P_1 and P_2 is omitted for simplicity.)

We shall prove now that K(t) monotonously decreases in the course of time. If we take into account the normalization condition (10) then

$$\int \frac{\partial P_1}{\partial t} dx = 0 \tag{15}$$

and the time derivative of K(t) is given by

$$\frac{dK}{dt} = \int \left(\frac{\partial P_1}{\partial t} \ln \frac{P_1}{P_2} - \frac{\partial P_2}{\partial t} \cdot \frac{P_1}{P_2} \right) dx . \tag{16}$$

Bearing in mind that P_1 and P_2 satisfy the Fokker-Planck equation we take $\partial P_1 / \partial t$ and $\partial P_2 / \partial t$ from (5) and obtain, by partial integration, using the boundary condition (12),

$$\frac{dK}{dt} = \int \left\{ \left[M_i P_1 - \frac{1}{2} \frac{\partial}{\partial x_k} (M_{ik} P_1) \right] \frac{\partial}{\partial x_i} \left(\ln \frac{P_1}{P_2} \right) + \left[-M_i P_2 + \frac{1}{2} \frac{\partial}{\partial x_k} (M_{ik} P_2) \right] \frac{\partial}{\partial x_i} \left(\frac{P_1}{P_2} \right) \right\} dx . \tag{17}$$

After simple algebraic manipulation we finally get

$$\frac{dK}{dt} = -\frac{1}{2} \int \left[M_{ik} P_1 \frac{\partial}{\partial x_k} \left(\frac{P_1}{P_2} \right) \frac{\partial}{\partial x_i} \left(\frac{P_1}{P_2} \right) \right] dx . \tag{18}$$

If we make use of the positive nature of the diffusion matrix, M_{ik}, and take $P_1 \geq 0$ it follows from (18) that

$$\frac{dK}{dt} \leq 0 . \tag{19}$$

Therefore, K(t) decreases in time until the integrand in (18) is equal to zero. For the even more general case of arbitrary convex functions an equivalent theorem has been derived by Crell et al. (1978).

The condition, $\frac{dK}{dt} = 0$, is fulfilled only if

$$\frac{\partial}{\partial x_i} \left(\frac{P_1}{P_2} \right) = 0 \qquad i = 1, \ldots, f \tag{20}$$

because $P_2(x, u; t)$ and $M_{ik}(x; u)$ are not zero for all X. From (20) we see that

$$P_1(x, u; t) = N P_2(x, u; t) . \tag{21}$$

By normalization the constant N in (21) is 1, and therefore $P_1(x, u; t)$ and $P_2(x, u; t)$ must be identical. Thus we conclude that the normalizable solutions of the Fokker-Planck equation approach each other in the course of time. If we take in particular $P_2(x, u; t)$ as the stationary solution, $P^0(x, u)$, the proof remains true. It follows that an arbitrary solution approaches $P^0(x, u)$ for long times. We may conclude, also, that $P^0(x, u)$

must be unique. Let us assume that there exists a second, normalizable, stationary solution, $\tilde{P}^o(x,u)$. Then it follows from the above theorem, that $\tilde{P}^o(x,u) \to P^o(x,u)$. This would be in contradiction to the assumed stationarity of $\tilde{P}^o(x,u)$.

The above considerations demonstrate that if there exists a normalizable-stationary probability distribution then it will be unique, and it will be approached by all probability distributions independent of the initial distribution.

Under certain conditions this is true even if the time evolution of $P(x,u;t)$ is governed by the full master equation, i. e. all the coefficients in the Kramers-Moyal expansion are different from zero (Lebowitz, Bergmann, 1957; Graham, 1973; Ebeling, 1976).

Substituting (1) instead of (5) in the expression for dK/dt (16) Lebowitz and Bergmann (1957) have shown that

$$\frac{dK}{dt} = -\int dx \int dx_o\, P_2(x_o,u;t)\,\Phi(x,u;t)\,G(x|x_o;u)\left[\frac{\Phi(x_o,u;t)}{\Phi(x,u;t)} \ln \frac{\Phi(x_o,u;t)}{\Phi(x,u;t)} - \frac{\Phi(x_o,u;t)}{\Phi(x,u;t)} + 1\right]$$

where (22)

$$\Phi(x,u;t) = \frac{P_1(x,u;t)}{P_2(x,u;t)} \quad . \tag{23}$$

The property $dK/dt \leq 0$ follows from the inequality

$$\Psi \ln \Psi - \Psi + 1 = \int_{\Theta=1}^{\Psi} \ln \Theta \, d\Theta \geq 0 . \tag{24}$$

Again K(t) decreases until the integrand in (22) is zero. From

$$\frac{\Phi(x_o,u;t)}{\Phi(x,u;t)} \ln \frac{\Phi(x_o,u;t)}{\Phi(x,u;t)} - \frac{\Phi(x_o,u;t)}{\Phi(x,u;t)} + 1 = 0 \tag{25}$$

with the normalization condition (10) one finds that

$$P_1(x,u;t) = P_2(x,u;t) \tag{26}$$

and there is nothing further to prove.

But to guarantee that $K(t)$ reaches its lower limit of zero it is, in addition, required that all points in the state space are connected with each other by some sequence of transitions. More precisely the case

$$G(x|x_0;u)=0 \quad \text{if} \quad \frac{\phi(x_0,u;t)}{\phi(x,u;t)} \ln \frac{\phi(x_0,u;t)}{\phi(x,u;t)} - \frac{\phi(x_0,u;t)}{\phi(x,u;t)} + 1 \neq 0 \qquad (27)$$

has to be excluded. Therefore, provided only that $\mathcal{X}$ cannot be decomposed, then $P_1(x,u;t)$ and $P_2(x,u;t)$ approach each other. The stationary distribution, especially, will be unique and independent of the initial distribution only if there no non-interacting regions within the state space exist.

Investigation of the Stationary Distribution by the Methods of Catastrophe Theory

Let us now consider $P^0(x,u)$ as a smooth, real hyperface (probability surface) spanned over the product space $R = \mathcal{X} \times C$ of the dynamic system. This, itself, suggests the application of some concepts of catastrophe theory (Thom, 1969, 1975; Golubitsky, Guillemin, 1973; Callahan, 1974, 1977; Güttinger, 1974; Woodcock, Poston, 1974; Bröcker, Lander, 1975; Zeeman, 1976; Golubitsky, 1978) to the investigation of the local shape of the stationary probability distribution. This new application of Thom's theory was originally proposed by Haken (1977) and by ourselves (Ebeling, 1976, 1978; Ebeling, Engel-Herbert, 1977). We note that the global behaviour of the probability surface is determined by the symmetry properties of the dynamic system. A requirement of the unique nature of the stationary probability distribution is the invariance of all the symmetry transformations of the dynamic system because, whilst such transformations leave the Fokker-Planck equation unaltered, $P^0(x,u)$ would change (Graham, 1973).

The most important qualitative characteristics of the stationary probability distribution are the number of maxima (modality) and the existence, or non-existence, of degenerate extrema or turning points. A catastrophe may be represented as an abrupt change in the number of maxima in the probability surface as the parameters, U, vary continuously. Obviously these changes are connected with qualitative changes in the structure of the dynamic system itself because formerly probable states become improbable, or, new probable states arise.

A function, $P^o(x,u)$, which has only nondegenerate critical points cannot undergo a catastrophic change. Therefore we assume that $P^o(x,u)$ possesses degenerate extrema where

$$\frac{\partial P^o(x,u)}{\partial x_i}\Big|_{x^o} = 0, \quad \Delta(P^o(x^o,u)) = \mathrm{Det}\left(\frac{\partial^2 P^o(x,u)}{\partial x_i \partial x_k}\Big|_{x^o}\right) = 0 \quad i,k = 1,..,f \tag{28}$$

We define the hyperface $\mathfrak{M} \subset R = X \times C$ of the states of extremal probability as

$$\mathfrak{M} = \left\{(x,u): \frac{\partial P^o(x,u)}{\partial x_i} = 0 \qquad i = 1,\dots,f\right\}, \tag{29}$$

and the transition set, $\mathfrak{K} \subset C$, according to

$$\mathfrak{K} = \left\{u: \frac{\partial P^o(x,u)}{\partial x_i} = 0,\ \Delta(P^o(x,u)) = \mathrm{Det}\left(\frac{\partial^2 P^o(x,u)}{\partial x_i \partial x_k}\right) = 0 \quad i,k = 1,\dots,f\right\}. \tag{30}$$

From the above definition $\mathfrak{K}$ consists of all points of the control space where the number and the type of the states of extremal probability change. Geometrically $\mathfrak{K}$ is generated by the orthogonal projection $\mathfrak{M} \to C$.

If the dimensionality of the control space is limited by

$$\dim C = \lambda \leq 4 \tag{31}$$

we may claim, following Thom's well known theorem, that there exist precisely seven essentially different transitions of the dynamic system if U crosses some critical values. From catastrophe theory the canonical forms which describe the probability surface in the vicinity of the transition points are also known. These forms are Thom's so-called *universal unfoldings* of the elementary catastrophe. The universal unfoldings have the property of the simplest functions that exhibit all the characteristic features of the given catastrophe. More complicated functions are related to universal unfoldings by a diffeomorphism.

A similar classification can be carried out if $\lambda \leq 6$. In the case $\lambda = 5$ for example, four new singularities appear and thus, four new unfoldings. Arnold (1972) has classified the normal forms of functions near degenerate critical points up to at least 25 dimensions.

Originally, catastrophe theory was applied in the determinsitic picture to gradient systems (Zeeman, 1973; Güttinger, 1974; Blum, Mehra, 1975; Feinn, Ortoleva, 1977; Ebeling, 1978). If we assume that $g_{ik}(x,u) = \sqrt{D}\,\delta_{ik}$ with D = const the Langevin-equation (7) is

$$\frac{dx_i}{dt} = -\frac{\partial V(x,u)}{\partial x_i} + \sqrt{D}\,\xi_i(t) \tag{32}$$

and we can determine the exact stationary solution of the corresponding Fokker-Planck equation via (9), for any dimension, f, of X which has the form

$$P^{\circ}(x,u) = N^{-1} \exp\left[-\frac{2}{D} V(x,u)\right]. \tag{33}$$

It is apparanet that both $P^{\circ}(x,u)$ and $V(x,u)$ have the same transition set σ because from $\frac{\partial P^{\circ}(x,u)}{\partial x_i} = 0$ we get $\frac{\partial V(x,u)}{\partial x_i} = 0$ and from $\Delta(P^{\circ}(x,u)) = 0$ it follows that $\Delta(V(x,u)) = \sigma$. Thus for gradient systems unique relations exist between the qualitative behaviour of the dynamic system in the deterministic, and the stochastic picture, described above.

As shown by Graham (1973) and Haken (1978) there exists a second case where the exact solution of the stationary Fokker-Planck equation is knwon. If the dynamic system is of Hamiltonian type ($f = 2, x_1 \equiv x, x_2 \equiv y$), then

$$\frac{dx}{dt} = \frac{\partial H(x,y,u)}{\partial y}, \quad \frac{dy}{dt} = -\frac{\partial H(x,y,u)}{\partial x} - u_1 \frac{dx}{dt} + \sqrt{D}\,\xi(t) \tag{34}$$

(H-Hamiltonian, u_1 -friction coefficient) and the solution of the stationary Fokker-Planck equation is given by

$$P^{\circ}(x,y,u) = N^{-1} \exp\left[-\frac{2u_1}{D} H(x,y,u)\right]. \tag{35}$$

Evidently in this case the topological properties of $P^{\circ}(x,y,u)$ are given by the Hamiltonian $H(x, y, u)$.

Examples

These general ideas will now be applied to some examples. As a one dimensional dynamic system ($f = 1, x_1 \equiv x$) we consider an enzyme-catalyzed reaction for which the substrate kinetics are described, for suitable chosen unites, by the equation (Czajkowski, Ebeling, 1977):

$$\frac{dx}{dt} = -\frac{dV(x,u)}{dx} + \sqrt{D}\,\xi(t) = u_1 - u_2 x - \frac{x + u_3 x^2}{1 + x + u_4 x^2 + u_5 x^3} + \sqrt{D}\,\xi(t) \tag{36}$$

In (36) $x > 0$ is the substrate concentration, $u_1 > 0$ and $u_2 > 0$ represent the flow of the substrate into the system, and the substrate decay rate, respectively. The rate of the enzyme-catalyzed reaction is taken as a rational function of the substrate concentration not only from theoretical considerations, but also, from experimental investigations (Cleland, 1963 a, b, c; Walter, 1972) of enzyme reactions with substrate inhibition. The intensity of substrate inhibition is characterized by the positive parameters u_3, u_4 and u_5. Again D = const is related to the intensity of the Gaussian δ-correlated fluctuations $\xi(t)$.

The stationary probability distribution has the form (33)

$$P^0(x,u) = N^{-1} \exp\left[-\frac{2}{D} V(x,u)\right]$$

where V(x,u) is determined by (36). As was explained in part 3 for a gradient system, $P^0(x,u)$ and V(x,u) possess an equal transition set $\mathfrak{K}$ Moreover, in example (36) it is not necessary to determine $\mathfrak{K}$ from (30) because the investigation of the extrema of $P^0(x,u)$ is based on the existence of positive roots of the equation

$$\frac{dV(x,u)}{dx} = 0 . \tag{37}$$

This may be related directly to one of Thom's universal unfoldings provided suitable control parameters are used. Indeed, if we write $\frac{dV(x,u)}{dx}$ in the form

$$\frac{dV(x,u)}{dx} = \frac{\frac{d}{dx}\tilde{V}(x, A_1, \ldots, A_4)}{\frac{1}{u_2 u_5}(1 + x + u_4 x^2 + u_5 x^3)} \tag{38}$$

where
$$\frac{d\tilde{V}(x, A_1, \ldots, A_4)}{dx} = x^4 + A_4 x^3 + A_3 x^2 + A_2 x + A_1$$

and

$$A_1 = -\frac{u_1}{u_2 u_5}, \qquad A_2 = \frac{u_2 + 1 - u_1}{u_2 u_5}$$
$$A_3 = \frac{u_2 + u_3 - u_1 u_4}{u_2 u_5} \qquad A_4 = \frac{u_2 u_4 - u_1 u_5}{u_2 u_5} \tag{39}$$

then (37) is fulfilled only if

$$x^4 + A_4 x^3 + A_3 x^2 + A_2 x + A_1 = 0 . \tag{40}$$

In order to eliminate the cubic term in (40) we redefine the state variable, x, by

$$x = z - \frac{A_4}{4} \tag{41}$$

and obtain

$$\frac{d\tilde{V}(z,q,v,w)}{dz} = z^4 + qz^2 + vz + w \tag{42}$$

with

$$q = A_3 - \frac{3}{8} A_4^2$$
$$v = A_2 - \frac{1}{2} A_3 A_4 + \frac{1}{8} A_4^3 \tag{43}$$
$$w = A_1 - \frac{1}{4} A_2 A_4 + \frac{1}{16} A_3 A_4^2 - \frac{3}{256} A_4^4 .$$

$\tilde{V}(z,q,v,w)$ is the unfolding of the swallowtail catastrophe which describes, therefore, the qualitatative behaviour of the system in the $R = z \times q \times v \times w$ space. In Fig. 20 the two-dimensional sections of the three-dimensional transition set of the swallowtail catastrophe are presented in a qualitative manner (Thom, 1969, 1975; Callahan, 1974, 1977; Güttinger, 1974; Woodcock, Poston, 1974; Bröcker, Lander, 1975; Zeeman, 1976; Golubitsky, Guillemin, 1977; Golubitsky, 1978). The possible instabilities of the system are indicated, as well as the number and type of extrema of $\tilde{V}(z,q,v,w)$ in the corresponding areas of $C = q \times v \times w$.

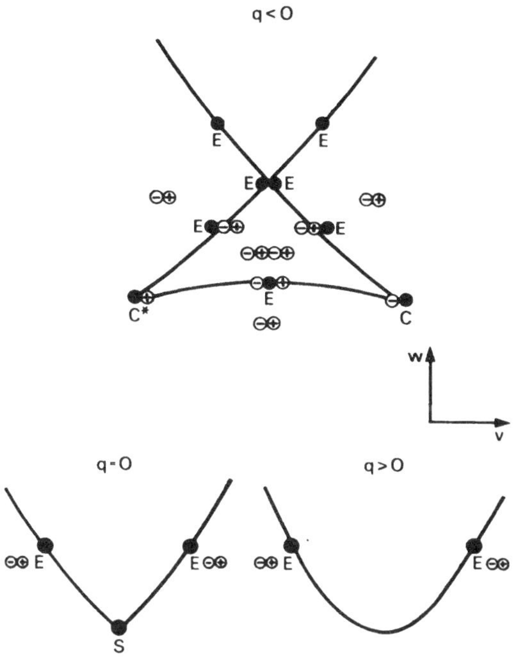

Fig. 20 The two-dimensional sections of the three-dimensional bifurcation set of the swallowtail catastrophe (Thom, 1975; Callahan, 1974, 1977) which describes the qualitative behaviour of the enzyme-controlled reaction (36) in different areas of the control space $C = q \times v \times w$
$\oplus$ - minimum, $\ominus$ - maximum, $\bigcirc$ - saddle point, $\bullet_E$, $\bullet_C$, $\bullet_{C^*}$, $\bullet_S$ degenerated extrema of end-type, cusp-type, dual cusp-type and swallow-tail-type respectively.

As an example of dynamic systems of second order ($f = 2$, $x_1 \equiv x$, $x_2 \equiv y$) we consider nonlinear oscillations under the influence of Gaussian white noise. The equation of motion for a particle of unit mass has the form

$$\frac{d^2x}{dt^2} + \gamma\left(x, \frac{dx}{dt}, u\right)\frac{dx}{dt} + \frac{dW(x,u)}{dx} = \sqrt{D}\,\xi(t) \tag{44}$$

where denotes the coordinate of the particle. $\gamma\left(x, \frac{dx}{dt}, u\right)$ represents the nonlinear friction of the particle during the motion in the external potential W(x, u). (44) is equivalent to the system of differential equations

$$\frac{dx}{dt} = y$$
$$\frac{dy}{dt} = -\gamma(x, y, u)y - \frac{dW(x,u)}{dx} + \sqrt{D}\,\xi(t) \tag{45}$$

which in general is not of gradient type. From (7, 8, 9) we obtain for (45) the following Fokker-Planck equation

$$\frac{\partial P(x,y,u,t)}{\partial t} = -y\frac{\partial P}{\partial x} + \frac{dW(x,u)}{dx}\frac{\partial P}{\partial y} + \frac{\partial}{\partial y}[\gamma(x,y,u)yP] + \frac{D}{2}\frac{\partial^2 P}{\partial y^2} . \quad (46)$$

First, let $\gamma(x,y,u) = const = u_1$. Then the stationary solution of (46) is the canonical distribution (35)

$$P^0(x,y,u) = N^{-1} exp\{-\frac{2u_1}{D}[\frac{y^2}{2} + W(x,u)]\} , \quad (47)$$

because, under these assumptions (45) is a special case of a Hamiltonian dynamic system (34) with $H(x,y,u) = \frac{y^2}{2} + W(x,y,u)$. The stationary probability distribution (47) exhibits local extrema (x^0, y^0) at $\frac{dW(x,u)}{dx}\Big|_{x^0} = 0$ and $y^0 = 0$. From

$$\Delta(P^0(x^0,y^0,u)) = -\frac{4u_1^2}{D^2} P^{02}(x^0,y^0,u)\frac{d^2W(x,u)}{dx^2}\Big|_{x^0} \quad (48)$$

and

$$\frac{\partial^2 P^0(x,y,u)}{\partial x^2}\Big|_{x^0,y^0} = \frac{2u_1}{D} P^0(x^0,y^0,u)\frac{d^2W(x,u)}{dx^2}\Big|_{x^0} \quad (49)$$

it follows that $P^0(x,y,u)$ has a maximum if $\frac{d^2W(x,u)}{dx^2}\Big|_{x^0} < 0$, and a saddle point if $\frac{d^2W(x,u)}{dx^2}\Big|_{x^0} > 0$.

The transition set K is obtained, according to (30), by eliminating x from

$$\frac{dW(x,u)}{dx} = 0 \quad \text{and} \quad \frac{d^2W(x,u)}{dx^2} = 0 . \quad (50)$$

Therefore, if for example the particle moves in the external potential

$$W(x,u) = x^6 + u_2x^4 + u_3x^3 + u_4x^2 + u_5x , \quad (51)$$

which corresponds to the universal unfolding of the butterfly catastrophe, four types of instabilities may occur. The four dimensional transition set contains a butterfly point, swallowtail lines, cusp surfaces and a fold hyperface. A two-dimensional section is given in Fig. 21.

Next we assume that $\gamma(x,y,u)$ depends on x as

$$\gamma(x,y,u) = u_1 + u_2x + u_3x^2 \quad (52)$$

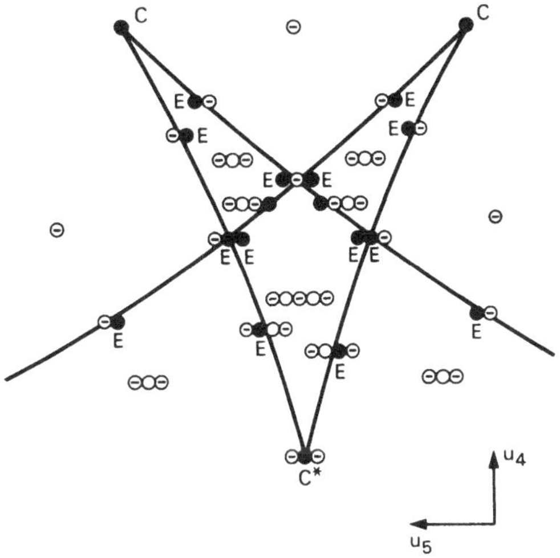

Fig. 21 The two-dimensional section of the four-dimensional bifurcation set of the butterfly catastrophe (Thom, 1975; Callahan, 1974, 1977) ($u_2 < 0$, $u_3 = \sigma$) which determines the topological properties of the stationary probability distribution $P^0(x, y, u)$ (47) in different areas of the control space $C = U_2 \times U_3 \times U_4 \times U_5$ if the external potential is chosen to be (51). Designations as in Fig. 20.

$$\gamma(x, y, u) = u_1 + u_2 x + u_3 x^2 \tag{52}$$

and consider, for simplicity, the motion of the system under potential $W(x, u) = x^2/2$. In this case the stationary solution of the Fokker-Planck equation (46) cannot be written simply in terms of quadratures as it was possible for $\gamma(x, y, u)$ = const.

However, if the parameters u_1, u_2, u_3 and D are small, i. e. if the system performs nonlinear oscillations under the influcence of weak frictional forces and weak fluctuations, we can apply a method developed by Stratonovich (1963) to determine $P^0(x, y, u)$. This consists of introducing the energy as

$$E = \frac{1}{2} y^2 + W(x, u), \tag{53}$$

and replacing the equations (45) for x and y by the adequate equations for x and E. Using the smallness of u_1, u_2, u_3 and D the two-dimensional Fokker-Planck equation for P(x, E, u;t) can be reduced, approximately, to a one-dimensional equation from which a stationary solution can be obtained in the usual way. Omitting all the details we give, here, the final result

$$P^0(x, y, u) = N^{-1} \exp\left[-\frac{2}{D} F(E, u)\right] \tag{54}$$

where

$$\frac{dF(E,u)}{dE} \quad \frac{\int_{R(E)} \gamma(x, \sqrt{2[E-W(x,u)]}, u)\sqrt{E-W(x,u)}\,dx}{\int_{R(E)} \sqrt{E-W(x,u)}\,dx} . \tag{55}$$

In (55) the integration has to be carried out over the region, R(E), where $W(x, u) < E$.

Another method has been proposed by Haken (1973 b). This is based on the proof that the problem in solving the high-dimensional stationary Fokker-Planck equation for a broad class of nonlinear dynamic systems can be reduced to solving a stationary Fokker-Planck equation whose dimension is given by the number of relevant constants of motion.

For the choice of $\gamma(x,y,u)$ (52) one obtains, from (53-55) with $W(x,u)=\frac{x^2}{2}$

$$P^o(x,y,u) = N^{-1}\exp\left\{-\frac{2}{D}\left[\frac{u_1}{2}(x^2+y^2)+\frac{u_3}{16}(x^2+y^2)^2\right]\right\} . \tag{56}$$

u_3 is assumed to be positive so as to guarantee that $P^o(x,y,u)$ is normalizable. The points of extremal probability appear at

$$x = 0 , \quad y = 0 \tag{57}$$

and

$$x^2 + y^2 = -\frac{4u_1}{u_3} . \tag{58}$$

From the identity

$$\Delta(P^o(x,y,u))\Big|_{x^2+y^2=-\frac{4u_1}{u_3}} \equiv 0 \tag{59}$$

we cannot directly determine the transition set, according to (30), by eliminating the state variables x and y from $\Delta(P^o(x,y,u)) = 0$ and $\partial P^o(x,y,u)/\partial x = 0$, $\partial P^o(x,y,u)/\partial y = 0$.

Nevertheless, if we make use of the symmetry property of $P^o(x,y,u)$ and introduce a new state variable r, such that

$$r^2 = x^2 + y^2 \tag{60}$$

then the transition set follows from (30) to be $u_1 = 0$. The transition from $u_1 > 0$ to $u_1 < 0$ corresponds to a special section of a cusp catastrophe set. The universal unfolding of a cusp catastrophe at $r = 0$,

$$V(v,a,b) = \frac{1}{4}v^4 + \frac{1}{2}av^2 + bv, \qquad (61)$$

is related to this case by $a = 4u_1/u_3$ and $b = 0$. The cusp catastrophe produces the well known bifurcation set

$$27b^2 + 4a^3 = 0. \qquad (62)$$

It also contains a cusp point, $a = b = 0$, and fold lines described by (62). In this case the cusp it truncated because there fold lines do not exist at $b = 0$. Consequently in this case sudden jumps in the behaviour of the dynamic system, due to the fold catastrophes along (62), are absent.

The situation can be better understood if we investigate the deterministic model, i. e. the case where $D = 0$. Stability analysis of the stationary state, $x = 0$, $y = 0$, shows that the system undergoes a Hopf-bifurcation if u_1 crosses zero. For $u_1 < 0$ the stationary state $x = 0$, $y = 0$ becomes unstable and a stable limit cycle appears. This is in complete agreement with the above result. $P^0(x, y, u)$ forms for, $u_1 < 0$, a probability crater with a minimum at $x = 0$, $y = 0$ and closed crest lines of equal height at $x^2 + y^2 = -4u_1/u_3$.

The location of the limit cycle may be determined, in the deterministic case, by applying the Pontrjagin theorem (1934) for Hamiltonian-like two-dimensional dynamic systems. The result (Andronov et al., 1967) is in complete agreement with (58). Moreover the parameter u_2 does not here either, enter into the deterministic bifurcation set.

Concluding remarks

It has been shown in this paper that for the stochastic processes, obeying Fokker-Planck or master equations, a function exists, the so-called K-function or gain of information, which monotonously decreases in the course of time and disappears for the stationary state:

$$K(P, P^{\circ}) = \begin{cases} > 0, & P \neq P^{\circ} \\ = 0, & P = P^{\circ} \end{cases} ; \quad \frac{dK}{dt} \leq 0 .$$

The K-function may be considered as a Lyapounov function of the stochastic process (Schlögel, 1971 a, b, 1972, 1973). The theorem, $dK/dt \leq 0$, asserts that the stationary state, $P^{\circ}(x, u)$ will indeed be an attractor in the function space spanned by the set of all probability distribution. This means that an arbitrary, initial distribution, $P_0(x, u)$, will approach the stationary distribution, $P^{\circ}(x, u)$, in the course of time, provided the state space is not decomposable.

Catastrophe theory appears to be a useful tool for the classification of the local form and structural changes of $P^{\circ}(x, u)$ with respect to changes of the control parameters u. Stochastic transitions caused by instabilities in the system are defined as topological changes in the local form of $P^{\circ}(x, u)$ due to the change of external parameters u. Catastrophe theory has been shown to allow general statements about, for example, the maximal number of stochastic transitions if the dimensionality of the space of external parameters does not exceed four. The method developed here can be also applied to transitons during the evolution of the probability distribution in time, if time is considered as an additional parameter.

Originally, catastrophe theory was applied to gradient systems. If there are two or more state variables the assumption that the dynamic system is of gradient type will not hold for most relevant cases, for example in chemical kinetics. There is no doubt that the class of nonlinear dynamic systems with unique, stationary, probability distribution is larger than the class of gradient systems. Therefore the application of catastrophe theory to the investigation of $P^{\circ}(x, u)$ is more general than the application to gradient systems. Both methods lead to identical results for constant intensity of the fluctuations. This was demonstrated here for an enzyme-catalyzed reaction system. Another solvable system treated here is the class of Hamiltonian systems.

The stochastic treatment of a nonlinear dynamic system which exhibits limit cycles remains a difficult problem. It consists in obtaining exact, or sufficiently good approximate, solutions of the stationary Fokker-Planck equation. Here, one example capable of approximate solution has been discussed.

Several authors (see Presnov and Zotina, 1978) have suggested that biological evolution and growth are stochastic processes underlying a criterion of evolution. Possibly the results given here could be useful in the elaboration of these concepts.

Acknowledgement

The authors would like to thank P. Alberti, T. Bröcker, R. Feistel and A. Uhlmann for stimulating discussions.

IV. The Problem of Phenomenological Nonequilibrium Thermodynamics

Irreversible Thermodynamics far from Equilibrium

L. I. Rozonoer

An exhaustiv phenomenological description of nonequilibrium thermodynamic processes in isolated systems is usually made with the aid of a set of macroscopic quantities so chosen that their future values are unambiguously determined by their values at the present moment. The process is described by a closed system of differential equations. From the point of view of statistical physics the possibility of such a description implies that the probability distribution at each moment may be unambiguously parametrized (of course, approximately) by the instantaneous values of macroquantities, i. e. the state of the system is fully characterized by these values. The very possibility of such a description, and also what type of macroquantities determine the state of the system in the above-mentioned sense, are determined, in principle, by microscopical properties of the system (its Hamiltonian and the initial and boundary conditions for the probability distribution). However in practical applications it is, as a rule, impossible to carry out such an analysis and one is forced to assume that the phenomenological description is possible from the outset and, moreover, to set up, from purely phenomenological considerations, a characteristic set of macroquantities. For example in chemical kinetics concentrations (and corresponding chemical potentials) are taken as fundamental quantities; in transport theory the macroquantities that set up the internal energy and mass of parts of the system (and corresponding intensive quantities - temperature and chemical potentials) are taken as fundamentals.

Linear thermodynamics of irreversible processes constitutes a general scheme which permits a purely phenomenological description of such processes near equilibrium. But the limits of applicability of the phenomological description are wider: it is also applied when the whole system is far from equilibrium. However, there exists such a division into sub-systems, that

Thermodynamics and Kinetics of Biological Processes

each of these may be regarded as, practically, at equilibrium (local quasi-equilibrium hypothesis). In such cases the description of the system is already nonlinear. Nonlinear chemical kinetics, based on the mass action law, is an example. The local quasi-equilibrium hypothesis is thus a precondition for the applicability of the phenomenological description both in linear and nonlinear cases (see Gurov, 1978). However, in a linear case there exists a general phenomenological scheme - linear irreversible thermodynamics. Whereas in a nonlinear case there is no sufficiently adequate general scheme based on this hypothesis, although its construction seems quite possible. Moreover, one may hope that the limits of applicability of the phenomenological description will appear to be still wider: even in the absence of a local quasi-equilibrium cases may exist where the system is described by a standard (even nonequilibrium locally) probability distribution unambiguously characterized by a set of macroquantities.

Thus, in order to work out a general phenomenological scheme of irreversible thermodynamics far from the equilibrium it is necessary first to formulate, in unified terms, and in a sufficiently general form, the hypothesis of probability distribution which would specify and generalize the local quasi-equilibrium hypothesis. It should be emphasized that statistical physics is required here only to predict correctly, the framework of the phenomenological description. The question which macroquantities should be chosen in this or that specific case, and what concrete form the equations should have, is decided exclusively upon phenomenological considerations. Micro-theory should yield very little - only some general principles that enables us to establish the form of phenomenological equations (such as for establishing the Onsager relations, when only very general considerations of micro-reversibility are required).

Mixing in Phase Space and Markov's Model of Physical Kinetics

As is known, the process of establishing equilibrium in a closed system consists, from a statistical point of view, in mixing the initial probability distribution up to a roughly uniform one on the surface of single-valued integrals of motion in the phase space of the system. This is not a Markov random process. However, if we consider sufficiently large macroscopic time intervales (greatly exceeding the characteristic times of collisions i. e. of molecular interactions) and sufficiently large regions in the phase space (which contain a large number of points corresponding to the collisions), then the rough density relaxation process may be approximately described as a Markov process. Markov's description of relaxation processes in concrete systems was used successfully some time ago (Kac, 1959; Prigogine, 1963) but its generality and universality for systems with mixing was first realized, apparently, by Krylov (1950).

Let a set of n macroscopic quantities be given as

$$\xi_i = F_i(\omega) \qquad i = 1, \ldots, n$$

whose values are a function of the point, ω, of the phase space Ω. The probability density, $\rho(\omega, t)$, in the phase space gives rise to the probability distribution $W(\xi_1, .., \xi_n; t)$ in the space of values of macroquantities. Let us consider the division of the phase space, Ω, into the regions wherein macroquantities take values from intervals, $(\xi_i, \xi_i + \Delta\xi_i)$:

$$\Omega(\xi) = \{\omega : \xi_i \leq F_i(\omega) \leq \xi_i + \Delta\xi_i , \; i = 1, \ldots, n\}.$$

The phase volume (with ergodic measure on the surface of single-valued integrals of motion) will be designated by $\Gamma(\xi_1, \ldots, \xi_n)\Delta\xi_1 .. \Delta\xi_n$. The fundamental assumption for what follows is: that there exist time intervals, $\sim\tau$, much larger than the characteristic time of collision, Θ, but much smaller than the characteristic time, T, of change of the macroquantities $\xi_1, .., \xi_n$. Within the time intervals $\sim\tau$ the phase flux in each of the regions, $\Omega(\xi)$, does not in practice overstep its limits, but inside $\Omega(\xi)$ it is intensively mixed. Suppose that τ is large enough for a uniform rough density to be established within this time inside each region $\Omega(\xi)$. From this assumption it follows that in the consideration of the process within a time the rough density $\hat{\rho}(\omega, t)$ in the phase space may be regarded as depending on ω only through the values of $F_1(\omega), \ldots, F_n(\omega)$, i. e.

$$\hat{\rho}(\omega, t) = P(F_1(\omega), \ldots, F_n(\omega); t) \tag{1}$$

where $P(\xi_1, \ldots, \xi_n; t)$ is a rough density inside the region $\Omega(\xi)$. The probability density of the values of macroquantities has here the form

$$W(\xi, t) = P(\xi, t)\Gamma(\xi), \tag{2}$$

where ξ denotes the set $\xi_1, \ldots, \xi_n$ just to contract the notation.

The assumption made (the hypothesis of intermediate mixing) enables us to consider a random relaxation process within $t \gg \tau$ as a Markov process in the space of values, ξ, of macroquantities. Let us characterize the phase flux by the quantity, $\gamma(\xi, \eta)$, of the phase volume flowing per unit time from unit phase volume of the region, $\Omega(\eta)$, to unit phase volume of $\Omega(\xi)$ as a result of collisions. The present paper considers, for simplicity, relaxation processes only conditioned by sharp interactions (collisions) with a characteristic time, Θ, in the absence of slow mechanical processes whose characteristic times are of the order of T. It is not difficult also to account for slow mechanical processes (which are, in particular, responsible for convective fluxes) in the framework of Markov's scheme (Kogan, Rozonoer, 1964). That part of the microsystem of the statistical ensemble flowing per unit time from the region, $\Omega(\eta)$, into the region $\Omega(\xi)$ is determined by the formula

$$\Gamma(\xi)\Delta\xi\, \gamma(\xi|\eta)\Gamma(\eta)\Delta\eta\, P(\eta, t) = \Delta\xi\, \Gamma(\xi)\, \gamma(\xi|\eta)\, W(\eta, t)\, \Delta\eta .$$

That is the transition probability density per unit time for the Markov process under consideration is given by

$$A(\xi|\eta) = \Gamma(\xi)\gamma(\xi|\eta). \tag{3}$$

Hence,

$$\frac{\partial W(\xi, t)}{\partial t} = \int A(\xi|\eta)W(\eta, t)d\eta - \int A(\eta|\xi)W(\xi, t)d\eta \tag{4}$$

where integration is carried out over an n-dimensional space of values of macroquantities. From reversibility of collisions it follows that

$$\gamma(\xi_1,\dots,\xi_n|\eta_1,\dots,\eta_n) \equiv \gamma(\eta_1,\dots,\eta_n|\xi_1,\dots,\xi_n).$$

Reversibility of collision is understood here as the following property of the phase flux: if a certain point, $\omega \in \Omega(\eta)$, transported for the time, dt, to the point $\omega_2 \in \Omega(\xi)$, there exists such a point, $\omega_2' \in \Omega(\xi)$, which is transported to the point, $\omega_1' \in \Omega(\eta)$, for the same time. Reversibility of collisions is usually based on time-invariance of microscopic equations. In a number of cases other invariance properties exist which entail collison reversibility.

Introducing the function

$$\Gamma(\xi,\eta) = \Gamma(\xi)\gamma(\xi|\eta)\Gamma(\eta),$$

which characterizes the total phase volume that flows per unit time from $\Omega(\eta)$ into $\Omega(\xi)$ and taking into account, that because of collision reversibility

$$\Gamma(\xi,\eta) = \Gamma(\eta,\xi) \tag{5}$$

we obtain, from (4), the basic equation of system evolution.

$$\frac{\partial W(\xi,t)}{\partial t} = \int \Gamma(\xi,\eta)\left[\frac{W(\eta,t)}{\Gamma(\eta)} - \frac{W(\xi,t)}{\Gamma(\xi)}\right] d\eta. \tag{6}$$

The Gibbs entropy, $H = -k\int_\Omega \hat{\rho} \ln \hat{\rho}\, d\omega$, of the statistical ensemble can be written, with the aid of (1) and (2), in the form

$$H = -k\int W \ln \frac{W}{\Gamma}\, d\xi, \tag{7}$$

where k is a Boltzmann constant. It is easy to find that by virtue of (5) and (6)

$$\begin{aligned} \frac{dH}{dt} = \frac{1}{2}k\iint \Gamma(\xi,\eta)\left[\frac{W(\eta,t)}{\Gamma(\eta)} - \frac{W(\xi,t)}{\Gamma(\xi)}\right]\cdot \\ \left[\ln\frac{W(\eta,t)}{\Gamma(\eta)} - \ln\frac{W(\xi,t)}{\Gamma(\xi)}\right] d\xi\, d\eta \geq 0, \end{aligned} \tag{8}$$

i. e. the Gibbs entropy does not decrease in the process of relaxation.

Relations (5) to (8) are the basic equations of Markov's kinetic model.

Now for the particular case when, in collisions, some physical quantity, ξ, (collision invariant), which is a linear combination of $\xi_1, \ldots, \xi_n$ is retained

$$\xi = \sum_1^n c_i \xi_i .$$

(for example, if ξ_i is the energy of the i-th part of an isolated system, then $\xi = \sum_1^n \xi_i$). In this case the function $\gamma(\xi \mid \eta)$, the transition probability density, $A(\xi \mid \eta)$, and the function $\Gamma(\xi, \eta)$ are identical to zero beyond the linear manifold $\sum_1^n c_i \xi_i - \sum_1^n c_i \eta_i = 0$ and therefore $\Gamma(\xi, \eta)$ has the form

$$\Gamma(\xi, \eta) = \tilde{\Gamma}(\xi, \eta) \delta\left(\sum_1^n c_i \xi_i - \sum_1^n c_i \eta_i\right) \tag{9}$$

where $\tilde{\Gamma}(\xi, \eta)$ is again symmetric about ξ, η. The system may have several invariants for collision, $\xi^k = \sum_{i=1}^n c_i^k \xi_i$, $k = 1, \ldots, l$. The set of linear invariants will be called complete, if the functions ξ^k are linearly independent, and, that every other linear invariant is a linear combination of ξ^k, $k = 1, \ldots, l$.

Statistical Ensemble of Maximum Entropy and Equations for Average Values of Macroquantities

A phenomenological description of the relaxation process is realized through average values of macroquantities:

$$x_i(t) = \int \xi_i W(\xi, t) d\xi , \qquad i = 1, \ldots, n$$

In order that such a description be possible, it is necessary that, at each moment, t, the distribution, $W(\xi, t)$, should be unambiguously determined by instantaneous values of $x_1(t), \ldots, x_n(t)$. When expressing $W(\xi, t)$ through $x(t)$, we proceed from the following considerations. Let time intervals $\sim \tau'$ exist, which greatly exceed the time τ for which Markov's description becomes valid, but are much less than the characteristic time, T, for the change of macroquantities. The time hierarchy $\Theta \ll \tau \ll \tau' \ll T$ corresponds exactly to the hierarchy of relaxation times described by Bogolyubov (1946). One may expect then, that the entropy, H, which increases by virtue of

(8), grows for the time $\sim\tau'$ up to the maximal value compatible with the values of $x_1, \ldots, x_n$ which do not, in practice, change within this time. Thus, the distribution, W , is determined unambigously through x with the aid of the maximum Gibbs' entropy condition at fixed average values of macroquantities

$$-\int W \ln \frac{W}{\Gamma} d\xi \to \max$$
$$\int W(\xi) d\xi = 1, \quad \int \xi_i W d\xi = x_i, \qquad i = 1, \ldots, n. \tag{10}$$

The solution of this variational problem is given by the expression

$$W(\xi) = c\Gamma(\xi) \exp\left(-\frac{1}{k} \sum_1^n X_i \xi_i\right), \tag{11}$$

where c and $X_1, \ldots, X_n$ are expressed through $x_1, \ldots, x_n$ with the aid of the conditions (10); the Boltzmann constant k is introduced for convenience. The entropy (7) on distribution (11) takes the form

$$H = \sum_1^n x_i X_i - G, \qquad G = \ln c \tag{12}$$

and, as is readily seen with the aid of (10) to (12), the values of x and X are connected with the Legendre transformation.

$$X_i = \frac{\partial H}{\partial x_i}, \quad x_i = \frac{\partial G}{\partial X_i}, \qquad i = 1, \ldots, n. \tag{13}$$

To obtain a closed set of equations for the average, $x_i(t)$, we multiply (6) by ξ_i and integrate over ξ . Substituting in the right-hand side of (6) expression (11) for $W(\xi)$ we have

$$\dot{x}_i = \exp \frac{G(X)}{k} \iint \xi_i \Gamma(\xi, \eta) [\exp(-\frac{X\eta}{k}) - \exp(-\frac{X\xi}{k}) d\xi d\eta, \tag{14}$$

where $X.\eta$ and $X\xi$ define the scalar products $(ab = \sum a_i b_i)$. Equations (14) together with relations (13) make up a closed system of differential equations since the right-hand side of (14) is a function of $X_1, \ldots, X_n$, which in turn is determined through $x_1, \ldots, x_n$ by relations (13). By virtue of (14) the entropy $H(x_1, \ldots, x_n)$ does not decrease since

$$\frac{dH}{dt} = \sum_1^n X_i \dot{x}_i = \exp\frac{G}{k}\,\frac{1}{2}\iint \Gamma(\xi,\eta)(X\cdot\xi - X\cdot\eta)\cdot$$
$$\cdot\left[\exp\left(-\frac{X\cdot\eta}{k}\right) - \exp\left(-\frac{X\cdot\xi}{k}\right)\right] d\xi\, d\eta \geq 0. \tag{15}$$

This relation (which is derived from (14) with the use of (5)) is also obtained directly from (8) through (11).

Thus making use of the maximum entropy ensemble enables us to describe the relaxation process directly through the average values of macroquantities. Since its enunciation by Boltzmann and Gibbs, the maximum entropy principle has been used in different ways to describe both equilibrium and nonequilibrium states. The role of maximum entropy ensembles was particularly stressed by Janes (1963) (see also Zubarev, 1974, wherein the maximum entropy principle is discussed in detail). However we should emphasize the following characteristic feature of the application of this principle in our previous (Kogan, Rozonoer, 1964) and present papers: the maximized entropy is defined for a rough density, and the distribution (11), which corresponds to rough density in the phase space, evolves in time according to Markov's equation (6) but not to Liouville's equation. This accounts for the fact that virtue of (14) the entropy increases so that (14) describes irreversible processes. Whereas, on the corresponding ensemble, built for a *thin* density, the entropy on the solution of Liouville's equation does not change (see Zubarev, 1974).

Let us consider the properties of equations (14) in the presence of the collision invariant, $\xi = \sum_1^n c_i \xi_i$. From (14) one can easily obtain, by virtue of (9), that

$$\sum_1^n c_i \dot{x}_i = 0,$$

i. e. that the quantity, $z = \sum_1^n c_i x_i$, is the integral of motion of the system of equations (24). From (9) it is also obvious that if $X_i = c_i,\ i = 1..,n,$ the right-hand sides of (14) vanish. This means that the point, $x = (x_1, \ldots, x_n)$ for which

$$X_i(x_1,\ldots,x_n) = c_i \qquad i = 1,\ldots,n$$

is the equilibrium point of the systen of equations (14). If $\zeta^k = c^k \cdot \xi$, $k = 1, \dots, \ell$ is a full set of linear collision invariants, any other invariant $\zeta = c\,\xi$ can be represented in the form $\zeta = \sum_{k=1}^{\ell} \lambda_k \zeta^k$, whence $c_i = \sum_{k=1}^{\ell} \lambda_k c_i^k$. Therefore the equilibrium points of the system (14) can be found from the condition

$$X_i(x_1, \dots, x_n) = \sum_{k=1}^{\ell} \lambda_k c_i^k \qquad i = 1, \dots, n \tag{16}$$

where λ_k are arbitrary numbers.

If the initial conditions $x_i(t_0) = x_i^0$, $i = 1..n$ are set up for the system of equations (14) and $c^k x$ are the integrals of motion, then

$$\sum_{i=1}^{n} c_i^k x_i = \sum_{i=1}^{n} c_i^k x_i^0 \qquad k = 1, \dots, \ell \ . \tag{17}$$

The coordinates of the equilibrium point, x^*, (along with values of λ_k) can be obtained from (16) and (17). It is convenient to introduce inner coordinates, $\varkappa_r$, $r = 1, .., n-\ell$ on the linear manifold (17) (its dimension is equal to $n - \ell$):

$$x_i = x_i^* + \sum_{r=1}^{n-\ell} b_i^r \varkappa_r \qquad i = 1, \dots, n \tag{18}$$

where each vector b^r is orthogonal to each vector c^k

$$b^r \cdot c^k = 0 \quad , \qquad r = 1, \dots, n-\ell; \quad k = 1, \dots, \ell$$

If we express the entropy, $H(x)$, with the aid of (18), in terms of $\varkappa_1, \dots, \varkappa_{n-\ell}$ and define

$$K_r = \frac{\partial \hat{H}}{\partial \varkappa_r} \qquad r = 1, \dots, n-\ell$$

then

$$K_r = \sum_{i=1}^{n} X_i b_i^r$$

and, by virtue of (16) and the orthogonality relations $b^r \cdot c^k = 0$ in the equilibrium, $K_r = 0$. By making a substitution of variables in (18), one can represent (14) as an equation for the quantities $\varkappa_r$, K_r (see below).

Let us represent the system (14) in a form more convenient for the phenomenological description. Introduce a function of $2n$ variables $P_1, .., P_n, Q_1, .. Q_n$

$$\Phi(P,Q) = k \exp\frac{G(P+Q)}{k} \iint \Gamma(\xi,\eta) \exp\left(-\frac{1}{k}P\cdot\xi - \frac{1}{k}Q\cdot\eta\right) d\xi\, d\eta . \qquad (19)$$

By direct verification we can ensure that if (5) is taken into account, the system (14) may be written, with the aid of the function Φ, in the form

$$\dot{x}_i = \left(\frac{\partial\Phi}{\partial P_i} - \frac{\partial\Phi}{\partial Q_i}\right)_{\substack{P=X \\ Q=0}} , \qquad i = 1,\ldots,n \qquad (20)$$

where the suffixes on the brackets indicate that the values $P_i = X_i$, $Q_i = 0$ are to be substituted after differentiation has been performed.

Note that the description in terms of averages is sufficiently complete only when these averages are representative. Representation of the averages, $x(t)$, implies that the probability density, $W(\xi, t)$, is concentrated near the values $\xi = x(t)$. From (7) it follows, in this case, that $H(x) \approx k \ln \Gamma(x)$ i. e. the Gibbs entropy corresponds to the Boltzmann entropy definition and may be identified with the phenomenological thermodynamic entropy. To emphasize this, we shall designate entropy as $S(x)$ and write the connection between x and X as follows:

$$X_i = \frac{\partial S}{\partial x_i} \qquad i = 1,\ldots,n . \qquad (21)$$

Let us consider the properties of the function $S(x)$ and $\Phi(P,Q)$. If the macroquantities chosen for macrodescription are extensive, then the function $S(x)$ is usually strictly concave (convex upward). This may be proved for additive functions (Khinchin, 1943) and seems to be valid for a very wide class of functions, $F_i(\omega)$, that set up extensive macroquantities. Concavity of $S(x)$ may be violated due to phase transitions. The function,

$$G(X) = \sum_{i=1}^{n} \hat{x}_i X_i - S(\hat{x}),$$

is here strictly concave (the sign $\frown$ implies that the variables, x_i, are expressed through X with the aid of (21)). By virtue of the non-negative nature of $\Gamma(\xi,\eta)$ the function

$$\iint \Gamma(\xi,\eta) \exp\left(-\frac{1}{k}P\xi - \frac{1}{k}Q\eta\right) d\xi\, d\eta$$

in (19) is convex in the variables P, Q. Thus the function, $\Phi(P,Q)$, is convex (not strictly obligatory) on the manifolds $P+Q = \text{const}$. It is assumed for simplicity that the function $\Phi(P,Q)$ is convex over the entire space. The case, when $\Phi(P,Q)$ is convex on the manifolds $P+Q = \mathrm{const}$ only, can be studied in the same manner. As is seen from definition (19) $\Phi(P,Q)$ is symmetric about the variables P, Q :

$$\Phi(P_1,\ldots,P_n;Q_1,\ldots,Q_n) = \Phi(Q_1,\ldots,Q_n;P_1,\ldots,P_n).$$

If there exists in the system the collision invariant $\rho = c\xi$, then, from definition (20) and formula (9), it follows that

$$\Phi(P+c,Q) \equiv \Phi(P,Q+c).$$

If $\rho^k = c^k \cdot \xi$, $k = 1,\ldots,\ell$ is a complete set, then

$$\Phi(P+\sum_{k=1}^{\ell}\lambda_k c^k, Q) \equiv \Phi(P, Q+\sum_{k=1}^{\ell}\lambda_k c^k), \tag{22}$$

this equality being identically valid at arbitrary $\lambda_1,\ldots,\lambda_\ell$.

It can be shown that movement to the variables $\varkappa_r, K_r$ (see (18)) is realized as follows. Let us define a function, $\hat{\Phi}$, of $2(n-\ell)$ variables $\hat{P}_1,\ldots,\hat{P}_{n-\ell};\hat{Q}_1,\ldots,\hat{Q}_{n-\ell}$

$$\hat{\Phi}(\hat{P},\hat{Q};\varkappa) = k\exp\frac{\hat{G}(\hat{P}+\hat{Q})}{k}\iint\hat{\Gamma}(\hat{\xi},\hat{\eta};\varkappa)\cdot \tag{23}$$
$$\cdot\exp\left(-\frac{1}{k}\hat{P}\hat{\xi} - \frac{1}{k}\hat{Q}\hat{\eta}\right)d\hat{\xi}\,d\hat{\eta}$$

which depends, generally speaking, on $\varkappa_1,\ldots,\varkappa_r$ as parameters. Here $\hat{\Gamma}(\hat{\xi},\hat{\eta};\varkappa)$ is defined by the function $\Gamma(\xi,\eta)$ and by the manifold (18). The function $\hat{G}(K)$ is determined through $\hat{S}(K)$ by the Legendre transformation. Instead of equations (14) and (20) one can now write

$$\dot{\varkappa}_r = \left(\frac{\partial\hat{\Phi}}{\partial\hat{P}_r} - \frac{\partial\hat{\Phi}}{\partial\hat{Q}_r}\right)_{\substack{\hat{P}=K\\ \hat{Q}=0}} \qquad r = 1,\ldots,n-\ell. \tag{24}$$

If $\hat{\Gamma}(\hat{\xi},\hat{\eta};\varkappa)$ is assumed somewhere, not equal to zero, then the function, $\hat{\Phi}$, is strictly convex in $\hat{P},\hat{Q}$.

Basic Properties of Phenomenological Equations

In the phenomenological description relations (20) and (21), derived from the considerations of physical statistics, can be taken as the initial relations. The system, wherein irreversible processes proceed is characterized, unambigously, by some set of extensive macroscopic quantities, $x_1, \ldots, x_n$, and the functions, $S(x_1,\ldots,x_n), \Phi(P_1,\ldots,P_n;Q_1,\ldots,Q_n)$, so that the equations have the form:

$$x_i = \left(\frac{\partial\Phi}{\partial P_i} - \frac{\partial\Phi}{\partial Q_i}\right)_{\substack{P=X\\ Q=0}} \qquad i=1,\ldots,n \tag{I}$$

where the intensive variables X_i are defined by the relations

$$X_i = \frac{\partial S}{\partial x_i}, \qquad i=1,\ldots,n. \tag{II}$$

Let us call $S(x)$ a nonequilibrium entropy, $\Phi(P,Q)$ a kinetic function of the system (the physical dimension of Φ is entropy/time). The entropy function is supposed to be strictly concave and the kinetic function a convex function of its variables, $\Phi(P,Q)$, being symmetric under re-arrangement of the variables P and Q . Relations (I) and (II)*, supplemented with the above requirements imposed on S and Φ , make up a general scheme of irreversible thermodynamics in isolated systems.

Consider some properties of this scheme. Let us show, first of all, that by virtue of (I) and (II) the entropy, S , does not decrease with time. To this end we pay attention to the following inequality

$$\sum_1^n \left(\frac{\partial\Phi}{\partial P_i} - \frac{\partial\Phi}{\partial Q_i}\right)(P_i - Q_i) \geq 0. \tag{25}$$

This holds for any convex function, $\Phi(P,Q)$, symmetric with respect to P, Q , and, if Φ is a strictly convex function, then

$$\sum_1^n \left(\frac{\partial\Phi}{\partial P_i} - \frac{\partial\Phi}{\partial Q_i}\right)(P_i - Q_i) > 0, \text{ if } \exists i\; P_i \neq Q_i. \tag{26}$$

Inequalities (25) and (26) follow from the known inequality (see e. g. Rocafellar, 1970):

$$\sum_{j=1}^m \left[\left(\frac{\partial\Psi}{\partial z_j}\right)_{z=z^1} - \left(\frac{\partial\Psi}{\partial z_j}\right)_{z=z^2}\right](z_j^1 - z_j^2) \geq 0$$

---

* Roman numerals indicate relations which are taken as the basic postulates of the phenomenological scheme.

for convex functions, where $z^1 = (z_1^1, \ldots, z_m^1)$, $z^2 = (z_1^2, \ldots, z_m^2)$ are two arbitrary points. To obtain inequality (25), one should set, in this inequality, $\Psi = \Phi$, $m = 2n$, $z^1 = (P_1, \ldots, P_n; Q_1, \ldots, Q_n)$, $z^2 = (Q_1, \ldots, Q_n; P_1, \ldots, P_n)$ and then make use of the symmetry condition $\Phi(P, Q) = \Phi(Q, P)$. For strictly convex functions, $\Psi(z)$, the inequality is strict if $z^1 \neq z^2$ from which (26) follows. From inequlity (25) it follows immediately that the entropy does not decrease if P is put equal to X, and Q equal to zero, since $dS/dt = \sum_1^n \dot{x}_i X_i$ and $\dot{x}_i$ is as expressed from (I), hence

$$\frac{dS}{dt} = \sum_1^n \left(\frac{\partial \Phi}{\partial P_i} - \frac{\partial \Phi}{\partial Q_i} \right)_{\substack{P=X \\ Q=0}} \cdot X_i \geq 0 .$$

Consider the equilibrium points of the system of equations (I) and (II). The equilibrium position is evidently the point $x = (x_1, \ldots, x_n)$ (if it exists at all) for which $X_i = X_i(x_1, \ldots, x_n) = 0$; $i = 1, \ldots, n$. If the function Φ is strictly convex, this equilibrium position is singular since in this case from inequality (26), it follows that $dS/dt > 0$ at $X \neq 0$ and, therefore, $\dot{x} \neq 0$ at $X \neq 0$. In the general case the equilibrium properties of systems (I) and (II) are closely associated with the presence of the integrals of motion (conservation laws). Let the system have ℓ independent invariants, $z^k = c^k x$, $k = 1, \ldots, \ell$. Suppose, that in this case, the function $\Phi(P, Q)$ possesses the property (see formula (22)):

$$\Phi(P + \sum_1^\ell \lambda_k c^k, Q) \equiv \Phi(P, Q + \sum_1^\ell \lambda_k c^k); \qquad \text{(III)}$$

this equality being identical with respect to P, Q and λ_k. We shall show that when (III) holds, the quantities $c^k \cdot x$ are actually the integrals of motion of systems (I) and (II). Differentiate (III) with respect to λ_k and then put $\lambda_k = 0$, $k = 1, \ldots, \ell$, yields

$$\sum_{i=1}^n \left(\frac{\partial \Phi(P, Q)}{\partial P_i} - \frac{\partial \Phi(P, Q)}{\partial Q_i} \right) c_i^k \equiv 0 .$$

Setting $P = X$, $Q = 0$, multiplying the corresponding equation in (I) by c_i^k and summarizing over i we ensure that $c^k \cdot \dot{x} = 0$. We now show that when (III) is valid, the equilibrium points of the systems (I) and (II) are those for which

$$X_i = X_i(x_1, \ldots, x_n) = \sum_{k=1}^\ell \lambda_k c_i^k \qquad i = 1, \ldots, n \qquad (27)$$

with arbitrary λ_k. To this end we differentiate (III) with respect to P_i and then put $P = Y$, $Q = Y$. We obtain

$$\left(\frac{\partial \phi}{\partial P_i}\right)_{\substack{P=Y+\Sigma\lambda_k c^k \\ Q=Y}} \equiv \left(\frac{\partial \phi}{\partial P_i}\right)_{\substack{P=Y \\ Q=Y+\Sigma\lambda_k c^k}} . \tag{28}$$

Similarly, when differentiating (III) with respect to Q_i and setting $P = Y$, $Q = Y$ we obtain

$$\left(\frac{\partial \phi}{\partial Q_i}\right)_{\substack{P=Y+\Sigma\lambda_k c^k \\ Q=Y}} \equiv \left(\frac{\partial \phi}{\partial Q_i}\right)_{\substack{P=Y \\ Q=Y+\Sigma\lambda_k c^k}} \tag{29}$$

On the other hand, from the symmetry, $\phi(P,Q)$, the identity

$$\left(\frac{\partial \phi}{\partial P_i} - \frac{\partial \phi}{\partial Q_i}\right)_{\substack{P=X \\ Q=Y}} \equiv \frac{1}{2}\left[\left(\frac{\partial \phi}{\partial P_i}\right)_{\substack{P=X \\ Q=Y}} - \left(\frac{\partial \phi}{\partial P_i}\right)_{\substack{P=Y \\ Q=X}}\right] + \frac{1}{2}\left[\left(\frac{\partial \phi}{\partial Q_i}\right)_{\substack{P=Y \\ Q=X}} - \left(\frac{\partial \phi}{\partial Q_i}\right)_{\substack{P=X \\ Q=Y}}\right] \tag{30}$$

follows at any X, Y. Assuming here $X = Y + \sum_1^l \lambda_k c^k$ and taking account of (28) and (29), we obtain

$$\left(\frac{\partial \phi}{\partial P_i} - \frac{\partial \phi}{\partial Q_i}\right)_{\substack{P=Y+\Sigma\lambda_k c^k \\ Q=Y}} \equiv 0 \qquad i = 1, \ldots, n. \tag{31}$$

On comparing this relation, at $Y = 0$, with the right-hand side of (1) we see that each point, $x = (x_1, \ldots, x_n)$, for which (27) holds is actually an equilibrium point of systems (I) and (II).

Relation (31) implies that the function $\phi(P,Q)$, which meets requirement (III), is not strictly convex. Indeed, from (31) it follows that

$$\sum_{i=1}^{n}\left(\frac{\partial \phi}{\partial P_i} - \frac{\partial \phi}{\partial Q_i}\right)(P_i - Q_i) = 0 \qquad \text{if } P - Q = \sum_1^l \lambda_k c^k \neq 0$$

in contradiction with (26). Suppose that, in spite of non-strict convexity in $\phi(P,Q)$, the inequality

$$\sum_{i=1}^{n}\left(\frac{\partial \phi}{\partial P_i} - \frac{\partial \phi}{\partial Q_i}\right)(P_i - Q_i) > 0 \qquad \text{if } P - Q \neq \sum \lambda_k c^k \tag{IV}$$

is fulfilled, which supplements the convexity condition (25). From this assumption it follows immediately (at $P = X$, $Q = 0$) that

$$\frac{dS}{dt} = \sum_{i=1}^{n} \dot{x}_i X_i > 0 \qquad \text{if } X \neq \sum_1^l \lambda_k c^k \tag{32}$$

and this means that the points for which $X \neq \sum_1^l \lambda_k c^k$ are not equilibrium points of systems (I) and (II). Thus (IV) guarantees that only those points are at equilibrium for which (27) holds.

Let for systems (I) and (II) the initial conditions $x_i(t_0)=x_i^0,\ i=1..n$ be set up. When (III) is satisfied, the system has ℓ integrals of motion and

$$\sum_{i=1}^{n} c_i^k x_i = \sum_{i=1}^{n} c_i^k x_i^0 \qquad k=1,\dots,\ell \qquad (33)$$

Therefore, the trajectory of the system lies entirely on the linear manifold (33). Relation (27) implies, by virtue of (II), that the equilibrium points of the system that lie on the manifold (33) are stationary points of the function $S(x)$ on this manifold. But since $S(x)$ is strictly concave, it can have no other stationary points on a linear manifold but a single point of global maximum. Thus, it is shown that if the entropy maximum is attained at a certain point x^* of the manifold (33), this is a single equilibrium point of the system (I), (II) on this manifold. If, in addition, (IV) is valid, then from (28) it follows that the point x^* is globally stable and $x(t) \to x^*$ at $t \to \infty$.

By introducing the coordinates $\varkappa_1,\dots,\varkappa_{n-\ell}$ on the manifold (33)

$$x_i = x_i^* + \sum_{r=1}^{n-\ell} b_i^r \varkappa_r \qquad i=1,\dots,n;\ b^r \cdot c^k = 0 \qquad (34)$$

and the corresponding intensive quantities

$$K_r = \frac{\partial \hat{S}}{\partial \varkappa_r}, \qquad (\text{II a})$$

where $\hat{S}$ is the entropy expressed by means of (34) through $\varkappa_1,\dots,\varkappa_{n-\ell}$, one can represent the system (I) in the form

$$\dot{\varkappa}_r = \left(\frac{\partial \hat{\Phi}}{\partial \hat{P}_r} - \frac{\partial \hat{\Phi}}{\partial \hat{Q}_r}\right)_{\substack{\hat{P}=K \\ Q=0}} \qquad r=1,\dots,n-\ell \qquad (\text{I a})$$

where $\hat{\Phi}(\hat{P},\hat{Q},\varkappa)$ is a convex function of the $2(n-\ell)$ variables $\hat{P}_1,\dots,\hat{P}_{n-\ell};\ \hat{Q}_1,\dots,\hat{Q}_{n-\ell}$. Condition (IV) is equivalent to the requirement that the function $\hat{\Phi}$ be strictly convex. In this case system (I a) and (II a) has only one global stable equilibrium point at which

$$K_r(\varkappa_1,\dots,\varkappa_{n-\ell}) = 0 \qquad r=1,\dots,n-\ell.$$

We shall now consider the properties of the systems (I) to (III) near equilibrium and show that in the linear approximation

$$\Delta \dot{x}_i = \sum_{j=1}^{n} L_{ij} \Delta X_j, \quad L_{ij} = L_{ji} \qquad i = 1, \ldots, n, \tag{35}$$

where

$$\Delta x_i = x_i - x_i^*, \quad \Delta X_j = \sum_{k=1}^{n} \left(\frac{\partial^2 S}{\partial x_j \partial x_k} \right)_{x=x^*} \Delta x_k . \tag{36}$$

Since $S(x)$ is strictly concave, the matrix $\left\| \frac{\partial^2 S}{\partial x_j \partial x_k} \right\|$ is negative and definite. Let us briefly express the right-hand side of (I) through $f_i(X_1, \ldots, X_n)$. Linearization of the function f_i near equilibrium gives

$$f_i \approx \sum_{j=1}^{n} \left(\frac{\partial f_i}{\partial X_j} \right)_{X=X^*} (X - X_j^*),$$

where $X_j^* = (\partial S / \partial X_j)_{x=x^*}$. At the same time linearization of the function $X_j = \partial S / \partial x_j$ gives

$$X_j \approx X_j^* + \sum_{k=1}^{n} \left(\frac{\partial^2 S}{\partial x_j \partial x_k} \right)_{x=x^*} \Delta x_k .$$

Therefore proof of (35) and (36) comes down to establishing the symmetry relations

$$L_{ij} = \left(\frac{\partial f_i}{\partial X_j} \right)_{X=X^*} = \left(\frac{\partial f_j}{\partial X_i} \right)_{X=X^*} = L_{ji} . \tag{37}$$

To this end we shall use identity (31) rewriting it in the form

$$\left(\frac{\partial \Phi}{\partial P_i} - \frac{\partial \Phi}{\partial Q_i} \right)_{\substack{P=Y+X^* \\ Q=Y}} = 0 . \tag{38}$$

Included in this is that at the equilibrium point $X^* = \sum_{k=1}^{l} \lambda_k c^k$. Differentiation of (38) with respect to Y_j and then putting $Y = 0$ yields

$$\left(\frac{\partial^2 \Phi}{\partial P_i \partial P_j} + \frac{\partial^2 \Phi}{\partial P_i \partial Q_j} - \frac{\partial^2 \Phi}{\partial Q_i \partial P_j} - \frac{\partial^2 \Phi}{\partial Q_i \partial Q_j} \right)_{\substack{P=X^* \\ Q=0}} = 0, \quad i,j = 1, \ldots, n .$$

After exchanging i and j, it is readily seen that

$$\left(\frac{\partial^2 \Phi}{\partial P_i \partial Q_j} \right)_{\substack{P=X^* \\ Q=0}} = \left(\frac{\partial^2 \Phi}{\partial P_j \partial Q_i} \right)_{\substack{P=X^* \\ Q=0}} . \tag{39}$$

Calculating now $(\partial f_i / \partial X_j)_{X=X^*}$ and $(\partial f_j / \partial X_i)_{X=X^*}$ we ensure that, by virtue of (39), relation (37) holds. Thus, relations (35) and (36) are proved. It is easily seen, also, that (as a consequence of the convexity of the function Φ) the symmetric matrix $\| L_{ij} \|$ is non-negative, i. e.

$$\sum_{i,j=1}^{n} L_{ij} u_i u_j \geqslant 0 .$$

This is not positive and definite, since for any integral of motion $c \cdot x$ the equality $\sum_{j=1}^{n} L_{ij} c_j = 0$ is obeyed. However by changing to the variables $\varkappa_1, \ldots, \varkappa_{n-l}$, in accordance with (34), one can obtain

$$\dot{\varkappa}_r = \sum_{q=1}^{n-l} \hat{L}_{rq} K_q \ , \quad L_{rq} = L_{qr} \qquad r = 1, \ldots, n-l \tag{40}$$

$$K_q = \sum_{p=1}^{n-l} \left(\frac{\partial^2 \hat{S}}{\partial \varkappa_q \partial \varkappa_p} \right)_{\varkappa=0} \varkappa_p \ , \tag{41}$$

where the matrix $\| \hat{L}_{rq} \|$ is positive and definite if (IV) holds (or, which is the same, if the function $\hat{\Phi}$ in (I a) is strictly convex).

Concluding Remarks and Examples

When appliying schemes (I) and (II), one should bear in mind that different functions, $\Phi(P, Q)$, may lead to one and the same equation (I) (i. e. to obtain the equations of motion it is not necessary, in general, to set up the function Φ). Thus, $\Phi(P, Q)$ cannot be unabigously re-established from the equations of motion if they are already known. So, for equations of linear approximation the general form of $\Phi(P, Q)$ is the quadratic form

$$\Phi(P, Q) = \frac{1}{2} \sum_{i,k=1}^{n} A_{ik} (P_i P_k + Q_i Q_k) + \frac{1}{2} \sum_{i,k=1}^{n} B_{ik} (Q_i P_k + Q_k P_i) .$$

It is readily seen that the equations of motions are determined, not by the matrices, $\| A_{ik} \|$ and $\| B_{ik} \|$, taken separately, but only by their difference and, therefore, without loss of generality one can put $B_{ik} = 0$. The generality of the re-establishment of $\Phi(P, Q)$ by the equations of motion for the average values of macroquantities is a consequence of the fact that, as can be shown, $\Phi(P, Q)$ contains exhaustive information on a Markov process, i. e. information not only on the dynamics of average values, but also, on the dynamics of time correlations.

Bearing in mind this fact, we shall dwell, briefly, on applications of the scheme presented to transport phenomena and chemical kinetics. Extensive quantities such as the intrinsic energies of parts of the system and the amount of substance in them should be chosen for the initial macroquantities.

The entropy, S , is, in this case, the entropy of the system expressed through extensive variables. In the choice of the function Φ one should immediately take note of the presence of conservation laws ((III) must be satisfied). For example, to describe heat conductivity between two volumes, one may put

$$\Phi(P_1, P_2; Q_1, Q_2) = \Psi(P_1 + Q_2, P_2 + Q_1),$$

where $\Psi(u,v)$ is some strictly convex function, symmetric with respect to u and v . The variables x_1, x_2 are volume energies and X_1, X_2 are reciprocal temperatures. To obtain the mass action law in a system of chemical reactions

$$\sum_{i=1}^{n} \alpha_{ri} A_i \rightleftharpoons \sum_{i=1}^{n} \beta_{ri} A_i \qquad r = 1, \ldots, m$$

that proceed in a closed volume (under assumption that temperature and pressure are constant), the function Φ may be set up in the form

$$\Phi = c + \sum_{r=1}^{m} k a_r \left[\exp\left(-\frac{\sum \alpha_{ri} P_i + \sum \beta_{ri} Q_i}{k}\right) + \exp\left(-\frac{\sum \alpha_{ri} Q_i + \sum \beta_{ri} P_i}{k}\right)\right]$$

where k is the Boltzmann constant, a_r are constant quantities, c is an unessential additive constant. The quantity x_i stands for the number of moles of the i-th substance, the entropy being set up in the usual way

$$S = -k \sum_{i=1}^{n} x_i (\ln x_i + \varphi_i) ,$$

where the constants, φ_i , are determined by temperature and pressure.

The phenomenological scheme has been, hitherto, presented in applications to discrete systems. Its extension to continuous systems is obvious: the discrete index, i, is replaced by $(j, \vec{r})$, where j is a discrete index designating type of physical quantity (energy, mass etc.) and $\vec{r}$ is the radius-vector of the point under consideration. Instead of the extensive quantities, x_i , their densities $u_j(\vec{r})$ are set out. The function $\Phi(P, Q)$ is replaced by the corresponding function of $P_j(\vec{r}), Q_j(\vec{r})$. The entropy density $S(\vec{r})$, which depends on the densities $u_j(\vec{r})$ in the normal way, is substituted for the entropy. Using the designations $\frac{\delta}{\delta P_j(\vec{r})}, \frac{\delta}{\delta Q_j(\vec{r})}$ for functional derivatives one can write the phenomenological scheme in the form

$$\frac{\partial u_j(\vec{r}, t)}{\partial t} = \left(\frac{\delta \Phi}{\delta P_j(\vec{r})} - \frac{\delta \Phi}{\delta Q_j(\vec{r})} \right)_{\substack{P_j(\vec{r}) = X_j(\vec{r}) \\ Q_j(\vec{r}) = \sigma}} \qquad \text{(I b)}$$

$$X_j(\vec{r}) = \left(\frac{\partial S}{\partial u_j}\right)_{u_j = u_j(\vec{r})} . \qquad \text{(II b)}$$

For example, in the description of the process of heat conductivity, $u(\vec{r},t)$ is the energy density at the point $\vec{r}$. Φ may be written in the form

$$\Phi = a\int[(\vec{\nabla}P)^2 + (\vec{\nabla}Q)^2]\,dV,$$

where integration is carried out over a three-dimensional volume, $\vec{\nabla} = \frac{\partial}{\partial \vec{r}}$, and a is a constant. The quantity $X(\vec{r})$, in this case, is a reciprocal temperature at the point $\vec{r}$.

We should note that equation (I b) does not contain convective terms. They could have been accounted for from the outset (see above, p. 222) but the manner of their introduction is, in many cases obvious after writing equation (I b).

The following interesting possibility arises in the scheme represented by (I) and (II). In the normal description of relaxation processes the fluxes of heat, matter etc. are expressed at each moment through temperatures, chemical potentials etc. Scheme (I) and (II) allows one to take into account inertial properties of the fluxes, i. e. to include these quantities into the system of equations as independent quantities, the usual connection between fluxes and the forces which induce them is established as a consequence of the equations only in some time (Kogan, Rozonoer, 1964). A detailed description, with independent fluxes, was first suggested by Grad (1949) in application to the theory of rarified gas (see also Kogan, 1965).

In conclusion some remarks about the thermodynamics of non-isolated systems. Non-isolated means here those systems which can exchange extensive quantities (energy, impulse, moment of impulse, mass) with the environment. Statistical consideration of irreversible processes in non-isolated systems is difficult since no standard statistical description exists for this case. Therefore, equations for non-isolated systems are more conveniently obtained in a phenomenological way by using, as the initial scheme, the phenomenological description for isolated systems. To this end the procedure for opening the system consists, in most cases, in setting up altered boundary conditions for the initial equations of an isolated system. For example, extending the heat-conductivity equation to an isolated system requires the assumption that the temperature or its gradient be non-zero on the boundaries. The procedure of opening may be understood, formally, in a broader sense, i. e. as any distortion of the initial equations of an isolated system that takes into account the physical processes of exchange of the extensive quantities with the environment. For instance, a system of equations of chemical kinetics may be opened by assuming that the concentrations of some substances participating in the reaction are known. The equations that involve derivatives of these concentrations being simply omitted from the initial system of equations. This way of opening systems takes into account the physical process of removing some reaction products from the reaction vessel. It is obvious that opening may change, qualitatively, the behaviour of the system, e. g. may lead to the appearance of dissipative structures in space and time (see Glansdorff, Prigogine, 1971) which were inadmissible in the initial isolated scheme. Standardization of the opening procedure, i. e. isolation of such peculiar classes of distortion of the initial equations which do correspond to real physical processes is an interesting and important problem. The general phenomenological scheme for describing non-isolated (in particular, open) systems may be discussed only after such a standardization has been carried out. One can show that the same scheme, (I) to (III), with the non-symmetrical functions, $\Phi(P, Q)$, describes a wide class of open systems.

On the Structuring of Thermodynamic Fluxes:
A Direct Implementation of the Dissipation Inequality

J. Bataille, D. G. B. Edelen, J. Kestin

The generally accepted principles on nonequilibrium thermodynamics (Meixner, Reik, 1959; Prigogine, 1967; de Groot, Mazur, 1969) start with an expression for the rate of entropy production, $\dot{\Theta}$, for the processes and material media under investigation. An examination of the mathematical form of the entropy production, together with the physical meaning of the various symbols that occur, leas to a representation of $\dot{\Theta}$ in a bilinear form:

$$\dot{\Theta} = X \cdot J \qquad (1)$$

by means of a choice of an n-dimensional vector X, whose components, $X_1, X_2, \ldots, X_n$, are identified with the thermodynamic forces, and a resulting identification of J as an n-dimensional vector whose components, $J_1, J_2, \ldots, J_n$, are the conjugate thermodynamic fluxes.

Although the choice of the X and the J in representation (1) is not necessarily unique, it must satisfy certain specific requirements.

(a) The first of these, is that the vector of thermodynamic forces is chosen so that X = 0 in every equilibrium state. We assume, for the purposes of this paper, that this requirement has been met.

(b) The second requirement, is that the vector, J, of thermodynamic fluxes, can be expressed as a vector-valued function of the thermodynamic forces, X, and, of the state variables of the accompanying thermostatic equilibrium state (Bataille, Kestin, 1976). Let the state variabels of the accompanying thermostatic equilibrium state be represented by an r -dimensional vector ω . We may then write

$$\mathfrak{J} = j(X, \omega) \tag{2}$$

where the vector-valued function $j(X, \omega)$ will be assumed throughout this paper to be continuous in X and have continuous first partial derivatives with respect to the entries of X for each admissible vector ω of the accompanying thermostatic equilibrium state.

(c) The third requirement, is that the fluxes vanish with the forces; that is $j(X, \omega)$ satisfies

$$j(0, \omega) = 0 \tag{3}$$

for each admissible ω , unless the forces have been artificially constrained (such as by the introduction of an anticatalyst, in which case the corresponding term in (4) vanishes).

(d) The fourth requirement, which is an expression of the second part of the second law, consists of the requirement that the functions $j(X, \omega)$ be such that

$$\dot{\Theta}(X, \omega) = X \cdot j(X, \omega) \geq 0 \tag{4}$$

for all posssible processes of the material system under consideration. In fact, it is possible to make the more exclusive assertion that

$$X \cdot j(X, \omega) > 0 \tag{5}$$

for all $X \neq 0$ and all admissible ω , (if we do not wish to include the unrealizable case of reversible processes).

It must be clearly understood that the four requirements just stated are not restricted to the case where the functions $j(X, \omega)$ are linear in the arguments X, for this would relegate nonequilibrium thermodynamics to the status of a discipline confined solely to the domain of validity of such linear approximations. Any such confinement is clearly artificial. In order to see this, it is only necessary to note that the thermodynamic fluxes usually define specific terms in the equations that govern the space and time evolution of the physical system. Nothing leads to the conclusion that these equations of evolution will have solutions such that the system will always remain in the domain of validity of the linear approximation of $\dot{j}(X, \omega)$ as a function of X. Further, there are known physical phenomena for which restriction of the fluxes to the domain of validity of the linear approximation gives an intrinsically incorrect description of the evolutionary processes; a case in point being that in which $\dot{j}(X, \omega)$ is given by the equations of chemical kinetics (Anderson, Boyd, 1971; Bataille et al., 1978).

(e) This observation notwithstanding, the linear approximation

$$\dot{j}(X, \omega) = L(\omega) : X + O(X \cdot X) \tag{6}$$

should yield the Onsager reciprocity relations, i. e.

$$L(\omega)^T = L(\omega) \tag{7}$$

if all the entries of X are assumed to have the same parity under time reversal. Here $L(\omega)^T$ denotes the transpose of the n-by-n matrix $L(\omega)$ and $O(X \cdot X)$ denotes quadratic and higher order terms in X. Under these conditions, satisfaction of (7) for the approximation given by (6) constitutes a fifth requirement that the functions $\dot{j}(X, \omega)$ must satisfy in order to fall within the practices of nonequilibrium thermodynamics.

The basic question to which this paper is addresses may be stated as follows: To what extent do the five requirements (a) to (e) of nonequilibrium thermodynamics determine the functions $\dot{j}(X, \omega)$ that specify the thermodynamic flows? A general, but specific answer to this question will provide not only a fuller understanding of linear nonequilibrium thermodynamics, but also dispense, once and for all, with certain historical prejudices that have arisen out of the intrinsic simplicity associated with the structure of linear phenomenological force-flux relations.

Reformulation of the Basic Problem

We have assumed that requirements (a) and (b) have been met and hence $j(X, \omega)$ has continuous, first partial derivatives with respect to the components of the vector X for each admissible ω. Under these conditions, we can show that satisfaction of requirement (c) follows from satisfaction of requirement (d). The precise statement is as follows.

If $j(X, \omega)$ is a continuous and differentiable vector-valued function such that (4) holds, then (3) is satisfied.

The proof is quite straightforward. We first note that $\dot{\Theta}(X, \omega)$ is continuous, differentiable, non-negative and vanishes at X = 0 by (4) so that $\dot{\Theta}(X, \omega)$ attains its minimum at X = 0;

$$\nabla_X \dot{\Theta}(X, \omega)\big|_{X=0} = 0 \tag{8}$$

Here ∇_X denotes the gradient with respect to X at fixed ω. We shall usually write ∇ for ∇_X where there is no need to designate the variable vector X explicitly. However, (4) also gives

$$\frac{\partial \dot{\Theta}(X, \omega)}{\partial X_k} = j_k(X, \omega) + \sum_{i=1}^{n} X_i \frac{\partial j_i(X, \omega)}{\partial X_k},$$

and hence an evaluation at X = 0 and use of (8) yields $j(0, \omega) = 0$.

The importance of this result is that it reduces our list of requirements to only two. The following reformulation of the problem can thus be made.

If $j(X, \omega)$ is of class C^1 with respect to the entries of the vector X for each admissible ω, then the thermodynamic fluxes,

$$J = j(X, \omega),$$

are constant with nonequilibrium thermodynamics if, and only if,

(i) $$\dot{\Theta}(X, \omega) = X \cdot j(X, \omega) \geq 0$$

for all vectors X and all admissible ω, and

(ii) $$j(X, \omega) = L(\omega) : X + O(X \cdot X), \qquad L(\omega)^T = L(\omega),$$

for all admissible ω if all of the components of X have the same parity under time reversal.

A Preliminary Observation

Most authors, after Ziegler (1966, 1972, 1977) and Li (1962) - who give physical arguments for their choice - proposed nonlinear constitutive laws that are such that the differential form $j(X,\omega)dX$ admits an integrating factor $u(X,\omega)$. On the other hand, if $j(X,\omega)$ is any vector-valued function of X and of ω, that is of class C^1 in the entries of X, we may always write

$$j(X,\omega) = u(X,\omega)(\nabla\Psi(X,\omega) + y(X,\omega)), \tag{9}$$

where ∇ denotes the gradient with respect to the entries of the vector X and where

$$u(X,\omega) > 0, \qquad \forall X, \omega .$$

It is important to realize that the function $u(X,\omega)$ may be taken to be unity, without loss of generality. If, however, $j(X,\omega)$ happens to satisfy the symmetry relations

$$j_i\left(\frac{\partial j_k}{\partial X_m} - \frac{\partial j_m}{\partial X_k}\right) + j_k\left(\frac{\partial j_m}{\partial X_i} - \frac{\partial j_i}{\partial X_m}\right) + j_m\left(\frac{\partial j_i}{\partial X_k} - \frac{\partial j_k}{\partial X_i}\right) = 0, \quad i,k,m = 1,\dots,n, \tag{10}$$

but not the conditions

$$\frac{\partial j_i(X,\omega)}{\partial X_k} = \frac{\partial j_k(X,\omega)}{\partial X_i}, \qquad i,k = 1,\dots,n, \tag{11}$$

then we obtain $y(X,\omega)=0$ with $\nabla_X u(X,\omega) \neq 0$. Thus, in this case only, is there an advantage to be obtained by selection of a non-constant function $u(X,\omega)$. However, since $u \neq 1$ has been taken as the point of departure in the above-mentioned works (Ziegler, 1966, 1972; Li, 1962), we shall continue to include the function $u(X,\omega)$ as an arbitrary function subject to the sole constraint $u(X,\omega) > 0$.

In the general case, namely where neither of the two conditions (10) nor (11) is satisfied, the form given by (9) can be obtained directly from Darboux's theorem (Sternberg, 1964; Slebodzinski, 1970). Although we shall not make specific use of the Darboux theorem in this paper, it is of interest to note that this theorem yields a specific form for the vector-valued function $y(X,\omega)$ and states that $u(X,\omega)>0$. Further, the Darboux theorem makes no claim that the quantities $u(X,\omega)$, $\Psi(X,\omega)$ and $y(X,\omega)$ are uniquely determined for given $j(X,\omega)$, and, in fact, explicit non-uniqueness can be demonstrated (Edelen, 1977).

Solution of the Basic Problem

We now inquire into the structure that is demanded of the quantities $u(X,\omega)$, $\Psi(X,\omega)$ and $y(X,\omega)$ in order that $j(X,\omega)$ satisfies the conditions imposed by nonequilibrium thermodynamics that are stated in Section 1. A direct substitution of (9) into (4) leads to the condition

$$u(X,\omega)X\{\nabla\Psi(X,\omega)+y(X,\omega)\}\geq 0,$$

for all X and all admissible ω . Thus, since $u(X,\omega)>0$ for all X and all admissible ω , we have to find all sufficiently smooth solutions of the differential inequality (dissipation inequality):

$$X\cdot\{\nabla\Psi(X,\omega)+y(X,\omega)\}\geq 0, \tag{12}$$

for all X and all admissible ω .

In order to solve this problem, we introduce an arbitrary, smooth, scalar function, $p(X,\omega)$, such that

$$p(X,\omega)\geq 0,\ \forall X \text{ and } \omega;\ p(0,\omega)=0. \tag{13}$$

These conditions imply that the function $p(X,\omega)$ achieves its minimum of zero at X = 0, i. e.

$$\nabla p(X,\omega)\big|_{X=0}=0, \tag{14}$$

and that

$$p(\lambda X, \omega) = O(\lambda^2), \tag{15}$$

where $O(\lambda^2)$ denotes quadratic and higher order terms in λ . For any such function, $p(X, \omega)$, solving the equation

$$X \cdot \nabla \Psi(X, \omega) = f(X, \omega), \tag{16}$$

where

$$f(X, \omega) = p(X, \omega) - X \cdot y(X, \omega), \tag{17}$$

is equivalent to solving (12).

The general solution of (16) is given by

$$\Psi(X, \omega) = \Psi_0(X, \omega) + \int_0^1 f(\lambda X, \omega) \lambda^{-1} d\lambda, \tag{18}$$

where $\Psi_0(X, \omega)$ is the general solution of the homogeneous equation

$$X \cdot \nabla \Psi_0(X, \omega) = 0. \tag{19}$$

Indeed, we need only show that the second term on the right-hand side of (18) is a particular solution of (16), (in view of the linearity of (16) in the variable $\Psi(X, \omega)$). Now,

$$X \cdot \nabla \int_0^1 f(\lambda X, \omega) \frac{d\lambda}{\lambda} = \int_0^1 X \cdot \nabla_X f(\lambda X, \omega) \frac{d\lambda}{\lambda} =$$
$$= \int_0^1 X \cdot \nabla_{\lambda X} f(\lambda X, \omega) d\lambda = \int_0^1 \frac{\partial}{\partial \lambda} f(\lambda X, \omega) d\lambda,$$

and hence, we obtain the identity

$$X \cdot \nabla \int_0^1 f(\lambda X, \omega) \frac{d\lambda}{\lambda} = f(X, \omega) - f(0, \omega). \tag{20}$$

However, since $f(0, \omega) = 0$ from (13) and (15), and the result is established.

What now remains is to determine $\Psi_0(X, \omega)$ so that (19) is satisfied. A direct application of Euler's theorem for homogeneous functions shows that (19) is satisfied by all functions $\Psi_0(X, \omega)$ that are homogeneous in X of degree zero for all ω :

$$\Psi_0(\lambda X, \omega) = \Psi_0(X, \omega) \tag{21}$$

for all $\lambda > 0$. A differentiation of (21) with respect to X leads to the relation

$$\lambda \nabla_{\lambda X} \Psi_0(\lambda X, \omega) = \nabla \Psi_0(X, \omega) . \tag{22}$$

If we take the limit of this relation as λ tends to zero, we obtain

$$0 = \nabla \Psi_0(X, \omega), \tag{23}$$

that is,

$$\Psi_0 = \Psi_0(\omega) \tag{24}$$

since it can readily be shown that $\nabla \Psi_0 |_{X=0}$ is given by

$$\nabla \Psi_0 |_{X=0} = \frac{j(0, \omega)}{u(0, \omega)} \tag{25}$$

and is therefore bounded. Indeed, it follows directly from (18) and (17) that

$$\nabla \Psi |_{X=0} = \nabla \Psi_0 |_{X=0} + \int_0^1 \{ \nabla_{\lambda X} p(\lambda X, \omega) - \nabla_{\lambda X} [\lambda X \cdot y(\lambda X, \omega)] \}_{X=0} d\lambda$$

and hence (10) and (14) show that

$$\nabla \Psi_0 |_{X=0} = \nabla \Psi |_{X=0} + y(0, \omega) = \frac{j(0, \omega)}{u(0, \omega)} .$$

A combination of (9), (18), (17) and (15), together with the fact that a function, $\Psi_0(\omega)$, of the thermostatic state variables may be discarded without loss of generality yields the following basic result.

Theorem A

A necessary and sufficient condition for a vector-valued function $j(X, \omega)$ of class C^1 to satisfy $X \cdot j(X, \omega) \geq 0$ is that it be of the form

$$j(X, \omega) = u(X, \omega) \{ \nabla [\Phi(X, \omega) - \int_0^1 X \cdot y(\lambda X, \omega) d\lambda + y(X, \omega) \} , \tag{26}$$

with

$$\Phi(X, \omega) = \int_0^1 p(\lambda X, \omega) \lambda^{-1} d\lambda , \tag{27}$$

where $y(X,\omega)$ is an arbitrary smooth vector-valued function and $p(X,\omega)$, $u(X,\omega)$ are arbitrary smooth scalar-valued functions such that

$$u(X,\omega) > 0, \tag{28}$$

$$p(X,\omega) \geq 0\,, \quad p(\lambda X,\omega) = O(\lambda^2) \tag{29}$$

for all X and all admissible ω.

Actually, the analysis given above only establishes necessity. It is, however, a trivial matter to check sufficiency directly from (26) to (29).

At this point we may choose to rewrite (26) to (29) in the equivalent form

$$j(X,\omega) = u(X,\omega)\{\nabla\Phi(X,\omega) + U(X,\omega)\}$$

with

$$U(X,\omega) = y(X,\omega) - \nabla\int_0^1 X\cdot y(\lambda X,\omega)\,d\lambda\,. \tag{30}$$

Now, it is easy to check that (30) implies the orthogonality condition

$$X\cdot U(X,\omega) = 0\,, \tag{31}$$

and differentiation of (31) followed by an evaluation at X = 0 yields

$$U(0,\omega) = 0\,. \tag{32}$$

We can therefore give the previous theorem an alternative formulation.

Theorem B

A necessary and sufficient condition for a vector-valued function $j(X,\omega)$ to satisfy the dissipation inequality is that it be of the form

$$j(X,\omega) = u(X,\omega)\{\nabla\Phi(X,\omega) + U(X,\omega)\} \tag{33}$$

with

$$\Phi(X,\omega) = \int_0^1 p(\lambda X,\omega)\lambda^{-1}\,d\lambda\,, \tag{34}$$

where $\mathcal{U}(X, \omega)$ is an arbitrary smooth vector-valued function that satisfies the orthogonality condition

$$X \cdot \mathcal{U}(X, \omega) = 0 \tag{35}$$

and $u(X, \omega)$, $p(X, \omega)$ are arbitrary smooth scalar-valued functions such that

$$u(X, \omega) > 0,$$
$$p(X, \omega) \geq 0, \quad p(\lambda X, \omega) = O(\lambda^2).$$

The essential difference between Theorem A and Theorem B is that Theorem A leaves $y(X, \omega)$ arbitrary while Theorem B imposes the orthogonality condition $X \cdot \mathcal{U}(X, \omega) = 0$ which, in turn, implies $\mathcal{U}(0, \omega) = 0$ (see (32)). The equivalence of Theorems A and B means, however, that the class of smooth solutions of

$$X \cdot \mathcal{U}(X, \omega) = 0$$

may be constructed from an arbitrary smooth vector field $y(X, \omega)$ by

$$\mathcal{U}(X, \omega) = y(X, \omega) - \nabla \int_0^1 X \cdot y(\lambda X, \omega)\, d\lambda \, . \tag{36}$$

There is, thus, no difficulty in securing the orthogonality condition (35) in Theorem B.

Generalized Dissipation Potentials and Nondissipative Fluxes

From now on, we shall confine the discussion to the results given by Theorem B and assume that an arbitrary, but definite, choice of a positive function $u(X, \omega)$ has been made. Thus any system of thermodynamic fluxes is given by (33) and (34), for some choice of the two generating functions $p(X, \omega)$ and $\mathcal{U}(X, \omega)$ that satisfy (29) and (35). Moreover the associated entropy production is obviously given by

$$\dot{\Theta}(X,\omega) = X \cdot j(X,\omega) = u(X,\omega)p(X,\omega). \tag{37}$$

All that remains is to interpret the quantities that appear in (33).

If we substitute (33) into the defining equation, $\dot{\Theta}(X,\omega) = X \cdot j(X,\omega)$, and note that (35) gives $u(X,\omega)X \cdot U(X,\omega) = 0$, we obtain

$$\dot{\Theta}(X,\omega) = u(X,\omega)X \cdot \nabla\Phi(X,\omega). \tag{38}$$

One part,

$$j_N = u(X,\omega)U(X,\omega), \tag{39}$$

of $j(X,\omega)$ thus makes no contribution to entropy production and we consequently refer to it as the *nondissipative part* of the thermodynamic flux vector. Further, since $u(X,\omega) > 0$, (38) shows that we may refer to

$$j_0 = u(X,\omega)\nabla\Phi(X,\omega), \tag{40}$$

as its *dissipative part* and to $\Phi(X,\omega)$ as a *generalized dissipation potential*. The latter may actually be evaluated directly in terms of $\dot{\Theta}(X,\omega)$ and $u(X,\omega)$. Indeed, substitution for $p(X,\omega)$ from (37) into (34) leads to:

$$\Phi(X,\omega) = \int_0^{\Lambda} \frac{\dot{\Theta}(\lambda X,\omega)}{\lambda u(\lambda X,\omega)} d\lambda = \int_0^{\Lambda} \frac{X \cdot j(\lambda X,\omega)}{\lambda u(\lambda X,\omega)} d\lambda. \tag{41}$$

The Linear Approximation and Onsager Reciprocity Relations

The fact that the nondissipative part,

$$j_N(X,\omega) = u(X,\omega)U(X,\omega) \qquad X \cdot j_N(X,\omega) = 0 \tag{42}$$

of any vector of thermodynamic fluxes makes no contribution to the entropy production function shows, directly, that any restrictions on the choice of $j_N(X,\omega)$ must come from outside the structure of the dissipative mechanisms of macroscopic, phenomenological, nonequilibrium thermodynamics. Perhaps the best known example of this is Onsager's derivation of his symmetry relations for linear force-flux relations from the requirement that the underlying laws of the microscopic system be invariant under time reversal. Since this case provides some useful insights, it is worth examining in detail.

Noting that $j(0,\omega)=0$, $u(0,\omega)=0$, by (3) and (32), so that $j_D(\lambda X,\omega)=O(\lambda)$, $j_N(\lambda X,\omega)=O(\lambda)$, (39) and (40) give

$$j_D(\lambda X,\omega)=\lambda u(0,\omega)L(\omega):X+O(\lambda^2), \tag{43}$$

$$j_N(\lambda X,\omega)=\lambda u(0,\omega)M(\omega):X+O(\lambda^2) \tag{44}$$

where $L(\omega)$ and $M(\omega)$ are square matrices and the colon denotes matrix multiplication. However, the orthogonality condition $X\cdot j_N(X,\omega)=0$ and (44) shows that $M(\omega)$ necessarily satisfies

$$M(\omega)^T=-M(\omega), \tag{45}$$

whilst (40) and (43) show that $L(\omega)$ necessarily satisfies

$$L(\omega)^T=L(\omega). \tag{46}$$

Here, the superscript T denotes matrix transposition. It is now a simple matter to see that

$$\begin{aligned} j(\lambda X,\omega) &= j_D(\lambda X,\omega)+j_N(\lambda X,\omega)= \\ &= \lambda u(0,\omega)(L(\omega)+M(\omega)):X+O(\lambda^2) \end{aligned} \tag{47}$$

will satisfy the Onsager reciprocity relation

$$(L(\omega)+M(\omega))^T=L(\omega)+M(\omega), \tag{48}$$

in the linear approximation if and only if, $M(\omega)=0$. A vector of thermodynamic fluxes, $j(X,\omega)$, will satisfy Onsager's reciprocity relations in the domain of validity of the linear approximation if, and only if, the nondissipative part, $j_N(X,\omega)$, satisfies

$$j_N(\lambda X,\omega)=O(\lambda^2). \tag{49}$$

It should be noted that (49) is, indeed, a new condition, for we previously had $j_N(\lambda X,\omega)=O(\lambda)$, not $j_N(\lambda X,\omega)=O(\lambda^2)$.

We may thus finally replace Theorem B in the following manner.

Theorem C

For any given smooth function $u(X,\omega)$, such that $u(X,\omega) > 0$, the requirements (a) to (e) of thermodynamics of irreversible processes structure any vector $j(X,\omega)$ of thermodynamic fluxes so that $j(X,\omega)$ is given by the sum to two fluxes

$$j(X,\omega) = j_D(X,\omega) + j_N(X,\omega)$$

defined by

$$j_D(X,\omega) = u(X,\omega)\nabla\Phi(X,\omega)$$

and

$$j_N(X,\omega) = u(X,\omega)\,U(X,\omega)\,, \qquad X \cdot j_N(X,\omega) = 0$$

which are, respectively, the dissipative and nondissipative parts of $j(X,\omega)$ associated with the particular choice of $u(X,\omega)$. The scalar-valued function $\Phi(X,\omega)$ is the generalized dissipation potential of the form

$$\Phi(X,\omega) = \int_0^1 p(\lambda X,\omega)\lambda^{-1}\, d\lambda\,,$$

and the generating function $p(X,\omega)$ and the vector $U(X,\omega)$ are arbitrary smooth quantities which must satisfy

$$p(X,\omega) = O(\lambda^2)\,, \quad U(X,\omega) = O(\lambda^2)\,, \quad X \cdot U(X,\omega) = 0$$

Further, the entropy production $\dot{\Theta}(X,\omega)$ is given by

$$\dot{\Theta}(X,\omega) = u(X,\omega)\,p(X,\omega)$$

and Φ is determined in terms of $\dot{\Theta}(X,\omega)$ or $j(X,\omega)$ by

$$\Phi(X,\omega) = \int_0^1 \frac{\dot{\Theta}(\lambda X,\omega)}{\lambda u(\lambda X,\omega)}\, d\lambda = \int_0^1 \frac{X \cdot j(\lambda X,\omega)}{u(\lambda X,\omega)}\, d\lambda\,.$$

Again, we take particular note of the fact that the non-negative function $u(X,\omega)$ that occurs throughout this formulation is arbitrary, and may be set equal to unity without loss of generality. The latter choice would lead to the equations obtained previously (Edelen, 1973 a, 1974, 1975) by an analysis based on a general decomposition theorem for C^1 vector-valued functions ($j = \nabla\Phi + U\,,\ X \cdot U = 0$) .

The Existence of Symmetry Relations

There is a clear advantage to be obtained, both theoretically and computationally, in extending Onsager's reciprocity relations into the regime of nonlinear phenomenological relations. A number of suggestions, which have been summarized by Bataille and Kestin (1977) have been put forward for such extensions. They essentially consist of the requirement that the differential form, $j \cdot dX$, admit an integrating factor, which is equivalent to the requirement that

$$j(X, \omega) = u(X, \omega) \nabla \Phi(X, \omega). \tag{50}$$

In view of Theorem C, this amounts to the requirement that the nondissipative part of the fluxes vanishes. In fact, satisfaction of (50), which is really an inaccessibility condition analogous to Caratheodory's principle precludes the possibility of finding a path, $X(\lambda)$, such that $X(0) = 0$, $X(1) = X_1$, where X_1 can be chosen arbitrarily and for which

$$j(X(\lambda), \omega) \cdot \frac{dX(\lambda)}{d\lambda} = 0$$

for all λ such that $0 \leq \lambda \leq 1$. The full significance of this observation is not clear at this juncture. It is given here in the expectation that such an alternative view may provide the reader with additional insight into the physical basis underlying the requirement that the nondissipative part of the fluxes vanishes.

It would seem, however, that in the general case with $u \neq 0$, recognition of a system of identically satisfied symmetry relations may prove helpful. If we use Theorem C, we obtain

$$j(X, \omega) - j_N(X, \omega) = u(X, \omega) \nabla \Phi(X, \omega). \tag{51}$$

Thus, since $u(X, \omega) > 0$ we obtain

$$u(X, \omega)^{-1} \{ j(X, \omega) - j_N(X, \omega) \} = \nabla \Phi(X, \omega). \tag{52}$$

It thus follows directly that the following symmetry relations are incidentally satisfied:

$$\frac{\partial}{\partial X_i} \left\{ \frac{j_k(X, \omega) - j_{N_k}(X, \omega)}{u(X, \omega)} \right\} = \frac{\partial}{\partial X_k} \left\{ \frac{j_i(X, \omega) - j_{N_i}(X, \omega)}{u(X, \omega)} \right\}. \tag{53}$$

These symmetry relations can be written in the equivalent form

$$\frac{\partial}{\partial X_i}\left\{\frac{\dot{j}_{Dk}(X,\omega)}{u(X,\omega)}\right\} = \frac{\partial}{\partial X_k}\left\{\frac{\dot{j}_{Di}(X,\omega)}{u(X,\omega)}\right\} \tag{54}$$

since the dissipative part of the flux, $\dot{j}_D$, is given by $\dot{j}-\dot{j}_N$. We note that for $u(X,\omega)=1$, the symmetry relations (54) reduce to those given in (Edelen, 1973 b, 1974, 1975).

Acknowledgement

Part of the work reported in this paper was supported under National Science Foundation Grant ENG 74-13217 to Brown University. One of us (J. B.) wishes to express his thanks to the National Science Foundation and to Brown University for making it possible for him to spend the summer of 1977 in Providence.

Axiomatic Models of Phenomenological Thermodynamics

E. V. Presnov, S. N. Malyghin

Considerable progress has lately been made in constructing a system of nonequilibrium, phenomenological thermodynamics (Ziegler, 1963; Biot, 1970; Gyarmati, 1970; Glansdorff, Prigogine, 1971; Day, 1972; Truesdell, 1972; Oster, Perelson, 1974; Presnov, 1978; Bataille, Kestin, 1979). An unexpected analogy with black hole physics has been detected (Frolov, 1978; Davies, 1978). New applications to biology have appeared (Eigen, 1971; Zotin, 1972; Oster et al., 1973; Nicolis, Prigogine, 1977; Schnakenberg, 1977; Lamprecht, Zotin, 1978). At the same time axiomatic principles of classical thermodynamics were being developed (Boyling, 1972; Hermann, 1973; Jauch, 1975; Edelen, 1976; Zeleznik, 1976). However, in the axiomatic approach there is no connection between classical and irreversible thermodynamics. Thus, one of the most complicated problems of irreversible thermodynamics is the existence of uniquely determined nonequilibrium entropy (Delgado Dominges et al., 1974). In some of the recent studies the concept of entropy has not been used (Meixner, 1969, 1973; Truesdell, 1972).

In the present paper we shall briefly describe only a few modern approaches based in axiomatic principles of equilibrium, phenomenological thermodynamics, and suggest a formal construction of irreversible thermodynamics for arbitrary dissipative systems. Our aim is a unified presentation of equilibrium and nonequilibrium phenomenological thermodynamics, classical mechanics and dynamics of chemical reactions.

Thermodynamics and Kinetics of Biological Processes

Classical Thermodynamics

First axiomatic theories of equilibrium thermodynamics are associated with the work by Schiller (1902); Caratheodory (1909); Afanasyeva-Ehrenfest (1928) and others. In this section we consider a new interpretation of Caratheodory's axiomatic approach (Hermann, 1973) and present Kulakov's principle of phenomenological theory (Kulakov, 1969).

Definition 1

(see also Arnold, 1978). Let M be a manifold of odd dimensionality, $2n+1$. A contact structure of the manifold M is determined by setting up the Pfaffian form, α , of the constant class $2n+1$. In this case, a differential 1-form, α , is called a contact form set up on the manifold M .

Definition 2

Let N be an n-dimensional manifold. The map $\Phi : N \to M$ is called a *smooth equilibrium thermodynamic system* if $\Phi \in C^{\infty}$ and

$$\Phi^{*}(\alpha) = 0 . \tag{1}$$

The contact manifold M is called, in this case, the Gibbs state space and the differential 1-form, d, the Gibbs form.
To develop definition 2 and to have its concrete physical interpretation we shall consider some more concepts.

Definition 3

The system of local coordinates (z, x_i, y_i) on the contact manifold (M, α) is called canonical if

$$\alpha = dz - \sum_{i=1}^{n} y_i \, dx_i .$$

The functions (y_i) and (x_i) are called conjugate with respect to the form α and interpreted as thermodynamic parameters of a smooth equilibrium thermodynamic system.

Definition 4

Let $z' = \sum_{i=1}^{n} y_i x_i - z$. Then, the functions $\Phi^{*}(z)$ and $\Phi^{*}(z')$ are said to be connected by a Legendre transformation in a smooth equilibrium thermodynamic system.

Definition 5

The set of functions $(g_1, \ldots, g_{2n})$ set up on the contact manifold, (M, α), is called a canonical set of functions if $d\alpha = dg_{n+1} \wedge dg_1 + \ldots + dg_{2n} \wedge dg_n$.

It is readily seen that a canonical set of functions is functionally independent. Now we are in a position to introduce ordinary thermodynamic concepts.

Definition 6

The function F, on the contact manifold (M, α), is called a potential function of a canonical set of functions $(g_1, \ldots, g_{2n})$ if

$$dF = \alpha + g_{n+1}\, dg_1 + \ldots + g_{2n}\, dg_n .$$

Thus it is clear, that a Legendre transformation makes it possible to find thermodynamic potentials in different canonical coordinate systems.

The first law of thermodynamics, in the ordinary form, os obtained if the map $\Phi : N \to M$ is considered to be parametrized in such a way that the functions $x_1, \ldots, x_n$ of the canonical coordinate system on the contact manifold (M, α) are also a system of local coordinates on the manifold N. Then (1) is the first law of thermodynamics

$$0 = dz - \sum_{i=1}^{n} y_i\, dx_i \qquad (2)$$

where $z = z(x)$ and $y_i = y_i(x)$.

It should be emphasized that in such a definition of thermodynamics the entropy occurs as one of the thermodynamic variables and temperature is a parameter conjugate to entropy.

Kulakov's theory of physical structures is a generalized concept of fundamental physical theories (Kulakov, 1969). Let us formulate a particular case of the Kulakov principle of phenomenological symmetry which determined the physical structure on a manifold of thermodynamic states, i. e. the Gibbs state space M .

Proceeding from the basic equation of thermodynamics (2), in traditional variables,

$$du = TdS + \sum_{j=1}^{n} y_j \, dx_j \tag{3}$$

consider the quantity $A^{(S)}_{im} = \int_i^m \sum_{j=1}^{n} y_i dx_i \Big|_{S=const}$ which represents the work performed by the system under an adiabatic transition from the state, i, to the state, m, and also regard the quantity $A^{(T)}_{mk} = \int_m^k \sum_1^n y_i dx_i \Big|_{T=const}$ which, in its turn represents the work performed by the system under an isothermal transition from the state, m, to the state, k, (here $i, m, k \in M$). Then, from equation (3), there follows the final equation of thermodynamics

$$a_{ik} = (T_i + T_k)(S_i - S_k) + 2(u_k - u_i),$$

where $a_{ik} = A^{(ST)}_{ik} + A^{(TS)}_{ik}$ (here $A^{(ST)}_{ik} = A^{(S)}_{im} + A^{(T)}_{mk}$; $A^{(TS)}_{ik} = A^{(T)}_{il} + A^{(S)}_{lk}$).

It can be verified that for any five thermodynamic states $i, j, k, l, m \in M$ there holds the relation

$$\begin{vmatrix} 0 & 1 & 1 & 1 & 1 & 1 \\ -1 & 0 & a_{ij} & a_{ik} & a_{il} & a_{im} \\ -1 & -a_{ij} & 0 & a_{jk} & a_{jl} & a_{jm} \\ -1 & -a_{ik} & -a_{jk} & 0 & a_{kl} & a_{km} \\ -1 & -a_{il} & -a_{jl} & -a_{kl} & 0 & a_{lm} \\ -1 & -a_{im} & -a_{jm} & -a_{km} & -a_{lm} & 0 \end{vmatrix} = 0. \tag{4}$$

On the other hand, for the quantities $S_{ik} = A^{(ST)}_{ik} - A^{(TS)}_{ik}$, we obtain

$$S_{ik} = -(T_i - T_k)(S_i - S_k)$$

and, therefore, for any four arbitrary thermodynamic states $i, j, k, l \in M$ the condition

$$\begin{vmatrix} 0 & 1 & 1 & 1 & 1 \\ 1 & 0 & S_{ij} & S_{ik} & S_{il} \\ 1 & S_{ij} & 0 & S_{jk} & S_{jl} \\ 1 & S_{ik} & S_{jk} & 0 & S_{kl} \\ 1 & S_{il} & S_{jl} & S_{kl} & 0 \end{vmatrix} = 0 \qquad (5)$$

is met.

Thus, as a consequence of the basic thermodynamic equation (1), we have two phenomenologically invariant relations, (4) and (5), which are the final form of the basic equation. These relations determine two forms of the spatial-type physical structure admissible by classical thermodynamics, i. e. the essence of thermodynamics may be formulated as follows: a manifold of thermodynamic states - the Gibbs state space M - possesses two types of phenomenological symmetry, namely, symmetry of rank 4 with a symmetric matrix and symmetry of rank 5 with an skew-symmetric matrix.

Phenomenological Representation of Dissipative Systems

As follows from the previous discussion, classical thermodynamics possesses a closed formal theory. Nonequilibrium, phenomenological thermodynamics, to be discussed here, is not yet a fundamental physical theory. Therefore, formalization of irreversible thermodynamics, which is a model theory, will depend upon the following assumption.

Let the time evolution of a dissipative system Ξ be given by differential equations

$$\dot{\xi} = f(\xi) \qquad (6)$$

on a suitable manifold W, where $\xi \in W$ is a vector field over the physical space V, the components of which are, at each point, the parameters, describing the local thermodynamic equilibrium state (Ruelle, 1973).

Example 1: Classical mechanics (Arnold, 1978). As a physical space V we should take, here the configuration, space $V = \{(q_1, \ldots, q_n)\}$, of our mechanical system, and, as the manifold W we should take the set of vector fields on V, i. e. the cotangent bundle $T^*V = \{(p, q)\}$ which admits introduction of a symplectic structure set up by the symplectic form $\omega^2 = dp \wedge dq$. Equations of motion of such a mechanical system are referred to as Hamilton equations for a certain Hamilton function $H: T^*V \to \mathbb{R}$ and are written in the form

$$\dot{p} = -\frac{\partial H}{\partial q} ; \qquad \dot{q} = \frac{\partial H}{\partial p} .$$

Example 2: Classical thermodynamics. The Gibbs state space M should be considered, here, as the physical space V. The constant Hamilton function, $H: T^*M \to \mathbb{R}$, $H = const$, is considered on the cotangent bundle $T^*M = \{(z, x, y, \hat{z}, \hat{x}, \hat{y})\}$ taken for the manifold W. In this case the Hamilton equations are

$$\dot{z} = 0 ; \quad \dot{x} = 0 ; \quad \dot{y} = 0 ; \quad \dot{\hat{z}} = 0 ; \quad \dot{\hat{x}} = 0 ; \quad \dot{\hat{y}} = 0 .$$

Suppose that the right-hand side of the auntonomous system of differential equations (6) may be represented in the form

$$\dot{\xi} = \Lambda (\nabla_{\xi} G) \tag{7}$$

where the nonlinear operator $\Lambda: \mathbb{R}^n \to \mathbb{R}^n$, $\Lambda \in C^0$ depends on the gradient of some potential $G: \mathbb{R}^n \to \mathbb{R}$, $G \in C^1$. For examples 1 and 2, such a representation has already been given. In the case of chemical kinetics the evolution equations describing the change of the vector, ξ, in time in the concentration space can also be written in the form (7) (Oster, Perelson, 1974).

We shall call the potential, G, admissible, and the dynamic model (7) a phenomenological G-representation of the dissipative system $\{\Xi, f\}$. Correspondingly $\dot{\xi} = J$ is a generalized flux (independent of the choice of G-representation), $\nabla_{\xi} G \equiv X^G$ is a generalized force (dependent on the choice of G-representation).

After the formal procedure, we shall speak of phenomenological equations of the dissipative system $\{\Xi, f\}$ which are induced by the initial dynamic model (6) and the phenomenological G-representation (7)

$$\mathfrak{I} = \Lambda(X^{G}). \tag{8}$$

It is readily seen that for any dissipative system, $\{\Xi, f\}$, there always exists a canonical representation realized by the potential $G_0 = \frac{1}{2}\sum_{1}^{n} \xi_i^2$, with the generalized forces $X^{G_0} = \nabla_\xi G_0 = \xi$ and the phenomenological equations $\mathfrak{I} = f(X^{G_0})$.

An analogue of the thermodynamic concept of entropy production in a dissipative system arises automatically.

Definition 7: A derivative of the function $G(\xi)$ along the vector field $f(\xi)$ (or the Lie derivative)

$$L_f G = f_1 \frac{\partial G}{\partial \xi_1} + \ldots + f_n \frac{\partial G}{\partial \xi_n} = \sum_{1}^{n} \mathfrak{I}_i X_i^{G}$$

is called production of the dissipative system $\{\Xi, f\}$ in G-representation.

For a rough classification of phenomenological representations of dissipative systems we shall use the following theorem of decomposition of phenomenological equations.

Theorem (Edelen, 1973; Presnov, 1973; 1978 a; Merkin, 1975, 1976; Presnov et al., 1977; Bataille et al., pp.239 of the present book). For the phenomenological equations (8) there exists a unique decomposition

$$\mathfrak{I} = \operatorname{grad}_{X^G} \Phi + U,$$

where the fluxes U are called non-dissipative on the strength of the identiy,

$$\sum_{i=1}^{n} X_i^{G} U_i = 0, \tag{9}$$

and the dissipative potential is determined by the formula

$$\Phi(X^G) = \int_0^1 \sum_{1}^{n} X_i^{G} \Lambda_i(\tau X^{G})\, d\tau.$$

Let us consider concrete models of some physical systems.

Conservative systems. The Hamilton equations set up on the cotangent bundle, $T^*V = \{(p,q)\}$, and written in the form

$$\dot{\xi} = I \nabla_\xi H$$

(here $\xi = (p,q) \in T^*V$ and $I = \begin{pmatrix} 0 & -E_n \\ E_n & 0 \end{pmatrix}$ is a symplectic matrix) are linear phenomenological equations, $J = I X^H$, with skew-symmetric reciprocity relations, $I = -I^T$. The generalized fluxes J are non-dissipative fluxes, and the conservation law (9) is $\sum_{i=1}^{n} X_i^H J_i = 0$. Production and dissipative potential in H-representation will be zero.

Nonlinear phenomenological equations of conservative systems occur, evidently, in the case when the dynamic model (6) possesses a first integral which is different from the Hamiltonian and is an admissible potential. For example, a dynamic model with the condition, $\mathrm{div}\, f = \mathrm{const}$, and the known vector field, $S(\xi)$, such that the Poisson bracket $[S, f]$ vanishes, has as first integral $E = \mathrm{div}\, S$ (González-Gascon, 1977). However, the condition, $\mathrm{div}\, f = \mathrm{const}$, alone is insufficient for the existence of a first integral (see below, the Lorenz model).

Canonical triplet (Obukhov, 1977). Euler's equations for a gyroscope in appropriate coordinates have the form

$$\dot{\xi} = A \nabla_\xi F .$$

Here $\xi = (\xi_1, \xi_2, \xi_3)$, $F(\xi) = \xi_1 \xi_2 \xi_3$ and $A = \begin{pmatrix} -1 & 0 & 0 \\ 0 & 1 & 0 \\ 0 & 0 & -1 \end{pmatrix}$.

The production and dissipative potential of the canonical triplet in F-representation are equal to $L_f F = -\xi_1^2 + \xi_2^2 - \xi_3^2$ and $\Phi(X^F) = \frac{1}{2} L_f F$, respectively. It is obvious that non-dissipative fluxes are absent in this case and, therefore, one can speak of a purely dissipative F-representation of a canonical triplet.

Volterra equations (Obukhov, 1977). The system described by these equations admits, besides a phenomenological representation of the general form (e. g. canonical), non-dissipative and purely dissipative representations. Indeed, at an appropriate substitution of variables the Volterra equations are reduced to the Hamiltonian form

$$\dot{\xi} = -\alpha(e^{\xi_2}-1) = -\frac{\partial H}{\partial \xi_2}\,; \qquad \dot{\xi}_2 = \gamma(e^{\xi_1}-1) = \frac{\partial H}{\partial \xi_1}\,,$$

where $H(\xi_1,\xi_2) = \gamma(e^{\xi_1}-\xi_1-1) + \alpha(e^{\xi_2}-\xi_2-1)$. On the other hand, assuming $R(\xi_1,\xi_2) = \gamma(e^{\xi_1}-\xi_1-1)-\alpha(e^{\xi_2}-\xi_2-1)$ one can write the Volterra equations in the form

$$\dot{\xi} = K\nabla_{\xi} R\,, \qquad K = \begin{pmatrix} 0 & 1 \\ 1 & 0 \end{pmatrix}.$$

This form of the Volterra equations allows us immediately to find a purely dissipative representation with the production, $L_f R = 2\frac{\partial R}{\partial \xi_1}\frac{\partial R}{\partial \xi_2}$, and the dissipative potential $\Phi(X^R) = \frac{1}{2} L_f R$.

These examples may be supplemented with the fact that one-dimensional dynamic models describe systems with only dissipative representations. The canonical representation of the gradient system, $\dot{\xi} = \mathrm{grad}_{\xi}\Psi(\xi)$, is purely dissipative.

Thus, an arbitrary dissipative system admits different phenomenological representations. Therefore, by analogy with the second law of thermodynamics, we can study phenomenological representations with non-negative production $L_f G \geq 0$.

It is easy to see that a non-negative phenomenological representation takes place for the systems described above, i. e. for a conservative, one-dimensional canonical triplet (with a potential $Q_K = -\left(\frac{\xi_1\xi_3}{\xi_2}\right)^{2K+1}$, $K=1,2,3,\ldots$) and also for linear systems. Non-negative representation is not unique.

The question of the existence of a non-negative phenomenological representation has characteristic features. From the general theory of first-order linear equations in partial derivatives there follows the existence, and uniqueness, of the local solution of an inhomogeneous equation (see e. g. Arnold, 1978)

$$L_f G = \sigma$$

(under condition that the functions $f_i, \sigma \in C^1$ do not vanish simultaneously). This means the local existence of a non-negative, phenomenological representation of a dissipative system with a given production, $\sigma(\xi) \geq 0$. But the question of the existence and uniqueness of the global solution of the inequality

$$L_{\xi} G \geq \sigma,$$

and the related question of the global existence and uniqueness of a non-negative phenomenological representation of an arbitrary dissipative system remains open.

In this connection we shall discuss two now popular models of dissipative systems which, evidently, have no non-negative global phenomenological representation.

Brusselator (Malek-Mansour et al., pp. 75 of this book)

$$\dot{\xi}_1 = A - B\xi_1 + \xi_1^2 \xi_2 ; \quad \dot{\xi}_2 = C\xi_1 - \xi_1^2 \xi_2 .$$

A non-negative local representation is set up, for example, by the potentials $G_1 = \frac{1}{2}\xi_2^2$ and $G_2 = -\frac{1}{2}\xi^2$ determined, correspondingly, in the regions

$$\mathcal{D}_1 = \{(\xi_1, \xi_2) : 0 \leq \xi_1 \xi_2 \leq C\} \text{ and } \mathcal{D}_2 = \mathbb{R}^2 - \mathcal{D}_1 .$$

Lorenz model (see e. g. Arnold, 1978).

$$\dot{\xi}_1 = -a\xi_1 + a\xi_2 ; \quad \dot{\xi}_2 = r\xi_1 - \xi_2 - \xi_1\xi_3 ; \quad \dot{\xi}_3 = -b\xi_3 + \xi_1\xi_2 ;$$
$$(a, b, r > 0).$$

These two models are of interest because, at some values of the parameters, they admit stochastic behaviour of phase trajectories.

In conclusion we shall describe one more possible *representation* of dissipative systems. Thus, for the canonical triplet that has a non-negative global representation (see above), the potential, $Q_0 = -\xi_1\xi_3/\xi_2$, determines the generalized forces, X^{Q_1} for which

$$L_{\xi} Q_0 = \sum_{i=1}^{3} J_i X_i^{Q_0} = \xi_3^2 + \left(\frac{\xi_1\xi_3}{\xi_2}\right)^2 + \xi_1^2 \geq \sigma .$$

The Jacobian of the transformation $\xi \to \nabla_{\xi} Q_0$ is, however, identical to zero, which means that this transformation cannot be inverted. Thus, the potential, Q_0 , is not admissible or, to put it differently, this potential cannot even be realized as a local phenomenological representation since one cannot find phenomenological equations $\dot{\xi} = \Lambda(\nabla_{\xi} Q_0)$. This property of phenomenological representations, evidently, shows some deficiency in the standard technique of phenomenological thermodynamics.

Axiomatic Approach to Irreversible Thermodynamics and Variational Principles

I. P. Virodov

Gyarmati (1970), who generalized the earlier work by Onsager (1931 b) and Prigogine (1967), has shown that there exists a variational principle formulated in terms of variations of the volume integral of entropy production and dissipative functions over generalized thermodynamic forces and fluxes. From this variational principle, Gyarmati (1970) obtained some particular results. In the paper by Vojty (1967) a variational principle has been postulated, within which the time integral of the thermodynamic Lagrangian density is varied.

In spite of some progress these authores did not formulate the general variational problem in irreversible thermodynamics. The heuristic approach used in these, and other, papers was reduced either to derivation of the differential transfer and hydrodynamic equations or to specific generalizations of specific results and did not make it possible to complete an axiomatic construction of irreversible thermodynamics. The principles of Gyarmati and Vojta, by themselves, do not meet the conditions of invariance under Lorentz transformations as long as they are taken as the fundamental variational field principles in irreversible thermodynamics. In contemporary irreversible thermodynamics nobody has hitehrto considered, clearly, the question of which of the principles, besides those of classical thermodynamics, that are to lay the foundations of this science in such a way that its axiomatic construction may be completed. Since thermostatics is generally considered to be a particular, limiting case of irreversible thermodynamics, the latter must involve, in the description of dynamic systems, the parameters and thermodynamic potentials introduced in thermostatics as well as the concept of the thermodynamic state. With such an addition, all the formalism of classical thermodynamics is maintained in the thermodynamics of

Thermodynamics and Kinetics of Biological Processes

irreversible processes. In spite of the fact that the development of thermodynamics was somewhat delayed, its perspectives were laid in thermostatics. In fact, the additions formulated above make up the essence of the local equilibrium principle, which takes into account the dynamic properties of the systems. Evolution of systems which meet this requirement, does not need any additional principles since they are founded in thermostatics (the copy of which is a thermodynamic system with thermodynamic parameters and potentials autonomic in time and spatial coordinates). Thus we are led to a closed system of axioms in irreversible thermodynamics that have some limits of applicability. The use made of the local equilibrium principle and completeness of information contained therin thus seem to be exhausted. In effect, however, the local equilibrium principle, in combination with the essential property of additivity of thermodynamic potentials, leads to the establishment of the fundamental variational principles of irreversible thermodynamics within the context of the continual formalism of Lagrangian-Hamiltonian dynamics. The relativistically invariant nature of these principles is provided by invariance of four-dimensional space and the scalar character of the thermodynamic Lagrangian density $\mathcal{L} = \rho \dot{s}$. Indeed, let us write the expression for a complete change of entropy characterized by the field parameters η_k (more precisely, by deviations from their equilibrium values) in the local form

$$\Delta[s(\eta) - s(0)] = \Delta(\Delta(s)) = \delta(\Delta S) + \tfrac{1}{2}\delta^2(\Delta s) + \ldots \qquad (1)$$

But according to the local equilibrium and entropy additivity principle,

$$\int_V \delta(\Delta s)\, dV = \delta \int_V \Delta s\, dV = \delta(\Delta s) = 0\,. \qquad (2)$$

The total change of entropy, ΔS , may be considered here as a Hamiltonian action.

Introducing the substantional derivative

$$\frac{d\Delta S}{dt} = \int_V \rho(\Delta \dot{s})\, dV \qquad (3)$$

and expressing with it the action integral, we obtain the Hamiltonian-Ostrogradsky minimum action principle for the density $\mathcal{L} = \rho \dot{s} = \rho(\Delta \dot{s})$

$$\delta \int_{(V)} \int_{t_1}^{t_2} \rho \dot{s}\, dV dt = \delta \int_{(V)} \int_{t_1}^{t_2} (\sigma - \operatorname{div} J_s)\, dV dt = 0\,. \qquad (4)$$

Under such variation, the initial and final points of the phase space of field variables do not vary. Writing the action in a relativistically invariant representation we find a functional variation of the integral over the four-dimensional value, V_4 , vanishes

$$\delta \int_{V_4} \mathcal{L}(\eta_k, \eta_{k/i}, x^i)\, d^{(4)}x = 0 \qquad (5)$$

provided that the variations of the field functions on the boundary of the region V_4 vanish*

$$(\delta \eta_k)_{(V_4)} = 0 \qquad (6)$$

The fundamental principle in irreversible thermodynamics (5) is a covariant relativistic principle from which the Lagrange-Euler equations in field theory follow

$$\frac{\delta \mathcal{L}}{\delta \eta_k} = \frac{\partial \mathcal{L}}{\partial \eta_k} - \left(\frac{\partial \mathcal{L}}{\partial \eta_{k/i}}\right)_{/i} = 0 . \qquad (7)$$

These reflect nullification of the functional variation. Writing the variational principle in the form (4) enables us to conclude that for a non-varying entropy flux on the surface $(V_3) = (V)$ we have

$$\delta(\Delta S) = \delta \int_{(V)} \int_{t_1}^{t_2} \sigma\, dV dt - \delta \int_{(V_3)} \int_{t_1}^{t_2} (J_s\, d(V_3)) dt = \delta \int_{(V)} \int_{t_1}^{t_2} \sigma\, dV dt = 0 . \qquad (8)$$

The expression obtained is a generalization of the minimum entropy production principle suggested earlier by Prigogine. Principles (8) and (4) were considered in detail by Lavrentev and Hreshchik (1978).

In the case of linear phenomenological irreversible thermodynamics, $\operatorname{div} J_s = \Phi + \Psi$, where Φ and Ψ are scattering potentials represented in terms of thermodynamic fluxes and forces, respectively. Substituting them in (4) we obtain a generalization of Gyarmati's principle. In this case, the variation over two types of variables is based on the representation of $\mathcal{L}$ in the form $\mathcal{L} = \mathcal{L}(\eta_k, \eta_{k/i}, x^i)$. This form assumes the realization of δ-variations over two types of field variables: $J_m(\eta_k, \eta_{k/i})$, and $X_m(\eta_k, \eta_{k/i})$ which follow from the dual Legendre transformation. Separation of the terms Φ and Ψ is carried out with the aid of the Gibbs

* The extremal entropy conditions are involved viz. $\delta U = \delta V = \delta \eta_i = 0 \quad i = 1, .., m$

equation. The same equations are also used for a representation of entropy production through generalized fluxes and forces, as well as for constructing the thermodynamic formalism as a whole.

Construction of the formalism of irreversible thermodynamics involves ambiguity. First, one may introduce the density of a thermodynamic Hamiltonian, $\mathcal{H}$, with the aid of Legendre transformations in the general covariant form and thus, keep in thermodynamics, a relativistic four-dimensional formalism. Second, one may depart from this canon of field theory and introduce in the Legendre transformations either the derivatives of field functions with respect to spatial coordinates (Vojta, 1967) or, according to the canons of Hamiltonian theory, their time derivatives (Gyarmati, 1970) thus giving up the covariance of the four-dimensional formalism. In this paper we follow the latter way by presenting $\mathcal{H}$ in the form

$$\mathcal{H} = \sum_{(i)} \Pi_i \dot{\eta}_i - \mathcal{L} . \tag{9}$$

Then

$$\delta(\Delta S) = \delta \int_V \int_{t_1}^{t_2} [\Sigma \Pi_i \dot{\eta}_i - \mathcal{H}]\, dV dt , \tag{10}$$

where the quantities Π_i and η_i vary independently if the canonical Hamiltonian equations in the variational derivatives

$$\dot{\eta}_i = \frac{\delta \mathcal{H}}{\delta \Pi_i} , \qquad \dot{\Pi}_i = - \frac{\delta \mathcal{H}}{\delta \eta_i} , \tag{11}$$

hold. This must be in accordance with the mass balance equations

$$\rho \dot{\eta}_i = \sigma_i - \operatorname{div} J_i \tag{12}$$

when thermodynamic fluxes and forces are introduced into equations (9) to (11).

Taking into account that $\sigma - \operatorname{div} J_s = \sum_{(i)} \Pi_i \dot{\eta}_i - \mathcal{H} = \mathcal{L}$ and that $\operatorname{div} J_s$ depends on the parameters of the environment for a given thermodynamic system (in addition, both J_s and, $\operatorname{div} J_s$, contribute to the variation of the parameters of the thermodynamic system, although the converse is untrue), we come to $\frac{\partial \mathcal{L}}{\partial \dot{\eta}_i} = \frac{\partial \sigma}{\partial \dot{\eta}_i} = \Pi_i$, i. e. σ is a homogeneous function of $\dot{\eta}_i$ and Π_i viz.

$$\sigma = \sum_{(i)} \Pi_i \dot{\eta}_i = \sum_{(i)} X_i J_i \tag{13}$$

and

$$\mathcal{H} = \mathrm{div}\, J_s . \tag{14}$$

Thus, to represent entropy production through the generalized fluxes J_i and forces X_i it is necessary to introduce the additional quantitaties $\varkappa_i$ (for scalar η_i, the quantities, $\varkappa_i$, are vector functions) by formulae

$$J_i = \varkappa_i \dot{\eta}_i , \qquad X_i = \Pi_i / \varkappa_i . \tag{15}$$

Then the mass balance equations will involve not, J_i but $\varkappa_i$: i. e.

$$\rho \dot{\eta}_i = \sigma_i - \mathrm{div}\,(\varkappa_i \dot{\eta}_i). \tag{16}$$

The description of a thermodynamic system at $\mathcal{H}$, which does not depend explicitly on time ($H = \int_{(V)} \mathcal{H}\, dV = h = \mathrm{const}$), is of particular interest. Under this (and some other) additional conditions an isoperimetric Lagrange problem is derived within which the character of variation in field varibles and functions must be changed. In particular, in the mechanics of material points the virtual (possible) changes of parameters must be equal to the actual changes: $\delta(\;) = d(\;) = \bar{\Delta}t(\dot{\;})$; where $\bar{\Delta}t = \varepsilon(t)$ is some function of time, and the total asynchronous variation, $\bar{\Delta}(\;)$, is connected with the synchronous variation $\delta(\;)$ by the relation $\bar{\Delta}(\;) = \delta(\;) + \bar{\Delta}t(\dot{\;})$. This variation takes into account mobility (transversality) of the ends of competing paths along some curves (transversals). It is rather important to note that four interconnected variations, namely, substantial, local, functional and complete, are introduced in field theory. The question of the choice of variations under certain conditions imposed upon the system in this general case will be discussed below. Under the chosen formalism, the change in the variation character does not entail a change in the local necessary extremum conditions i. e. the Legendre-Weierstrass conditions and Lagrange-Euler equations. Nor do the conditions change at cusp points and fold lines. There only appears some additional transversality connecting variations of the terminal points of extremal path with the values of thermodynamic functions and some of their derivatives. For the isoperimetric Lagrange problem on transversals $\bar{\Delta}\eta(r_1, t_1) = \bar{\Delta}\eta(r_2, t_2) = 0$, i. e.

$$(\bar{\Delta}\eta_k)_{(V_4)} = 0 . \tag{17}$$

In the Hamiltonian formalism the Lagrangian minimum action principle,

$$\bar{\Delta}\int_{(V)}\int_{t_1}^{t_2}\sum_{(i)}\Pi_i \dot{\eta}_i \, dV dt = 0 , \tag{18}$$

is valid at H = const. But since (see Leech, 1958)

$$\frac{d\mathcal{H}}{dt} = \frac{\partial\mathcal{H}}{\partial t} - \sum_{(i,j)}\frac{d}{dx_j}\left(\frac{\partial\mathcal{L}}{\partial\eta_{i/j}}\dot{\eta}_i\right) \tag{19}$$

then the variational principle (18) is realized under the additional condition that

$$\frac{d\mathcal{H}}{dt} + \sum_{(i,j)}\frac{d}{dx_j}\left(\frac{\partial\mathcal{L}}{\partial\eta_{i/j}}\dot{\eta}_i\right) = 0 \qquad \text{or} \qquad \frac{d\mathcal{H}}{dt} + \operatorname{div} J_{\mathcal{H}} = 0 , \tag{20}$$

which is introduced for short $J_{\mathcal{H}}$. The second equation follows from the mass balance equation (of type (12)) for the quantity $\mathcal{H}$ with account taken of $\sigma_{\mathcal{H}} = 0$. From (20) we find

$$\frac{dH}{dt} = \int_{(V)}\frac{d\mathcal{H}}{dt}dV = -\int_{(V)}\operatorname{div} J_{\mathcal{H}}\, dV = -\int_{(V_3)}(J_{\mathcal{H}}\, d(V_3)) = \frac{d}{dt}\int_{(V_3)} J_s\, d(V_3)) . \tag{21}$$

From this we discover that H = const. if $(J_{\mathcal{H}})_{(V_3)} = 0$ or $(J_s)_{(V_3)} =$ const. This case is also realized if

$$\sum_{(j)}\frac{\partial\mathcal{L}}{\partial\eta_{i/j}}\dot{\eta}_i\, d(V_3)_j = 0 \quad i = 1, 2, 3 \qquad \text{or} \qquad (J_{\mathcal{H}})_n = \sum_{(i)}\frac{\partial\mathcal{L}}{\partial\nabla\eta_i}\dot{\eta}_i = 0 \tag{22}$$

i. e. for a system isolated with respect to fluxes $(J_{\mathcal{H}})_n$. But

$$(J_{\mathcal{H}})_n = \sum_{(i)}\frac{\partial\mathcal{L}}{\partial\nabla\eta_i}\cdot\frac{J_i}{\varkappa_i} . \tag{23}$$

From this it follows that H = const(t), if, on the surface (V_3) , the dependence of $\mathcal{L}$ on the gradients of field functions vanishes, or, if the fluxes, $J_i = 0$. In this case, the isolation of the system with respect to $\mathcal{H}$-quantity leads to the minimum entropy production principle for the asynchronous variation

$$\bar{\Delta}\int_{(V)}\int_{t_1}^{t_2}\sum_{i}\Pi_i \dot{\eta}_i \, dV dt = \bar{\Delta}\int_{(V)}\int_{t_1}^{t_2}\sigma \, dV dt . \tag{24}$$

From relation (21) it also follows that

$$\left(\frac{dJ_s}{dt} + J_{\mathcal{H}}\right)_n = 0 . \tag{25}$$

Thus, if $(J_{\mathcal{H}}) = 0$, then either $(J_s)_n = 0$ (condition under which the system is adiabatic) or $(J_s)_n = \mathrm{const}$ (stationarity of the system with respect to entropy flux on the hypersurface of the system (V_3)). This relation resembles symmetry transformations in Hamiltonian mechanics under which the form-invariance of the Hamiltonian function is kept (in particular, this is reflected in the condition H = const.) if J_S is regarded as an infinitesimal generating function which is the constant of motion.

Since relativistic covariance is guaranteed by the Lagrange formalism, more general conservation laws may be obtained with the aid of symmetry transformation for a complete change of the thermodynamic Lagrangian density (see Schmutzer, 1973)

$$\begin{aligned}\bar{\Delta}\mathcal{L} = \frac{\delta\mathcal{L}}{\delta\eta_k}(\delta\eta_k + \Delta_s\eta_k - \eta_{k|i}\xi^i) + (\Pi^{kj}\delta\eta_k)_{|j} + \\ + [\Pi^{kj}\Delta_s\eta_k - (\Pi^{kj}\eta_{k|i} - \mathcal{L}g^j_i)\xi^i]_{|j} .\end{aligned} \tag{26}$$

$\bar{\Delta}\mathcal{L}$ is a complete variation equal to the sum of the mass and functional variations, $\Delta_s\mathcal{L}$ and $\Delta\mathcal{L}$, resp. which is realized with account taken of continuous Lorentz transformations

$$x^{i'} = x^i + \xi^i = x^i + \alpha^i_j x^j + \alpha^i \tag{27}$$

at infinitesimal α_i . The quantity $g^j_i = \delta^j_i$ in (26) is the Kronecker tensor. In accordance with the general aim of constructing a thermodynamic system for irreversible processes, all field functions and functionals of them are autonomic in space-time coordinates. Therefore, in the symmetry transformation

$$\bar{\Delta}\mathcal{L} = - I^j_{|j} , \tag{28}$$

which provides a relativistic covariance, the right-hand side of the above relation is zero at I = const.

$$\bar{\Delta}\mathcal{L} = 0. \tag{29}$$

Some arbitrariness in the choice of an infinitesimal function, I, is justified by the conservation of the field equations in the Lagrange-Euler form and is restricted to the conditions on transversals, and to the conservation laws. From (29) it follows that when the variational derivatives (7) vanish, the following relations (weak conservation law)

$$[\Pi^{kj}\bar{\Delta}\eta_k - (\Pi^{kj}\eta_{k|j} - \mathcal{L}g^j_i)\xi^i]_{|j} = 0 \tag{30}$$

must hold. These represent a conservation law in four-dimensional space. With the aid of the Ostrogradsky-Gauss theorem we can go from the volume integral to a hypersurface of four-dimensional space and obtain

$$\int_{(V_4)}[\Pi^{kj}\bar{\Delta}\eta_k - (\Pi^{kj}\eta_{k|i} - \mathcal{L}g^j_i)\xi^i]\,df_j = 0 \tag{31}$$

or, by virtue of the conditions (17),

$$\int_{(V_4)}(\Pi^{kj}\eta_{k|i} - \mathcal{L}g^j_i)\,df_j = 0. \tag{32}$$

Thus for the set of differential equations (7) we have found the first marginal problem, within which the value of the thermodynamic Lagrangian density on the surface (V_4) is given by the relation

$$g^j_i\,\mathcal{L} = \Pi^{kj}\eta_{k|i}\,. \tag{33}$$

But, since $\dfrac{\partial\mathcal{L}}{\partial\eta_{k|i}} = \Pi^{ki}$, in the relativistic covariant case

$$\mathcal{L} = \Pi^{ki}\eta_{k|i} - \mathcal{H}, \tag{34}$$

where, according to (33)

$$\mathcal{H} = \Pi^{ki}\eta_{k|i} - \tfrac{1}{4}\Pi^{ki}\eta_{k|i} = \tfrac{3}{4}\Pi^{ki}\eta_{k|i}\,. \tag{35}$$

On the other hand, if we introduce, by analogy with the mechanics of continuous media (or field theory), the tensor,

$$s^i_j = \Pi^{ki}\eta_{k|j} - g^i_j\,\mathcal{L}, \tag{36}$$

that possesses the dimensions of entropy production, which we will refer to as a canonical thermodynamic tensor of the entropy flux production-thermodynamic fluxes of forces (EFP-TFF tensor), then we find that the thermodynamic Hamiltonian density is

$$\mathcal{H} = \mathrm{Sp}\,s^i_j + 3\,\mathcal{L}\,. \tag{37}$$

By virtue of the independence of space-time translations and relations it follows, from relation (32), that in a covariant passage from the integral over the four-dimensional volume to the integral over the surface (V_4), on the hypersurface (V_4)

$$s_j^i = 0. \tag{38}$$

This means that at any t on the surface, (V_3), of the system $\mathcal{H} = 3\mathcal{L}$.

Although the addition to $\mathcal{L}$ of a four-dimensional divergent term is known not to change the form of differential field equations and the integral functions of a thermodynamic system, the form of local functions may vary. Symmetry transformations in the Lagrangian formalism lead to a form-invariance of the $\mathcal{L}$-functional which is subject to a functional and coordinate transformation. They are canonical transformations whose sequence is responsible for the process of system's motion. Hence, the fulfilment of (29) leads to $\bar{\Delta}\mathcal{H} = 0$ and = const. during the motion of the system if $\mathcal{L}$ = const.

In the general case the symmetry transformation (28) leads to a pair of differential conservation laws (Schmutzer, 1976)

$$(\Pi^{ki}\delta\eta_k + \delta I^i)_{|i} = 0, \tag{39}$$

$$(\Pi^{ki}\Delta_s\eta_k - s_j^i\xi^j + I\xi^i)_{|i} = 0 \tag{40}$$

which, with the aid of the chosen formalism of irreversible thermodynamics and Gibbs equations, allow us to express Π^{ki} and $\mathcal{L}$ in terms of the field parameters η_i and are, a system of differential equations in partial derivatives with respect to I and η_i. With the use of some additional data, the solution of this system will make it possible to determine the quantity The knowledge of this very important field characteristic may be used to establish the nature of variation in the main variational principle. And although from (1) and (2) we cannot determine, in advance, what type of variations exist in irreversible thermodynamics, the conditions of the autonomy of thermodynamic functions and field parameters with respect to x^i involves the realization of variations, with account taken of functional and coordinate transformations. The solution of this problem will not influence the form of the basic local differential field equations obtained when the variational derivatives of the Lagrangian density vanish. In this case, additional information on transversals will be obtained. This information should, evidently, be associated with the conditions of realization of the local equilibrium principle.

Note, in conclusion, that in the mechanics of material points, integral variational principles are derived from the basic Lagrange equation of motion, whereas in continuum mechanics and, in field theory, there are no such rigorous grounds for the minimum action principle. At the same time we should emphasize that the description of processes in the continuum is based on its physical properties, for example, in the continuity reflected in continuity equations etc. The phenomenological theory is formulated, in this case, with the aid of known analytical technique. In field theory the situation is somewhat different, although the continuum hypothesis can also be used here. In field theory its deductive foundation, based on variational principles, is laid from the outset.

The thermodynamics of irreversible processes is based on the idea of the existence of field parameters and field potentials (the usual thermodynamics (thermostatics) is a static field theory). Although, as a rule, the field concept is introduced to describe relativistic interactions, in the thermodynamics of irreversible processes the introduction of a field is dictated, not so much by disregard of long-range, and regard for short-range, character of interaction as by potentiality of the main thermodynamic functions corresponding to the local equilibrium. In this, and other works of the present author (Virodov, 1977, 1978, 1979), based on a continual-field approach, a unified method for a rigorous introduction of variational principles has been established and a means for further development of irreversible thermodynamics has been demonstrated.

Indeed, on introducing, according to Legendre transformations, generalized thermodynamic potentials in the energy- and entropy-respresentations

$$\Psi_k = u - \sum_{i=1}^{k} \eta^i \eta_i = \sum_{j=k+1}^{n} \eta^j \eta_j \;; \qquad \Phi_k = s - \sum_{i=1}^{k} \eta^{*i} \eta_i = \sum_{j=k+1}^{n} \eta^{*j} \eta_j \qquad (41)$$

we obtain, by analogy with (5), the following generalized variational principles for continual phenomenological thermodynamics of irreversible processes

$$\overset{\circ}{\delta}\int_{V_4} \varrho \Delta \dot{\Psi}_k d^4x = 0; \quad \eta^{\dot{j}} = \text{const}, \eta_j = \text{const} \quad (j = 1, \dots, k) \tag{42}$$

$$\overset{\circ}{\delta}\int_{V_4} \varrho \Delta \dot{\Phi} d^4x = 0; \quad \eta^{*j} = \text{const}, \eta_j = \text{const} \quad (j = 1, \dots, k). \tag{43}$$

At K = 2, from principle (42), it follows that

$$\overset{\circ}{\delta}\int_{V_4} \varrho \Delta \dot{g} d^4x = 0; \quad T, P, n_1, \dots, n_m = \text{const}. \tag{44}$$

At K = 1, from (42), it follows that

$$\overset{\circ}{\delta}\int_{V_4} \varrho \Delta \dot{f} d^4x = 0; \quad T, V, n_1, \dots, n_m = \text{const}. \tag{45}$$

At K = 0, from the principles (42) and (43), it follows that

$$\overset{\circ}{\delta}\int_{V_4} \varrho \Delta \dot{U} d^4x = 0; \quad S, V, n_1, \dots, n_m = \text{const}. \tag{46}$$

$$\overset{\circ}{\delta}\int_{V_4} \varrho \Delta \dot{S} d^4x = 0; \quad U, V, n_1, \dots, n_m = \text{const}. \tag{47}$$

The character of the variation, δ°, in the above variational principles is determined by the form of the variational principle (or, correspondingly, by a set of thermodynamic parameters for the system, i. e. by the character of the contact or isolation of the system from the surrounding systems). For perfect variation of the function, $I = \int_{V_m} \mathcal{L} d^n f$, we find

$$\Delta I = \int \frac{\delta \mathcal{L}}{\delta \eta_\lambda} \delta \eta_\lambda d^4x + i \int \left[\frac{\partial \mathcal{L}}{\partial \eta_{\lambda j}} (\eta_{\lambda j} \Delta x_j - \Delta \eta_\lambda) - \mathcal{L} \Delta x^j \right] df_j = 0 \tag{48}$$

Hence, when the Lagrange-Euler equations hold, the following transversality condition is met on the surface (V_4) :

$$(\mathcal{L} - \Pi^{\lambda j} \eta_{\lambda, j}) \Delta x_j + \Pi^{\lambda j} \Delta \eta_\lambda = 0. \tag{49}$$

For *free* boundaries, in perfect variations, this condition leads to the following three particular cases:

1. $\Delta \eta_\lambda = 0; \Delta x_j = 0$. This is a variation of the form $\Delta(\ ) \equiv \delta(\ )$;
2. $\Delta \eta_\lambda \neq 0, \Delta x_j \neq 0$. This is the case when

$$\mathcal{L} = \Pi^{\lambda j} \left(\eta_{\lambda j} - \frac{\Delta \eta_\lambda}{\Delta x_j} \right) = - \Pi^{\lambda j} \frac{\delta \eta_\lambda}{\Delta x_j}; \quad \mathcal{H} = \Pi^{\lambda j} \frac{\Delta \eta_\lambda}{\Delta x_j}; \tag{50}$$

3. $\Delta\eta_\lambda = 0,\ \Delta x_j \neq 0$. In this case

$$\mathcal{L} = \sigma = \Pi^{\lambda j}\eta_{\lambda,j} \quad , \quad \mathcal{H} = \sigma ; \tag{51}$$

and variation on (V_4) is substantially realized. This case is of interest because it contains the minimum entropy production principle, since the substantial character of variations leads to $\mathcal{L} = \sigma$ inside the range where the variational principle acts.

In conclusion we note that for an axiomatically closed character for the phenomenological thermodynamics of irreversible processes must be based on cellular additivity of thermodynamic potentials. This exhibits the manner of accounting for the processes of dissipation, when the local equilibrium principle holds. A statistical interpretation of thermodynamics with the aid of equilibrium statistical mechanics leads us to the necessity of introducing, by analogy with phase, an elementary cell regarded as being in an unambigous correspondence with a definite set of thermodynamic parameters which satisfy the state equation. It is in this sense that, from the state equation, the Gibbs equation follows in the local representation in nonequilibrium thermodynamics. It does not contain any information on the character of the formation of local equilibrium, nor on the limits of its stability beyond the elementary cell, where one can obtain information about the character of dissipation processes. In order to keep the extremal character of the thermodynamic potentials and the stability conditions in nonequilibrium thermodynamics and also to account for dissipative processes in the variational principles formulated above, it is necessary to derive the relation imposed on the thermodynamic parameter variations extending the system beyond the limits of the elementary cell. Let us restrict ourselves to the consideration of principle (47).

Let the entropy increment, $\Delta \check{S}$, have a definite sign (e. g. negative). Then for a region with varied thermodynamic parameters, the stability of thermodynamic equilibrium is determined by the inequality

$$\sum_{n \geq 2} \frac{1}{n!} \delta^n \check{S} < 0 . \tag{52}$$

Now, taking into account the state equation, $S = S(U, a_i, N_j)$, and expanding $\check{\Delta} S$ in a Taylor series we arrive at the conclusion that the state is stable only, under certain limiting *extension* variations $\check{\Delta}U, \check{\Delta}a_i, \check{\Delta}N_j$ which have to be independent in the establishment of the variations $\delta^n S$, i. e. $\check{\Delta}U = \delta U$, $\check{\Delta}a_i = \delta a_i$; $\check{\Delta}N_j = \delta N_j$. Therefore, the variations, $\delta^n S$, may be found by direct variation of the Gibbs equation taking into account that $\delta^n U = 0$, $\delta^n a_i = 0$, $\delta^n N_j = 0$, $n \geq 2$. Substituting this into inequality (52) and taking account of the thermodynamic state equation, we obtain the following inequality imposed on the variations of the thermodynamic parameters responsible for stability in the regions beyond elementary cells:

$$\sum_{(n)} \frac{1}{n!} \left\{ \sum_i \left[l a_i \delta^n (T^{-1}) + T^{-1} \delta^n A_i \right] \delta a_i - \sum_{(j)} T^{-1} \delta^n \mu_i \delta N_j \right\} > 0 \qquad (53)$$

$$n! = 2!, 3!, 4!, \dots ,$$

where la_i are the hidden heats of the change of the inner parameters a_i ; the quantities μ_j and A_i appear in the inequality,

$$T \check{\Delta} S - \Delta U - \sum_{(i)} A_i \check{\Delta} a_i + \sum \mu_j \check{\Delta} N_j < 0 , \qquad (54)$$

for *extension* variations.

The Second Lyapunov Method and The Variational Principle

V. N. Kapranov, A. I. Lopushanskaya

In investigations of stability in the linear approximation based on Lyapunov's method (in thermodynamics, for example, $\delta^2_m \varrho z$), we only obtain information about the real part of the spectrum of decrements. Meanwhile, proceeding from this approach, one can find conditions under which non-monotonous disturbances will not be realized and, what is more important, one can construct a variational formulation of the problem for monotonous disturbances. The former problem is, of course, the first to be solved.

If $\delta^2_m \varrho z$ is used as Lyapunov's function, the behaviour of the system in the domain of the state in question is described by the quantities

$$\delta_m P = \frac{1}{2}\frac{\partial}{\partial z}\delta^2_m \varrho z\,,\quad \delta^2_m \Pi;\quad P_m[\delta z] = \int \delta_m P\, dV,\quad \Pi_m[\delta z] = \int \delta_m \Pi\, dV$$

and for disturbance corresponding to the decrement

$$Re\,\tau = \tau_r = \frac{\delta_m P}{\delta^2_m \varrho z} = \frac{P_m[\delta z]}{\int \delta^2_m \varrho z\, dV} = \frac{P_m[\delta z]}{\delta^2_m z} \quad \text{and} \qquad (1)$$

$$Im\,\tau = \tau_{im} = \frac{\delta_m \Pi}{\delta^2_m \varrho z} = \frac{\Pi_m[\delta z]}{\delta^2_m z}\,. \qquad (2)$$

Thus, $\delta_m P$ and $P_m[\delta, z]$ characterize the stability of motion, while $\delta_m \Pi$ and $\Pi_m[\delta z]$ are associated with the imaginary part of the decrements. In addition if $\Pi_m[\delta z] = 0$, then the monotony (aperiodicity) principle holds (Glansdorff, Prigogine, 1970, 1971).

It is obvious that each time that linearized disturbed equations are considered, which describe x_k-deviations of x_k from their values, x_{k_0} , in the stationary state under investigation, for an arbitrary $\varphi(a_k) = \sum_k a_k |x|^2$ with time-independent coefficients a_k then

Thermodynamics and Kinetics of Biological Processes

$$\tau_r = \frac{\sum a_k(\dot{x}_k^* x_k + x_k^* \dot{x}_k)}{2\varphi(a_k)} = \frac{\delta_m P(\varphi)}{\varphi(a_k)} = \frac{P_m[\Phi]}{\Phi(a_k)} \quad (3)$$

$$\tau_{im} = i\,\frac{\sum a_k(\dot{x}_k^* x_k - x_k^* \dot{x}_k)}{2\varphi(a_k)} = \frac{\delta_m \Pi(\varphi)}{\varphi(a_k)} = \frac{\Pi_m(\Phi)}{\Phi(a_k)} . \quad (4)$$

Therefore, at $\varphi \neq 0$, $\Phi \neq 0$, we arrive at a sufficient local, and integral, condition of stability for $\vec{x} = 0$:

$$\varphi(a_k) > 0 \ (<0), \ \delta_m P(\varphi) < 0 \ (<0); \ \Phi(a_k) > 0 \ (<0), \ P_m[\Phi] < 0 \ (>0) \quad (5)$$

and to those of monotony of disturbances

$$\varphi(a_k) \neq 0; \ \delta_m \Pi(\varphi) = 0; \ \Phi(a_k) \neq 0; \ \Pi_m[\Phi] = 0. \quad (6)$$

If the choice of the coefficients, a_k , gives us $\Pi_m[\Phi(a_k)] = 0$, then relations (6) lead to sufficient conditions under which the monotony (aperiodicity) principle holds in the form

$$\varphi(a_k) \neq 0 \qquad \Phi(a_k) \neq 0 , \quad \text{and} \quad (7)$$

where $\varphi(a_k)$ and $\Phi(a_k)$ are defined for the aperiodic disturbances only. In this case, the quadratic form $\varphi(a_k)$ has no fixed sign, but if stability for the disturbances leading to $\varphi(a_k) > 0$ and those for which $\varphi(a_k) < 0$ is investigated separately, then $\varphi(a_k)$ may be chosen as Lyapunov's function.

When the choice of a_k does not enable us to obtain $\Pi_m[\Phi(a_k)] = 0$, then it is more convenient to work with a function which is sign-defined for all disturbances, and, for the isolation of aperiodic disturbances to impose an additional condition, namely that

$$\Pi_m[\Phi] = 0 . \quad (8)$$

This excludes the appearance of non-monotonous disturbances.

Let us consider the chemical system (see Glansdorff, Prigogine, 1971) described by the following equations

$$A \xrightarrow{k_1} X \qquad Y + 2X \xrightarrow{k_2} 3X$$
$$B + X \xrightarrow{k_3} Y + D \qquad X \xrightarrow{k_4} E \tag{9}$$

in which the concentrations of substances A, B, D, E and the temperature, T, are constant.

In the case of one-dimensional diffusion, when superimposition effects may be neglected, the kinetics of the chemical reactions (9) are described by the set of equations

$$\dot{X} = k_1 A + k_2 X^2 Y - (k_3 B + k_4) X + D_x \nabla_r^2 X,$$
$$\dot{Y} = k_3 B - k_2 X^2 Y + D_Y \nabla_r^2 Y, \tag{10}$$

which, for periodical boundary conditions, admits a spatially homogeneous stationary solution viz.

$$X_0 = \frac{k_1}{k_4} A, \quad Y_0 = \frac{k_3}{k_2} \frac{B}{X_0}. \tag{11}$$

The disturbed problem in the linear approximation can be written in the form

$$\dot{x} = (k_3 B - k_4) x + k_2 X_0^2 y + D_x \frac{\partial^2}{\partial r^2} x \quad \text{, and}$$
$$\dot{y} = - k_3 B - k_2 X_0^2 Y_0 y + D_Y \frac{\partial^2}{\partial r^2} y. \tag{12}$$

Let us consider the functions

$$\varphi(a_1, a_2) = a_1 |x|^2 + a_2 |y|^2, \; \Phi(a_1, a_2) = \int \varphi(a_1, a_2) dr. \tag{13}$$

Then

$$2 P_m [\Phi(a_1, a_2)] = \int dr \{ k_2 [2a_1 (X_0 Y_0 - \frac{k_4}{k_2}) |x|^2 - 2a_2 X_0^2 |y|^2 +$$
$$+ X_0 (a_1 X_0 - a_2 Y_0)(x^* y + y^* x) - 2a_1 D_x |\nabla_r x|^2 - 2a_2 D_Y |\nabla_r y|^2 \} \tag{14}$$

$$2\Pi_m = ik_2 \int dr\, X_0 (a_1 X_0 + a_2 Y_0)(y^* x - x^* y) \tag{15}$$

follows.

From (15) it is seen that no choice of a_1, a_2 of the same sign can make the quantity $\Pi_m[\Phi]$ equal to zero, i. e. the monotony principle, generally speaking, does not hold. However if $a_1 = 1, a_2 = -X_0/Y_0$, then $\Pi_m[\Phi] = 0$ and, therefore, the sufficient conditions for realization of monotonous disturbances only have the form

$$\varphi(1, -\frac{X_0}{Y_0}) \neq 0; \quad \Phi(1, -\frac{X_0}{Y_0}) \neq 0 \tag{16}$$

If the excess entropy $(a_1 = 1, a_2 = \frac{X_0}{Y_0})$ given by

$$\delta_m^2 S = -(|x|^2 + \frac{X_0}{Y_0}|y|^2) < 0 \tag{17}$$

is used as Lyapunov's functions which is negatively definite for all disturbances, the sufficient conditions for asymptotic stability of the state (11) with respect to arbitrary disturbances will be given by

$$P_m[\delta S] = \int dr \{(k_4 - k_2 X_0 Y_0)|x|^2 + D_x |\nabla_r x|^2 + \frac{X_0}{Y_0} D_y |\nabla_r y|^2 + \frac{k_2 X_0^3}{Y_0}|y|^2\} > 0. \tag{18}$$

A substitution of the disturbances, $a_i(r,t) = a_i(t)\, e^{kr}$, transforms (18) into

$$\int dr \{(k_4 + k^2 D_x - k_2 X_0 Y_0)|x|^2 + \frac{X_0}{Y_0}(k_2 X_0^2 + k^2 D_y |y|^2\} > 0. \tag{19}$$

From this it follows that at

$$B \leq \frac{k_4 + k^2 D_x}{k_3} \tag{20}$$

then state (11) is asymptotically stable with respect to any disturbances $(|x|^2 \gtrless \frac{X_0}{Y_0}|y|^2)$.

a) Let $|x|^2 = \frac{X_0}{Y_0}|y|^2$ (the necessary condition for the appearance of oscillatory disturbances). In this case, (19) transforms into

$$\int [k_4 - k_2 X_0 Y_0 + k^2(D_x + D_y) + k_2 X_0^2]|x|^2 dr > 0$$

and state (11) is asymptotically stable at

$$B < \frac{1}{k_3}[k_4 + k_2 X_0^2 + k^2(D_x + D_y)], \tag{21}$$

and unstable if

$$B > B_{cr} = \frac{1}{k_3}\left[k_4 + k_2 X_0^2 + k^2(D_x + D_y)\right], \tag{22}$$

i. e. for the disturbances satisfying the condition $|x|^2 = X_0/Y_0 \cdot |y|^2$ the excess entropy $\delta_m^2 S$ is a necessary and sufficient Lyapunov function.

b) Let $|x|^2 \neq X_0/Y_0 \cdot |y|^2$. In such cases, oscillatory disturbances cannot be realized, and the quadratic form, $\varphi(1, -X_0/Y_0) \neq 0$, is convenient to use as Lyapunov's function, so that the sufficient conditions for asymptotic stability of (11) take the form

$$\int dr \left\{ (k_3 B - k_4 - k^2 D_x) x^2 + \frac{X_0}{Y_0}(k_2 X_0^2 + k^2 D_y) y^2 + 2xy\, k_2 X_0^2 \right\} > 0 \ (<0).$$

From this relation it follows that if $B \geq \frac{k_4 + k^2 D_x}{k_3}$, and $B > \frac{(k_4 + k^2 D_x)(k_2 X_0^2 + k^2 D_y)}{k_3 k^2 D_y}$

the disturbances, for which $\varphi(1, -\frac{X_0}{Y_0}) < 0$, are not dangerous, which demonstrates after noting (20), that they do not lead to a violation of the stability of the state (11). On the contrary, with respect to perturbations leading to $\varphi(1, -\frac{X_0}{Y_0}) > 0$ at

$$B > \frac{(k_4 + k^2 D_x)(k_2 X_0^2 + k^2 D_y)}{k_3 X^2 D_y} \tag{23}$$

the stationary state (11) is unstable.

Comparison with the results obtained on the basis of kinetic stability theory (Glansdorff, Prigogine, 1971) shows that condition (23) is necessary and sufficient for the appearance of monotonous instability, and equation (22) is associated with oscillatory instability, so that the condition, $|x|^2 = \frac{X_0}{Y_0}|y|^2$, is evidently necessary and sufficient for the realization of non-monotonous disturbances in the given problem (correspondingly, the condition $\Phi(1 - \frac{X_0}{Y_0}) \neq 0$ is necessary and sufficient for the existence of aperiodic disturbances).

Now we shall investigate the stability of the mechanical equilibrium

$$\vec{V}^{st} = 0,\ (\hat{\Pi})^{st}_{ik} = \left(\frac{\partial V_i}{\partial x_k} + \frac{\partial V_k}{\partial x_i}\right)^{st} = 0,\ \nabla T^{st} \vec{\lambda} = -\beta_1$$
$$\nabla c^{st} \cdot \vec{\lambda} = -\beta_2,\ \rho^{st} = \rho^{st}(z),\ P^{st} = P^{st}(z) \tag{24}$$

of a non-contractible binary mixture which fills some volume in a field of gravity. Except for viscosity depending, for example, on temperature, all the properties of the liquid are constant.

If the distance, the time, the velocity, the temperature, the concentration and the pressure are measured, respectively, in the units

$$L,\ \frac{L^2}{\nu^+},\ \frac{\varkappa}{L},\ \beta_1 L,\ \beta_2 L\frac{\varkappa}{D},\ \rho_0\frac{\nu^+\varkappa}{L^2}$$

then the solutions for the disturbed velocity, $\vec{u}$, temperature, Θ, pressure, ξ, and light-component concentration, γ, in the linear approximation are described by the equations (Gershuni, Zhukhovitsky, 1972)

$$\begin{aligned}
&\nabla\vec{u} = 0,\\
&S_c\frac{\partial\gamma}{\partial t} = \vec{u}\cdot\vec{\lambda} + \nabla^2\gamma + b\nabla^2\Theta,\\
&\frac{\partial\vec{u}}{\partial t} = -\nabla\xi + \nabla\left[\nu(T^{st})\hat{\pi}\right] + (R\Theta + Rd\gamma)\vec{\lambda}\\
&P_1\frac{\partial\Theta}{\partial t} = \vec{u}\cdot\vec{\lambda} + (1+a)\nabla^2\Theta + \frac{a}{b}\nabla^2\gamma,\\
&(\hat{\pi})_{ik} = \frac{\partial u_i}{\partial x_k} + \frac{\partial u_k}{\partial x_i}.
\end{aligned} \quad (25)\ \text{and}$$

Here $\vec{V}$ is the velocity, T is the temperature, C is the concentration of a lighter component, $\vec{\lambda}$ is a unit vector antiparallel to gravity, ρ is density, $S_c = \nu^+/D$ is the Schmidt number, $P_1 = \nu^+/\varkappa$ is the Prandtl number, $R = \frac{\alpha_1 L^4 g}{\nu^+\varkappa}\beta_1$ is the Rayleigh number, $Rd = \frac{\alpha_2 L^4 g}{\nu^+ D}\beta_2$ is the the diffusion Rayleigh number, ν^+ is some specific viscosity taken for the unit of measurement, $\varkappa$, D are the thermal conductivity and the diffusion coefficients, resp., α_1 and α_2 are the thermal- and the "concentration-expansion" coefficients, resp., $\beta_1, \beta_2 > 0$ when $\nabla T^{st}\vec{\lambda} < 0$, $\nabla C^{st}\cdot\vec{\lambda} < 0$, a, b are parameters characterizing the thermal diffusion and diffusion heat conductivity, resp.. It is necessary to add boundary conditions to equations (25), which are described below.

Consider the quadratic form of the equations; viz.

$$\varphi(A_1, P_1A_2, S_cA_3) = A_1|\vec{u}|^2 + P_1A_2|\Theta|^2 + S_cA_3|\gamma|^2. \quad (26)$$

Then, taking into account (25), we obtain

$$2P[\Phi(A_1, P_1A_2, S_cA_3)] = -\int dV\{A_1\nu(T^{st})|\hat{\pi}|^2 + 2A_2(1+a)|\nabla\Theta|^2 +$$
$$+ 2A_3|\nabla\gamma|^2 - (A_1Rd + A_3)(\vec{u}^*\gamma + \gamma^*\vec{u})\cdot\lambda - (A_1R + A_2)(\vec{u}^*\Theta + \Theta^*\vec{u})\vec{\lambda} +$$
$$+ (A_3b + A_2\tfrac{a}{b})(\nabla\gamma^*\nabla\Theta + \nabla\gamma\nabla\Theta^*) + \nabla[A_1(\vec{u}\hat{\pi} + \vec{u}\hat{\pi}^* - \xi\vec{u} + \xi^*\vec{u}) + \tag{27}$$
$$+ A_3(\gamma^*\nabla\gamma + \gamma\nabla\gamma^*) + (1+a)A_2(\Theta^*\nabla\Theta + \Theta\nabla\Theta^*) +$$
$$+ A_2\tfrac{a}{b}(\Theta^*\nabla\gamma + \Theta\nabla\gamma^*) + A_3b(\gamma^*\nabla\Theta + \gamma\nabla\Theta^*)\}$$

and

$$2\Pi[\Phi(A_1, P_1A_2, S_cA_3)] = i\int dV\{(A_3 - A_1Rd)(\vec{u}^*\gamma - \vec{u}\gamma^*)\cdot\vec{\lambda} +$$
$$+ (A_2 - A_1R)(\vec{u}^*\Theta - \Theta^*\vec{u})\cdot\vec{\lambda} + \nabla[A_1(\vec{u}\hat{\pi}^* - \hat{\pi}\vec{u}^* - \xi^*\vec{u} + \xi\vec{u}^*) + \tag{28}$$
$$+ A_2(1+a)(\Theta\nabla\Theta^* - \Theta^*\nabla\Theta) + A_3(\gamma\nabla\gamma^* - \gamma^*\nabla\gamma) + A_3b(\gamma\nabla\Theta^* - \gamma^*\nabla\Theta) +$$
$$+ \tfrac{a}{b}A_2(\Theta\nabla\gamma^* - \Theta^*\nabla\gamma)] + (\tfrac{a}{b}A_2 - bA_3)(\nabla\gamma\cdot\nabla\Theta^* - \nabla\gamma^*\cdot\nabla\Theta)\}.$$

Consider some simple problems with boundary conditions such that the surface integrals in (27) and (28) vanish.

Let R and Rd be independent and a = b = 0. From (28) it follows that for the realization of aperiodical disturbances it is only necessary that

$$A_1 = 1\ ,\quad A_2 = R\ ,\quad A_3 = Rd \tag{29}$$

so that at $R, Rd > 0$ the monotony principle holds since $\varphi(1, P_1, R, Sc, Rd) = -\delta_m^2 z = -\delta^2 z < 0$ for all disturbances, and, the sufficient integral conditions of asymptotical stability take the form

$$P[\delta z] = \int\{\tfrac{1}{2}\nu(T^{st})\hat{\pi}^2 + R(\nabla\Theta)^2 + Rd(\nabla\gamma)^2 - 2\vec{\lambda}\vec{u}(R\Theta + Rd\gamma)\}dV > 0\ . \tag{30}$$

From expression (30) it is seen that mechanical equilibrium will be not stable for any arbitrary values of the temperature and concentration gradients. Indeed, for example, for rather large values of R expression (30) will be replaced by

$$R\int\{(\nabla\Theta)^2 - 2\vec{\lambda}\vec{u}\Theta\}dV > 0,$$

i. e. by the same condition as in a one-component liquid. At a certain value of R the state of mechanical equilibrium in such systems becomes un-unstable (Glansdorff, Prigogine, 1971; Gershuni, Zhukhovitsky, 1972). Hence, we find that

$$R_{cr} = \frac{\int\{\frac{1}{2}\nu(T^{st})\hat{\pi}^2 + Rd(\nabla\gamma)^2 - 2\vec{\lambda}\vec{u}\gamma\}dV}{\int\{2\vec{\lambda}\vec{u}\Theta - (\nabla\Theta)^2\}dV} \; . \tag{31}$$

This is defined by the condition $P(\delta z) = 0$ such that at $R > R_{cr}$ state (24) will not be realized.

If $R, Rd < 0$ then

$$\delta_m^2 z - (|\vec{u}|^2 + P_1|R||\Theta|^2 + S_c|Rd||\gamma|^2) < 0 \text{ and}$$
$$P_m(\delta z) = \int dV\{\frac{1}{2}\nu(T^{st})|\hat{\pi}|^2 + |R||\nabla\Theta|^2 + |Rd||\nabla\gamma|^2\} > 0, \tag{32}$$

i. e. state (24) is not violated at any values of the temperature and concentration gradients. At the same time,

$$\Pi_m[\delta z] = i\int dV\{|R|(\vec{u}^*\Theta - \Theta^*\vec{u}) + |Rd|(\vec{u}^*\gamma - \gamma^*\vec{u})\}\cdot\vec{\lambda}$$

and, therefore, non-monotonous disturbances are also possible. In order to exclude them it is necessary to impose an additional condition, that, $\Pi_m[\delta z] = 0$. If we choose

$$A_1 = 1, \; A_2 = -|R|, \; A_3 = -|Rd|, \; \Pi_m[\Phi(1, -P_1|R|, -S_c|Rd|)] = 0$$

the sufficient conditions for the asymptotical stability of (24) will be as follows

$$\Phi(1, -P_1|R|, -S_c|Rd|) > 0 \quad (< 0)$$
$$P[\Phi(1, -P_1|R|, -S_c|Rd|)] = -\int dV\{\frac{1}{2}\nu(T^{st})\hat{\pi}^2 - |R|(\nabla\Theta)^2 - \tag{33}$$
$$- |Rd|(\nabla\gamma)^2 + 2\vec{\lambda}\vec{u}(|R|\Theta + |Rd|\gamma)\} < 0 \qquad (> 0).$$

In this case, the requirement $\Phi(1, -P_1|R|, -S_c|Rd|) \neq 0$ is met automatically, so that $\varphi(1, -P_1|R|, -S_c|Rd|) \neq 0$ is defined for monotonous disturbances only. Thus at $R, Rd < 0$ the stability of (24) relative to aperiodic disturbances is defined either by (32) under the condition $\Pi_m[\delta z] = 0$ or by (33).

When R, Rd are of different signs, i. e. define $\mathcal{R} > 0$ then:

$$\Pi_m[\delta z] = i|Rd|\int dV(\vec{u}^*\gamma - \gamma^*\vec{u})\cdot\vec{\lambda} \tag{34}$$

$$P_m[\delta z] = \int dV\{\tfrac{1}{2}\nu(T^{\Delta t})|\hat{\pi}|^2 + R|\nabla\Theta|^2 + |Rd||\nabla\gamma|^2 - R\vec{\lambda}(\vec{u}^*\Theta + \Theta^*\vec{u})\} \tag{35}$$

$$\Phi(1, P_1R, -S_c|Rd|) = \int\{|\vec{u}|^2 + P_1R|\Theta|^2 - S_c|Rd||\gamma|^2\}dV \tag{36}$$

$$\Pi_m[\Phi(1, P_1R, -S_c|Rd|] = 0 \tag{37}$$

$$P[\Phi(1, P_1R, -S_c|Rd|] = -\int\{\tfrac{1}{2}\nu(T^{\Delta t})(\hat{\pi})^2 + R(\nabla\Theta)^2 - |Rd||\nabla\gamma|^2 + 2\vec{\lambda}\vec{u}(|Rd|\gamma - R\Theta)\}\ . \tag{38}$$

From this the stability of (24) with respect to monotonous disturbances is given by

$$P_m[\delta z] > 0\ , \quad \Pi_m[\delta z] = 0 \tag{39}$$

when $\delta_m^2 z$ is used as Lyapunov's function, and, by

$$\Phi(1, P_1R, -S_c|Rd|) > 0\ (<0)\ , \quad P[\Phi(1, P_1R, -S_c|Rd|)] < 0\ (>0)\ , \tag{40}$$

when the quadratic form $\varphi(1, P_1R, -S_c|Rd|) \neq 0$ is chosen as Lyapunov's function. The critical temperature gradients can be found from the conditions $P_m[\delta z]$ at $\Pi_m[\delta z] = 0$, or $P[\Phi(1, P_1R, -S_c|Rd|)] = 0$ and will be defined by the expressions

$$R_{cr} = \frac{\int\{\tfrac{1}{2}\nu(T^{\Delta t})|\hat{\pi}|^2 + |Rd||\nabla\gamma|^2\}dV}{\int[\vec{\lambda}\cdot(\vec{u}^*\Theta + \vec{u}\Theta^*) - |\nabla\Theta|^2]dV} \quad \text{at} \quad \Pi_m[\delta z] = 0 \tag{41}$$

and

$$R_{cr} = \frac{\int\{\tfrac{1}{2}\nu(T^{\Delta t})\hat{\pi}^2 - |Rd|(\nabla\gamma)^2 + 2|Rd|\vec{\lambda}\cdot\vec{u}\gamma\}dV}{\int\{2\vec{\lambda}\cdot\vec{u}\Theta - (\nabla\Theta)^2\}dV} \tag{42}$$

This may constitute a means to consider other problems. The problems under investigation may be also solved by Lyapunov's first method. Moreover, in the definition of critical values, for example the Rayleigh number, one must use Lyapunov's first method since the eigenvalues are determined by the solutions of the problem. It is remarkable that, with the aid of the quantities $\delta_m^2 z$, $P_m[\delta z]$ and $\Pi_m[\delta z]$ (or corresponding to them Φ, $P(\Phi)$ and $\Pi(\Phi)$), leading to sufficient integral conditions of asymptotic stability and monotony of disturbances, one can also derive a variational formulation of the corresponding problem for monotonous disturbances.

When a = b = 0, the problem is described by the set of equations (25) at a = b = 0, which, after the substitution of the disturbances, $a_i(\vec{r},t) = a_i(\vec{r}) \exp \tau t$, is reduced to

$$\begin{aligned} &\nabla \cdot \vec{u} = 0 \\ &\tau S_c \gamma = \vec{u} \cdot \vec{\lambda} + \nabla^2 \gamma \\ &\tau \vec{u} = -\nabla \xi + \nabla \cdot [\nu (T^{st}) \hat{\pi}] + (R\Theta + Rd\gamma)\vec{\lambda} \\ &\tau P_1 \Theta = \vec{u} \cdot \vec{\lambda} + \nabla^2 \Theta . \end{aligned} \tag{43}$$

At $R, Rd > 0$ aperiodic disturbances alone are possible, and (43) are Euler-Lagrange type equations for the variational problem

$$P[\delta z] = extr, \tag{44}$$

under the additional conditions

$$\delta^2 z = const \quad \text{and} \quad \nabla \cdot \vec{u} = 0 . \tag{45}$$

Indeed, if we introduce the Lagrange multipliers, τ, and, 2ξ, and, make the first variation of the function equal to zero then

$$\mathfrak{F} = P[\delta z] + \tau \delta^2 z + \int 2\xi \nabla \cdot \vec{u} \, dV \quad \text{and}$$

we arrive at (43).

Then utilizing the Cauchy-Bunjakovsky inequality,

$$\int \vec{\lambda} \cdot \vec{u} a_i \, dV \leq [\int \vec{u}^2 dV \int a_i^2 dV]^{1/2},$$

we find that

$$P[\delta z]=\int\{\tfrac{1}{2}\nu(T^{st})\hat{\pi}^2+R(\nabla\Theta)^2+Rd(\nabla\gamma)^2-2\vec{\lambda}\vec{u}(R\Theta+Rd\gamma)\}dV$$

has an infimum. Thus, since $\delta^2 z$ is negative, the highest eigenvalue τ_0, is equal to

$$\tau_0 = max\ \frac{P[\delta z]}{\delta^2 z} \tag{46}$$

for all arbitrary $\vec{u}, \Theta$ being rather smooth, which meets the condition of non-contractibility and, such boundary conditions under which the surface integrals in (27) and (28) vanish. Since the eigenfunctions pertaining to different levels of the spectrum satisfy the orthogonality condition

$$\int\{\vec{u}_n\vec{u}_m + P_1 R\Theta_m\Theta_n + S_c Rd\gamma_n\gamma_m\}dV = M\delta_{nm}, \tag{47}$$

then the subsequent τ can be found from (46) with account taken of (47).

When $P[\delta z] = 0$, (44) transforms into

$$\begin{aligned}
&\nabla\cdot\vec{u} = 0\\
&\vec{u}\cdot\vec{\lambda} + \nabla^2\gamma = 0\\
&-\nabla\xi + \nabla[\nu(T^{st})\cdot\hat{\pi}] + (R_{cr}\Theta + Rd\,\gamma)\vec{\lambda} = 0\\
&\vec{u}\cdot\vec{\lambda} + \nabla^2\Theta = 0
\end{aligned} \tag{48}$$

i. e. in this problem the critical Rayleigh numbers R_{cr} (or Rd) are eigenvalues, and, when each of them is attained, the equilibrium becomes unstable with respect to one more form of disturbance. At different critical motions satisfy the orthogonality conditions

$$\int\vec{u}_m\vec{\lambda}\Theta_n dV = \int\nabla\Theta_m\cdot\nabla\Theta_n dV = \int\nu(T^{st})\hat{\pi}_m:\hat{\pi}_n dV = \int\vec{u}_m\vec{\lambda}\gamma_n dV = 0\ . \tag{49}$$

Equations (48) are obtained from

$$I_1 / I_2 = extr \tag{50}$$

at $\nabla\cdot\vec{u} = 0$.

According to (30) and (31) $P[\delta z] = I_1 - R I_2$ and, at a fixed R_{cr}, the difference between $P[\delta z]$, for an arbitrary motion, and the motion associated with the chosen R_{cr} is given by

$$\Delta P[\delta z] = \Delta I_1 - R_{cr} \Delta I_2 > 0,$$

since all the disturbances, except the critical one, for a given R_{cr} satisfy the stability condition. This gives

$$\Delta (I_1 / I_2) > 0,$$

so that the smallest critical Rayleigh number, R^{o}_{cr}, is equal to

$$R^{o}_{cr} = \min I_1 / I_2 , \qquad (\nabla \cdot \vec{u} = 0),$$

and all the subsequent R_{cr} are to be found from (50) under the orthogonality condition (49).

For the conditions $R > 0$ and $Rd < 0$ then the problem presented is most interesting, since it is one of the simplest non-self-conjugate problems containing instability. In this case, (43) can be rewritten in the form

$$\begin{aligned} &\nabla \cdot \vec{u} = 0 \\ &\tau S_c \gamma = \vec{u} \cdot \vec{\lambda} + \nabla^2 \gamma \\ &\tau \vec{u} = -\nabla \xi + \nabla \cdot [v(T^{st}) \hat{\pi}] + \vec{\lambda}(R\Theta - |Rd|\gamma) \\ &P_1 \tau \Theta = \vec{u} \cdot \vec{\lambda} + \nabla^2 \Theta \end{aligned} \tag{51}$$

Let the principle of the exchange of stabilities hold. Then, if one uses the excess entropy, $\delta^2_m z$, as Lyapunov's function, the problem (51) is equivalent to the variational problem,

$$P_m[\delta z] = extr, \tag{52}$$

$$\text{at } \delta^2_m z = const; \quad \Pi_m[\delta z] = 0; \quad \nabla \cdot \vec{u} = 0. \tag{53}$$

Indeed, introducing the Lagrange multipliers $2\tau, i, 2\xi, 2\xi^*$ and nullifying the first functional variation

$$\mathcal{F} = P_m[\delta z] + 2\tau \delta^2_m z - i \Pi_m [\delta z] + 2\int (\xi^* \nabla \vec{u} + \xi \nabla \vec{u}^*) dV$$

after integrating by parts and noting that the surface integrals vanish, then, making the expressions of the independent variations, $\delta\xi^*$, $\delta\vec{u}^*$, and $\delta\Theta^*$, $\delta\gamma^*$ be equal to zero, we are led to equations (51). The fact that the expressions for $\delta\xi$, $\delta\vec{u}$, $\delta\Theta$, $\delta\gamma$ are equal to zero results in expressions for the quantities complex-conjugate to $\vec{u}$, Θ, γ, ξ. It is readily seen that the function $P_m[\delta z]$ is restricted from below, so that, under the additional condition (53), the highest eigenvalue, τ_0, is equal to

$$\tau_0 = \max \frac{P_m[\delta z]}{\delta_m^2 z} . \tag{54}$$

When the quadratic form, $\varphi(1, P_1 R, -S_c|Rd|) \neq 0$, is chosen as Lyapunov's function, the stability of (24) can be investigated separately for disturbances satisfying $\varphi > 0$ ($\Phi > 0$) and those for which $\varphi < 0$ ($\Phi < 0$). So, in the case of $\varphi(1, P_1 R, -S_c|Rd|) > 0$ the sufficient integral conditions for asymptotic stability of (24) have the form

$$\begin{aligned} P[\Phi(1, P_1 R, -S_c|Rd|] = -\int dV \{\tfrac{1}{2}\nu(T^{st})\hat{\pi}^2 + \\ + R(\nabla\Theta)^2 - |Rd|(\nabla\gamma)^2 + 2\vec{\lambda}\cdot\vec{u}(|Rd|\gamma - R\Theta)\} < 0 . \end{aligned} \tag{55}$$

It is obvious that at fixed P_1, S_c and Rd one can always choose R such that sufficient stability conditions, like in the case of one-component liquid, are specified by the expression

$$\int \{\tfrac{1}{2}\nu(T^{st})\hat{\pi}^2 + R(\nabla\Theta)^2 - 2R\vec{\lambda}\cdot\vec{u}\Theta\} dV > 0 .$$

Therefore, for sufficiently large R the integral in (55) will be negative, so that for the disturbances leading to $\varphi > 0$ the principle of the exchange of stabilities holds. It is clear that if Φ is limited, $P[\Phi]$ has a maximum, i. e. for $\varphi > 0$, if the liquid is not contractible, the highest eigenvalue, τ_0, is

$$\tau_0 = \max \frac{P[\Phi(1, P_1 R, -S_c|Rd|]}{\Phi(1, P_1 R, -S_c|Rd|} .$$

Similarly one can give a variational formulation of the problems for chemical reactions when $a \neq 0$, $b \neq 0$ etc. In all these cases, if the monotony principle holds, then the problem is reduced to the solution of the problems

$$P_m[\delta z] = extr\,, \quad \delta_m^2 z = const\,, \quad \Pi_m[\delta z] = 0,$$

under some additional conditions inherent in each particular problem.

A thermodynamic theory of stability based on the use of the excess generalized entropy as Lyapunov's functions has been generally recognized. However, as is seen from this paper, the potential of this theory is much wider. In particular, it is essential to investigate the potential for obtaining a variational formulation of the problem on the basis of $\delta_m^2 \rho z$. But, although this formulation fits only for monotonous disturbances, it is also important since, if both types of instability, monotonous and oscillatory, are possible, an individual solution of the problem should be found for each type of instability. In addition, in a formal use of the variational technique used in this approach, we may consider decrements to be complex quantities. Therefore, in this case the results are the same as in the method of local potential. This method, however, yields no information on the general properties of the spectrum of disturbandes. This obviously makes the first approach advantageous.

Exploiting Lyapunov's functions defined for monotonous disturbances only is, of course, preferable since in addition to the possibility of reducing the problem to a variational problem it also gives some information about conditions under which aperiodic disturbances alone are realized. Such a situation is, however, rather rare. At the same time the additional condition $\Pi_m[\delta z] = 0$, which also excludes disturbances connected with complex decrements, may always be used.

It should be noted that the variational formulation of some problems including non-self-conjugate ones. was given in some papers (Sorokin, 1953; Sorokin, Sushkin, 1960; Nakoryakova, 1964) for the case of aperiodic disturbances. These papers formulate the variational principle by proceeding directly from the expression for τ_r obtained for $\tau_{im} = 0$. But since this condition is achieved by the choice of the quadratic form of the amplitudes of disturbances, such an approach is based, practically, on the method of Lyapunov's functions defined only for aperiodic disturbances and, considered at the initial moment.

V. Ecology and Populations

The Rate of Correlation Decay in One-Dimensional Ecological Models

L. A. Bunimovich, Ja. G. Sinay

In recent years the models of population number dynamics with discrete time have attracted much attention. Such models describe the evolution of an isolated population with non-overlapping generations. The dynamics of one-dimensional ecological models with continuous time is trivial (Smith, 1974). At the same time, simple models with discrete time turn out to demonstrate a wide range of different types of behaviour: from the presence of a globally stable single equilibrium state, or a limit cycle up to essentially stochastic behaviour, when correlations in the system decay in the course of time (in this case the system is said to possess the property of mixing). These topics, as well as some other models, have been discussed in a review by May (1974).

From the point of view of application, one of the most important characteristics of a system which possesses the property of mixing, is the velocity of correlation decay within it. This paper is devoted to obtaining such an estimate for the so-called logistic model.

The most general one-dimensional model of population number dynamics is described by the equation

$$x_{n+1} = x_n f(x_n),$$

where x_n and x_{n+1} are, respectively, the population numbers in the n-th and (n+1)-the generations, and f is some function. The population numbers may acquire values from zero to infinity, i. e. the phase space of the dynamic system under consideration is a semiaxis,

Thermodynamics and Kinetics of Biological Processes

If $f(x) = const$, we are dealing either with an exponentially propagating (Malthusian) population, or with an exponentially degenerating one, or $x_{n+1} = x_n$.

None of these situations corresponds to the behaviour of real populations in which numbers vary with time. The effect of competition is not taken into account in this case, whereas it is predominant when the population numbers are not small.

Therefore, the following requirements are usually imposed upon f : $f(x) \underset{x \to 0}{\to} const$, and $xf(x) \underset{x \to \infty}{\to} 0$. It is a natural requirement that the function $xf(x)$ should have a single maximum. If $f(x)$ depends linearly on x, then we can move to the logistic model

$$x_{n+1} = a x_n (1 - x_n) , \qquad (1)$$

where x_n now designates the population density (i. e. the phase space here is the integral $[0, 1]$, a is the parameter characterizing the propagation velocity, $0 \leq a \leq 4$).

The ergodic properties of this model have been studied in theoretical and quantitative treatments. We draw attention to the paper by Ruelle, 1977, who was the first to show that there exists a value of the parameter $a < 4$, at which the transformation $S_a : x_n \to x_{n+1}$ possesses an invariant measure absolutely continuous with respect to the Lebesgue measure. Yakobson (1978) has recently shown that there exists a continuum in the values of the parameter a at which the transformation S_a possesses an absolutely continuous invariant measure. Pianigiani (1977) proved the existence of an invariant measure for S_a at a countable number of values of the parameter a.

However, the methodology used to prove the existence of an absolutely continuous invariant measure does not make it possible to estimate the rate of the correlation decay. To obtain this estimate we make use of the method first suggested for the estimation of the velocity of correlation decay for expanding transformations of the integral (Sinay, 1970). This method also simultaneously permits proof of the existence of an absolutely continuous invariant measure for the transformations of an interval of form similar to (1) (see Bunimovich, 1970; Pianigiani, 1976).

Some Remarks on the Properties of Mapping S_a

The mapping S_a at different values of the parameter may be presented graphically. It is known (May, 1974) that at $0 < a < 3$, S_a has a globally attracting stationary point, at $3,526 \geq a \geq 3$; a globally attracting cycle of the second order, at $3,692 \geq a \geq 3,526$; a globally stable cycle of order 4 and, finally, at $3,692 \leq a \leq 4$ there exist cycles of all periods. Thus, we shall consider the parameter a from the latter diapason.

It has been shown (Pianigiani, 1977) that there exists a countabie number of values of the parameter a at which the following conditions are met

(i) there exists $p > 1$ such that

$$S_a^p\left(\tfrac{1}{2}\right) = 1 - S_a^{2p}\left(\tfrac{1}{2}\right), \quad S_a^p\left(\tfrac{1}{2}\right) < \tfrac{1}{2};$$

(ii)

$$S_a^p(I) = I, \quad S_a^k(I) \cap \left(\tfrac{1}{a}, 1 - \tfrac{1}{a}\right) = \phi$$

at $k = 1, \ldots, p-1$, where $I = \left(S_a^p\left(\tfrac{1}{2}\right), S_a^{2p}\left(\tfrac{1}{2}\right)\right)$. For such values a_p, $p = 1, 2, \ldots$ of the parameter a the existence of an absolutely continuous invariant measure, μ, for the transformation S_a was proved.

It is easy to see that from condition (i) it follows that $S_a^{p+1}(\frac{1}{2})$ is a repulsing periodic point of the period p. Indeed, we have

$$S_{a_p}^{p+1}\left(\tfrac{1}{2}\right) = S_{a_p}^{2p+1}\left(\tfrac{1}{2}\right) = S_{a_p}^p\left(S_{a_p}^{p+1}\left(\tfrac{1}{2}\right)\right).$$

The density $\rho_{a_p}(x)$ of the invariant measure is not a bounded function but has singularities at the points $S_{a_p}(\frac{1}{2}), S_{a_p}^2(\frac{1}{2}), \ldots, S_{a_p}^{2p}(\frac{1}{2})$, of the form $\rho_{a_p}(x)_{x \to S_{a_p}^k(\frac{1}{2})} \sim \dfrac{const}{\sqrt{x - S_{a_p}^k(1/2)}}$, if $k = 2, 3, \ldots, p+1$ and

$$\rho_{a_p}(x) \underset{x \to S^k_{a_p}(\frac{1}{2})}{\sim} \frac{const}{\sqrt{S^k_{a_p}(\frac{1}{2}) - x}},$$

if = p + 2, p + 3, ..., 2p, 1

Estimation of the Rate of Correlation Decay

The main result of the present work is the following theorem.

Theorem. Let f(x) and g(x) be continuously differentiable bounded functions defined on the interval [0,1]. Then at the values of the parameter a satisfying conditions (i) and (ii), the following inequality

$$\rho_{f,g}(n) = \left| \int f(S^n_a(x)) g(x) \rho_a(x) dx - \int f \rho_a dx \cdot \int g \rho_a dx \right| \leq \leq C_1 \exp(-C_2 n^{1/3}), \tag{2}$$

holds where C_1, C_2 are positive constants, the integral is calculated over the entire interval [0,1] and $p_a(x)$ is the density of an absolutely continuous invariant measure for the transformation S_a.

From (2) it follows that at such values of the parameter a the transformation S_a possesses the property of mixing and, consequently, that nonequilibrium distributions absolutely continuous with respect to the Lebesgue measure over the interval [0,1] under the action of dynamics converge to the equilibrium distribution (given by the density $\rho_a(x)$).

Proof. First of all we shall show that $S^k_a(I)$ and $S^\ell_a(I)$ do not overlap at $1 \leq k < \ell \leq p-1$. Indeed, if $S^k_a(I) \cap S^\ell_a(I) \neq \phi$, then

$$S^{p-\ell+k}_a(I) \cap S^p_a(I) \neq \phi.$$

The latter inequality contradicts the condition (ii), since

$$S^p_a(I) \subset (\tfrac{1}{a}, 1-\tfrac{1}{a}), \quad S^{p-\ell+k}_a(I) \cap (\tfrac{1}{a}, 1-\tfrac{1}{a}) = \phi.$$

Thus, the support of the invariant measure set up by the density $\rho_a(x)$ is the collection of the segments $I, S_a(I), S_a^2(I), \dots, S_a^{p-1}(I)$, which transform into one another by iterative action of the transformation S_a. In this case S_a gives a one-to-one mapping, $S_a^k(I)$ on $S_a^{k+1}(I)$ if $1 \le k \le p-1$. When I is transformed on $S_a(I)$, each point of the interval $S_a(I)$ (except for the point $S_a(\frac{1}{2})$ has exactly two prototypes. Let us designate $\bigcup_{k=0}^{p-1} S_a^k(I)$ through A_I. It is clear that $\int_A f(S_a^n x) g(x) \rho_a(x) dx = 0$.

Consider first the transformation S_a^p acting on the interval I. According to (ii), S_a^p transforms the interval I into itself. It is clear that the transformation $S_a^p | I$ has an absolutely continuous invariant measure ν, set up by the density $\bar{\rho}_a(x)$. The connection between the measures μ and ν is given by the relation

$$\mu = (\nu + \nu \cdot S_a^{-1} + \dots + \nu \cdot S_a^{-p+1})/p.$$

In what follows we shall write T instead of S_a^p and omit the index a in all the transformations and functions considered below.

As has been shown above, any point $x \in T$ (except $T(\frac{1}{2})$ has exactly two prototypes $x_1 = T^{-1}x$ and $x_2 = T^{-1}x$, symmetric about the point 1/2. Therefore, the interval, I, can be divided into two intervals $\Delta_1^{(1)}$ and $\Delta_2^{(1)}$, such that $T\Delta_1^{(1)} = I$ and $T\Delta_2^{(1)} = I$. It is evident that $\Delta_1^{(1)}$ and $\Delta_2^{(1)}$ are half the interval I. Each of the intervals $\Delta_1^{(1)}$ and $\Delta_2^{(1)}$ may, in their turn, be divided into two intervals $\Delta_{11}^{(2)}$ and $\Delta_{12}^{(2)}$, $\Delta_{21}^{(2)}$ and $\Delta_{22}^{(2)}$ such that $T\Delta_{i_1 i_2}^{(2)} = \Delta_{i_2}^{(1)}$, where $i_1, i_2 = 1, 2$. This process may be continued and as a result for any k we obtain 2^k intervals of the k-th rank, $\Delta_{i_1 i_2 \dots i_k}^{(k)}$ and $T\Delta_{i_1 i_2 \dots i_k}^{(k)} = \Delta_{i_2 \dots i_k}^{(k-1)}$. If $\Delta_i^{(k)}$ is an arbitrary interval of the k-th rank, then $T^k \Delta_i^{(k)} = I$, $i = (i_1, i_2, \dots, i_k)$.

Let us designate the function which sets up this mapping through $\varphi_{\Delta_i^{(k)}}$. Note that

$$\varphi_{\Delta_i^{(k)}}(x) = \underbrace{\varphi(\varphi(\dots\varphi(x))\dots)}_{k\text{-times}}.$$

Hence, if $y = T^k x$, dx is the Lebesgue measure on $\Delta_i^{(k)}$, then $dy = \varphi'(T^{k-1}x)\cdot\varphi'(T^{k-2}x)\ldots\cdot\varphi'(x)\,dx$. Consequently, the Lebesgue measure on $\Delta_i^{(k)}$ induces, by means of the mapping, $\varphi_{\Delta_i^{(k)}}(x)$ the measure on I with the density

$$\Psi_{\Delta_i^{(k)}}(y) = \frac{1}{\ell(\Delta_i^{(k)})\cdot\varphi'(T^{k-1}x)\cdot\varphi'(T^{k-2}x)\cdot\ldots\cdot\varphi'(x)},$$

where $\ell(\Delta_i^{(k)})$ is the length of the interval $\Delta_i^{(k)}$ (normalized to the length of I). It is easily seen that $\Psi_{\Delta_i^{(k)}}(y)$ is defined for all the points of the interval I, except for its boundaries. Whence

$$\int_I f(T^k dx)\,dx = \sum_{i=1}^{2^k}\int_{\Delta_i^{(k)}} f(T^k dx)\,dx = \sum_{i=1}^{2^k}\ell(\Delta_i^{(k)})\int f(y)\Psi_{\Delta_i^{(k)}}(y)\,dy.$$

Thus, the transformations T^k generate a sequence of densities $\varphi_k(y)$ which have the form

$$\varphi_k(y) = \sum_{i=1}^{2^k}\ell(\Delta_i^{(k)})\Psi_{\Delta_i^{(k)}}(y). \tag{3}$$

It can be shown that this sequence $\{\varphi_k(y)\}$ converges integrally to the function p(y) on I which is the density of the invariant measure ν. This fact has been proved in a similar way as by Bunimovich (1970) with the use of the results obtained by Pianigiani (1976).

In addition, from the recent work of Pianigiani (1976, 1977) it follows that the transformation T is conjugated, with the aid of an absolutely continuous mapping, F, to the transformation R of the interval $[0,1]$, which is an expanding interval. More precisely $T = F^{-1}\cdot R\cdot F$, where $F(x) = \int_0^x h(y)dy$, $h(y) = \frac{c}{\sqrt{|(T(\frac{1}{2})-y)(T^2(\frac{1}{2})-y)|}}$, $c>0$ is a normalizing constant providing the equality $\int_I h(y)dy = 1$ and the transformation R satisfies the following conditions:

1. there exists partition of the interval 0,1 into two, such that the restriction on R for each of the intervals obtained is a function of class C^1;

2. the function $1/R'(x)$ is absolutely continuous over each of these intervals;

3. $\inf|R'(x)| \geq \lambda > 1$.

It has even been proved (Pianigiani, 1977) that $\lambda \geq \sqrt{2}$, but we shall not use this directly.

It is readily seen that the measure ν, invariant under the transformation T, and the measure $\tilde{\nu}$, invariant under the transformation R, are related as $\nu = \tilde{\nu} \cdot F$. Whence, if $\tilde{\nu}$ is absolutely continuous, then ν is absolutely continuous, too, and vice versa.

From this it follows that the set of intervals, $\Delta_i^{(k)}$, such that $T^k \Delta_i^{(k)} = I$ corresponds to $F\Delta_i^{(k)} = \tilde{\Delta}_i^{(k)}$, such that $R^k \tilde{\Delta}_i^{(k)} = [0,I]$ and $R\tilde{\Delta}^k_{i_1 i_2 \ldots i_k} = \tilde{\Delta}^{k-1}_{i_2 \ldots i_k}$. Similarly, the sequence of densities φ_k corresponds to the sequence of densities $\tilde{\varphi}_k$.

Now consider the expression

$$J = \int f(T^n x) g(x) \hat{\rho}(x) dx = \int_I f(T^n x) g(x) \tilde{\rho}(x) dx =$$
$$= \sum_{\Delta_i^{(m)}} \int_{\Delta_i^{(m)}} f(T^n x) g(x) \tilde{\rho}(x)\, dx ,$$

where f and g are functions of class C^1 defined on the interval $[0,1]$. We have

$$J = \sum_{\Delta^{(m)}} \int_{\Delta^{(m)}} f(T^n x) g(x) \tilde{\rho}(x) dx = \sum (g\tilde{\rho})_{\Delta^{(m)}} \int_{\Delta^{(m)}} f(T^n x) dx + A ,$$

where $(g\tilde{\rho})_{\Delta^{(m)}}$ is the mean (with respect to the Lebesgue measure) value of the function $g(x)\tilde{\rho}(x)$ over the interval $\Delta^{(m)}$. Therefore, $\sum_{\Delta^m} (g\tilde{\rho})_{\Delta^{(m)}} \ell(\Delta^{(m)}) = \int_I g(x)\tilde{\rho}(x) dx$, where $\ell(\Delta^{(m)})$ is the length of the interval $\Delta^{(m)}$. Let us now estimate the value of A.

From condition (3) for the transformation R it immediately follows that $\max_{\tilde{\Delta}_i^{(k)}} \ell(\tilde{\Delta}_i^{(k)}) < \lambda^{-k}$. Whence $\max_{\Delta_i^{(k)}} \ell(\Delta_i^{(k)}) < c_1 \lambda^{-k/2}$.

Finally, from (3) and the expression for the function h it follows that $A < c_2 \lambda^{-k/6}$.

In fact, the function f(x) is bounded from above and from below, g(x) and its first derivative is also bounded from above and from below and, finally, for the function $\rho(x)$ we have $\tilde{c}_1 h(x) \le \tilde{\rho}(x) \le \tilde{c}_2 h(x)$. Let us divide the interval I into two regions I_1 and I_2 such that in the region I_2 the estimate $|h'(x)| < \lambda^{k/3}$ is valid for the derivative h'(x). It is clear that $\nu(I_1) \le \tilde{c}_3 \lambda^{-k/6}$. Let us now represent the sum in the expression for J as two sums: over the intervals $\Delta^{(m)}$ for I_1 and I_2. Correspondingly, the remainder, A, is represented in the form $A = A_1 + A_2$. We have $|A_1| \leq \tilde{C}_4 \lambda^{-k/6}$, $|A_2| \leq \tilde{C}_5 \lambda^{-k/6}$.

We transform the integrals under the summation

$$\int_{\Delta^{(m)}} f(T^n x)\,dx = \sum_{\Delta^{(n)} \subset \Delta^{(m)}} \int_I f(x)\left(\varphi^{-1}_{\Delta^{(n)}}(x)\right)' dx,$$

and make a change of variables, setting $y = F(x)$.

Then

$$\sum_{\Delta^{(n)} \subset \Delta^{(m)}} \int_I f(x)\left(\varphi^{-1}_{\Delta^{(n)}}(x)\right)' dx =$$

$$= \sum_{\tilde{\Delta}^{(n)} \subset \tilde{\Delta}^{(m)}} \int f(F^{-1}(y))\left(\tilde{\varphi}^{-1}_{\tilde{\Delta}^{(n)}}(y)\right)' \frac{1}{h(\tilde{\varphi}^{-1}_{\tilde{\Delta}^{(n)}}(y))} =$$

$$= \sum_{\tilde{\Delta}^{(n)} \subset \tilde{\Delta}^{(m)}} \int \frac{f(F^{-1}(y))}{h(\tilde{\varphi}^{-1}_{\tilde{\Delta}^{n}}(y))}\, dy\, (\tilde{\varphi}^{-1}_{\tilde{\Delta}^{(n)}}(y) = \ell(\tilde{\Delta}^{(m)}) \sum_{\tilde{\Delta}^{(n)} \subset \tilde{\Delta}^{m}} \frac{\ell(\tilde{\Delta}^{(n)})}{\ell(\tilde{\Delta}^{(m)})} \int G(y) D_{\tilde{\Delta}^{(n)}}(y)\,dy,$$

where $G(y)$ stands for the function $f(F^{-1}(y))/h(\tilde{\varphi}^{-1}_{\tilde{\Delta}^{(n)}}(y))$, $D_{\tilde{\Delta}^{(n)}}(y) = (\tilde{\varphi}^{-1}_{\tilde{\Delta}^{(n)}}(y))'/\ell(\tilde{\Delta}^{(n)})$.

These latter considerations are analogous to Sinay (1970). The difference between our latter expression and a similar one obtained by Sinay (1970) is only that his function, G(y), was continuously differentiable and the derivative of our G(y) has singularities of the type $1/\sqrt{y}$, $y \to +0$ at the boundaries of the interval [0,1].

Let us fix r and transform the latter expression

$$\sum_{\tilde{\Delta}^{(n)} \subset \tilde{\Delta}^{(m)}} \frac{\ell(\tilde{\Delta}^{(n)})}{\ell(\tilde{\Delta}^{(m)})} \int G(y)\left(\tilde{\varphi}^{-1}_{\tilde{\Delta}^{(n)}}(y)\right)' dy =$$

$$= \sum_{\tilde{\Delta}^{(n)} \subset \tilde{\Delta}^{(m)}} \frac{\ell(\tilde{\Delta}^{(n)})}{\ell(\tilde{\Delta}^{(m)})} \sum_{\tilde{\Delta}^{(r)}} (G)_{\tilde{\Delta}^{(r)}}\, q_n(\tilde{\Delta}^{(r)} \mid \tilde{\Delta}^{(m)}) + B,$$

where $q_n(\tilde{\Delta}^{(r)} \mid \tilde{\Delta}^{(m)}) = \sum_{\tilde{\Delta}^{(n)} \subset \tilde{\Delta}^{(m)}} \frac{\ell(\tilde{\Delta}^{(n)})}{\ell(\tilde{\Delta}^{(m)})} \int_{\tilde{\Delta}^{(r)}} G(y)\,dy$,

and B is the remainder term that occurred under the replacement of the function G(y) by its mean value $(G)_{\tilde{\Delta}^{(r)}}$. Let us estimate the value of B.

As in the estimation of the value of A, we use here the singularity of the function G(y) and the fact that the length of the interval $\tilde{\Delta}^{(r)}$ is exponentially small (with respect to r).

Further estimates coincide completely with analogous estimates carried out by Sinay (1970) and, therefore, we shall only clarify the basic points of the remaining part of the proof.

Let us consider, for any two r-th rank intervals, $\tilde{\Delta}_1^{(r)}$ and $\tilde{\Delta}_2^{(r)}$, the following expression

$$\tilde{\pi}^{(n)}_{\tilde{\Delta}_1^{(r)},\tilde{\Delta}_2^{(r)}}(\tilde{\Delta}^{(m)}) = \frac{\sum_{\tilde{\Delta}^{(n-r)}\subset\tilde{\Delta}^{m}} \ell(\tilde{\Delta}^{(n-r)})\int_{\tilde{\Delta}_1^{(r)}} D_{\tilde{\Delta}^{(n-r)}}(y)\,dy \int_{\tilde{\Delta}_2^{(r)}} D_{\tilde{\Delta}^{(n)}}\,dy}{\sum_{\tilde{\Delta}^{(n-r)}\subset\tilde{\Delta}^{(m)}} \ell(\tilde{\Delta}^{(n-r)})\int_{\tilde{\Delta}_1^{(r)}} D_{\tilde{\Delta}^{(n-r)}}(y)\,dy} .$$

This expression is to be regarded as an element of the matrix $\Pi^{(m)}(\tilde{\Delta}^{(m)})$. This is a stochastic matrix. Indeed, all its elements are non-negative, and it is evident that

$$\sum_{\tilde{\Delta}_2^{(n)}} \pi^{(n)}_{\tilde{\Delta}_1^{(n)},\tilde{\Delta}_2^{(n)}}(\tilde{\Delta}^{(m)}) = 1 .$$

Now consider the vector $q_n(\tilde{\Delta}^{(m)})$, whose set of components represents the probability distribution over the intervals of rank r:

$$q_n(\tilde{\Delta}^{(m)})|_{\tilde{\Delta}^{(r)}} = \sum_{\tilde{\Delta}^{(n)}\tilde{\Delta}^{(m)}} \ell(\tilde{\Delta}^{(n)})\int_{\tilde{\Delta}^{(r)}} D_{\tilde{\Delta}^{(n)}}(y)\,dy,$$

$$q_n(\tilde{\Delta}^{(m)}) = \Pi^{(n)}(\tilde{\Delta}^{(m)})q_{n-r}(\tilde{\Delta}^{(m)}) = \Pi^{(n)}(\tilde{\Delta}^{(m)})\cdot\Pi^{(n-r)}(\tilde{\Delta}^{(m)})q_{n-2r}(\tilde{\Delta}^{(m)})$$

$$= \dots = \Pi^{(n)}(\tilde{\Delta}^{(m)})\cdot\Pi^{(n-r)}(\tilde{\Delta}^{(m)})\dots\Pi^{(n-kr)}(\tilde{\Delta}^{(m)})q_{n-(k+1)r}(\tilde{\Delta}^{(m)}) .$$

This expression implies that $n-(k+1)r \geq m$.

Let two probability distributions $q' = \{q'_{\tilde{\Delta}}\}$, $q'' = \{q''_{\tilde{\Delta}}\}$ over the intervals $\tilde{\Delta} = \tilde{\Delta}_{i_1 \dots i_r}$ of rank r be given. We shall introduce a metric in the space of all such distributions, putting $d(q',q'') = \sum |q'_{\tilde{\Delta}} - q''_{\tilde{\Delta}}|$. In a similar manner to Bunimovich (1970) it can be shown that all the elements of the matrix $\Pi^{(n)}(\tilde{\Delta}^{(m)})$ are positive. Hence,

$$d(\Pi^{(n)}(\tilde{\Delta}^{(m)})q', \Pi^{(n)}(\tilde{\Delta}^{(m)})q'') \leq (1-\alpha)d(q',q''),$$

where $0<\alpha<1$.

Let $q_0(\tilde{\Delta}^{(m)})$ be the eigenvector of the matrix $\Pi^{(r)}(\Delta^{(m)})$, i. e. a stationary distribution corresponding to a Markovian chain with the transition matrix $\Pi^{(r)}(\tilde{\Delta}^{(m)})$. Now consider the set, 0, of probability distributions, q, for which $d(q,q_0)\leq\lambda^{-r/4}$. As has been shown (Bunimovich, 1970; Sinay, 1970), we obtain at $m<r\leq n$:

$$\left| \frac{\pi^{(n)}_{\Delta'',\Delta'}(\tilde{\Delta}^{(m)})}{\pi^{(r)}_{\Delta'',\Delta'}(\tilde{\Delta}^{(m)})} - 1 \right| \leq c_5 \lambda^{-r/2} \tag{4}$$

From (4) it follows that if r is sufficiently large, then $\Pi^{(n)}(\tilde{\Delta}^{(m)})0=0$ (Bunimovich, 1970; Sinay, 1970). Assuming that $r=m=[n^{2/3}]$ and substituting for all these estimates, we obtain

$$\left|\sum_{\tilde{\Delta}^{(r)}} (f)_{\tilde{\Delta}^{(r)}} q(\tilde{\Delta}^{(r)}|\tilde{\Delta}^{(m)}) - \sum_{\tilde{\Delta}^{(r)}} (f)_{\tilde{\Delta}^{(r)}} q_0(\tilde{\Delta}^{(r)})\right| \leq c_6\lambda^{-r/4}.$$

From the foregoing it also follows that

$$\left|\sum_{\tilde{\Delta}^{(r)}} (f)_{\tilde{\Delta}^{(r)}} q_0(\tilde{\Delta}^{(r)}) - \int_I f(x)\tilde{\rho}(x)dx\right| \leq c_7\lambda^{-r/4}.$$

Finally we obtain

$$\left|\int_I f(T^n x)g(x)\tilde{\rho}(x)dx - \int_I f(x)\tilde{\rho}(x)dx \cdot \int_I g(x)\tilde{\rho}(x)dx\right| \leq$$

$$\leq \mathrm{const}\,\lambda^{-r/4} = c_7 \exp(-c_8 n^{1/3}).$$

Thus we have obtained the required estimate of the rate of correlation decay but now only for the transformation $T=S_a^p$ defined over the interval I. An analogous estimate (of course, with other constants) also proves valid for the transformation S which acts upon A_I. In fact,

$$\int_{A_I} f(S_a^n(x))g(x)\rho(x)dx = \sum_{k=0}^{p} \int_{S_a^k I} f(S_a^n x)g(x)\rho(x)dx.$$

Now over each interval $S_a^k I$ we shall make a change of variables with the aid of the mapping, S_a^{p-k}. As a result, each term of the latter sum takes the form

$$\int_{S_a^k I} f(S_a^n x) g(x) \rho(x)\, dx = \int_I f(S^{n-k-p} y) g(S_a^{k-p} y) \rho(S^{k-p} y)(S^{k-p}(y))' dy.$$

Thus we may consider the corresponding set of intervals $\bar{\Delta}^{(k)}_{i_1 \dots i_k}$, such that if $\bar{\Delta}^{(k)}_{i_1 \dots i_k} \subset S^j I$, $0 < j < p$, then $S_a^{p-j} \bar{\Delta}^{(k)}_{i_1 \dots i_2} = \Delta^{(k)}_{i_1 \dots i_k}$ and make analogous considerations for them. As is easily seen, everything proceeds automatically, since the mapping S_a^{p-j} transforms $S^j I$ onto I in a one-to-one manner and is a mapping of class C^1.

Our statement now proceeds from the fact that $c_9 < |(S_a^{p-k}(y))'| < c_{10}$ at all k and also from the expression of the measure $d\nu = \tilde{\rho}(y) dy$ through $d\mu = \rho(x) dx$.

A better estimate $\rho_{f,g}(n) \leq \text{const} \exp(-\text{const}\, n)$, evidently holds, though at present we are not able to prove it.

From the proof of the theorem it follows that the proof is also valid for the case when f and g are characteristic functions of connected subsets of the interval [0,1] . Therefore with the aid of our estimate of the correlation decay we can estimate the probability of the fact that in an n-th generation the density of population will belong to the subset A, if in the zero generation it belonged to the subset B. To do this it is sufficient to put in (2) $f = \chi_B$, $g = \chi_A$.

Kinetics and Nonequilibrium Processes and Ecology

V. V. Alekseev

A typical example of an irreversible process in ecology is succession in which co-operation of organisms replace one another in a given region. The phenomenon of ecological succession is characterized by the following features: 1) it is a directed process; 2) it is a result of the changes made in the environment by the co-operation itself; 3) succession ends in the formation of a stable climacteric biocenosis. For instance, eutrofication of lakes due to accumulation of precipitates determines a rather well studied succession series. The lake is initially deep, rich in oxygen, its water is transparent but its productivity is low. Gradually, it grows shallow, the concentration of biogens in the water increases, and the lake becomes highly productive. Along with changes in the physico-chemical conditions, the specific composition of biogenocenosis also changes.

To predict the succession process·in a real biotope, one needs to know the specific composition of biogenocenoses surrounding the given biotope, since succession is determined, e. g. in a reservoir, on the one hand, by an increase of biogen concentration and, on the other hand, by the possibility of putting in the reservoir some biological species from a neighbouring ecological system. Knowing the list of species, one can draw a trophic scheme of their interaction and investigate it from the point of view of stability.

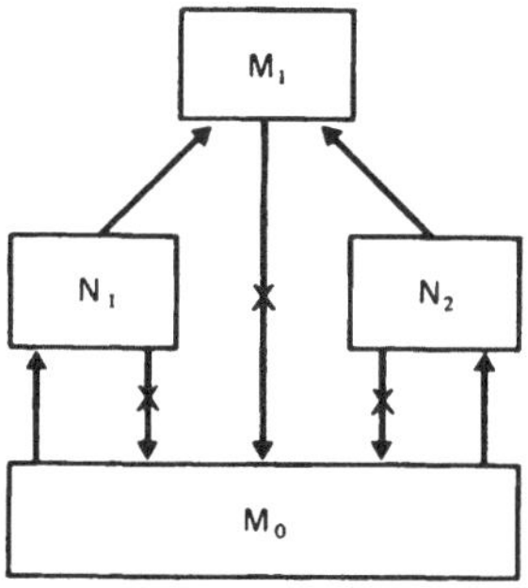

Fig. 22 Trophic scheme of biogenocenosis of three species.

Thermodynamics and Kinetics of Biological Processes

Figure 22 presents a trophic scheme for closed biocenosis which consists of three populations: two autotrophic, N_1 and N_2, and one heterotrophic population M_1 (N_1, N_2, M_2 are the amounts of the limiting biogenic elements in the corresponding population, M_o is the mass of the limiting biogenic element in the environment). As in previous papers (Alekseev, 1976; Alekseev et al., 1978), arrows designate pair interaction between species and crossed arrows represent reactions leading to disappearance, the rates of which are proportional to the amount of the corresponding biomass.

The set of equations describing the dynamics of this process in a given model has the form

$$\frac{dN_1}{dt} = -\varepsilon_1 N_1 - \gamma_1 N_1 M_1 + \beta_1 N_1 M_0$$

$$\frac{dN_2}{dt} = -\varepsilon_2 N_2 - \gamma_2 N_2 M_1 + \beta_2 N_2 M_0 \tag{1}$$

$$\frac{dM_1}{dt} = -\varepsilon M_1 + \tilde{\gamma}_1 M_1 N_1 + \tilde{\gamma}_2 M_1 N_1 .$$

Here M is the total mass of the limiting biogenic element in the system: $M = M_o + N_1 + N_2 + M_1$.

Let us investigate the stability of possible stationary states at different values of M. From the set of equations (1) it is clear that the stationary state, corresponding to the absence of living organisms, will be stable when $\varepsilon_1/\beta_1 > M$, while for the existence of one species the following inequalities should hold

$$\frac{\varepsilon_1}{\beta_1} < M < \frac{\varepsilon_1}{\beta_1} + \frac{\varepsilon}{\tilde{\gamma}_1} .$$

To investigate the stability of the stationary states

$$\bar{N}_1 = \frac{\varepsilon}{\tilde{\gamma}_1}, \quad \bar{M}_1 = \frac{M - \varepsilon_1/\beta_1 - \bar{N}_1}{1 + \gamma_1/\beta_1} \quad \text{and}$$

$$\bar{N}_2 = \frac{\varepsilon}{\tilde{\gamma}_2}, \quad \bar{M}_1' = \frac{M - \varepsilon_2/\beta_2 - \bar{N}_2}{1 + \gamma_2/\beta_2}$$

corresponding to the presence of two species in the system, we rewrite equations (1) in the form

$$\frac{1}{N_1}\frac{dN_1}{dt} = (\gamma_1+\beta_1)(\bar M_1 - M) + \beta_1(\bar N_1 - N) - \beta_1 N_2$$

$$\frac{1}{M_1}\frac{dM_1}{dt} = -\tilde\gamma_1(\bar N_1 - N) + \tilde\gamma_2 N_2 \tag{2}$$

$$\frac{dN_2}{dt} = (M\beta_2 - \varepsilon_2)N_2 - (\gamma_2+\beta_2)N_2 M_1 - \beta_2 N_2^2 - \beta_2 N_1 N_2 .$$

Multiplying the first equation in (2) by $-\tilde\gamma_1(\bar N_1 - N_1)$ and the second one by $-(\gamma_1+\beta_1)(\bar M_1 - M_1)$ and summing them, we obtain

$$\begin{aligned}\tilde\gamma_1\left(1-\frac{\bar N_1}{N_1}\right)\frac{dN_1}{dt} + (\gamma_1+\beta_1)\left(1-\frac{\bar M_1}{M_1}\right)\frac{dM_1}{dt} = \\ = -\beta_1\tilde\gamma_1(\bar N_1 - N_1)^2 + [\beta_1\tilde\gamma_2 N_1 - \tilde\gamma_2(\gamma_1+\beta_1)\bar M_1]N_2 + \\ + (\gamma_1+\beta_1)\tilde\gamma_2 N_2 M_1 - \beta_1\tilde\gamma_1 N_1 N_2 .\end{aligned} \tag{3}$$

Multiplication of equation (3) by $(\gamma_2+\beta_2)$, and the third equation in (2) by $(\gamma_1+\beta_1)\tilde\gamma_2$ and summing, yields

$$\begin{aligned}&\frac{d}{dt}[\tilde\gamma_1(\gamma_2+\beta_2)(N_1-\bar N_1 \ln N_1) + (\gamma_1+\beta_1)(\gamma_2+\beta_2)(M_1-\bar M_1 \ln M_1) + \\ &+ (\gamma_1+\beta_1)\tilde\gamma_2 N_2] = -\beta_1(\gamma_2+\beta_2)\tilde\gamma_1(\bar N_1 - N_1)^2 - \beta_2\tilde\gamma_2(\gamma_1+\beta_1)N_2^2 - \\ &-[\beta_1\tilde\gamma_1(\gamma_2+\beta_2)+\beta_2\tilde\gamma_2(\gamma_1+\beta_1)]N_1N_2 + [\tilde\gamma_2(\gamma_1+\beta_1)(M\beta_2-\varepsilon_2) + \\ &+ \beta_1\tilde\gamma_1(\gamma_2+\beta_2)\bar N_1 - \gamma_2(\gamma_1+\beta_1)(\gamma_2+\beta_2)\bar M_1]N_2 = \\ &= -\beta_1(\gamma_2+\beta_2)\tilde\gamma_1(\bar N_1 - N)^2 - \beta_2\tilde\gamma_2(\gamma_1+\beta_1)N_2^2 - \\ &-[\beta_1\tilde\gamma_1(\gamma_2+\beta_2)+\beta_2\tilde\gamma_2(\gamma_1+\beta_1)](N_1-\bar N_1)N_2 + \\ &+ \tilde\gamma_2\beta_1\beta_2\left(\frac{\gamma_1}{\beta_1}-\frac{\gamma_2}{\beta_2}\right)\left(M-\alpha-\frac{\varepsilon}{\tilde\gamma_1}\right)N_2 ,\end{aligned} \tag{4}$$

where

$$\alpha = \frac{\frac{\varepsilon_2}{\beta_2}-\frac{\varepsilon_1}{\beta_1}}{\frac{\gamma_1}{\beta_1}-\frac{\gamma_2}{\beta_2}} + \frac{\frac{\varepsilon_2}{\beta_2}\cdot\frac{\gamma_1}{\beta_1}-\frac{\varepsilon_1}{\beta_1}\cdot\frac{\gamma_2}{\beta_2}}{\frac{\gamma_1}{\beta_1}-\frac{\gamma_2}{\beta_2}} =$$

$$= \frac{\varepsilon_1}{\beta_1} + \left(1 + \frac{\gamma_1}{\beta_1}\right) \frac{\varepsilon_2/\beta_2 - \varepsilon_1/\beta_1}{\gamma_1/\beta_1 - \gamma_2/\beta_2}.$$

Within time derivative there is an expression which extends the quantity

$$G = \Sigma_1 \left(\frac{N_1}{\bar{N}_1} - \ln \frac{N_1}{\bar{N}_1}\right) + \Sigma_2 \left(\frac{M_1}{\bar{M}_1} - \ln \frac{M_1}{\bar{M}_1}\right)$$

which is used, as a rule, in the analysis of biological co-operations (Goel et al., 1971), to the case of zero stationary values. This expression, like G, has one minimum at $N = \bar{N}_1, M_1 = \bar{M}_1, N_2 = 0$. Near the stationary state, one may disregard small quantities of the second order, and the stability will take the form

$$\left(\frac{\gamma_1}{\beta_1} - \frac{\gamma_2}{\beta_2}\right)\left(M - \alpha - \frac{\varepsilon}{\tilde{\gamma}_1}\right) < 0. \qquad (5)$$

When $\gamma_1/\beta_1 < \gamma_2/\beta_2$, this condition turns out to be equivalent to $M > \varepsilon_1/\beta_1 + \varepsilon/\tilde{\gamma}_1$, since $\alpha < \varepsilon_1/\beta_1$. But when $\gamma_1/\beta_1 > \gamma_2/\beta_2$, the condition (5) implies $M < \alpha + \varepsilon/\tilde{\gamma}_1$

The stability condition of the stationary state $N_2 = \bar{N}_2$, $M_1 = \bar{M}_1'$, $N_1 = 0$ is found from the symmetry of N_1 and N_2, i. e. from (5) by permutation of indices 1 and 2

$$\left(\frac{\gamma_2}{\beta_2} - \frac{\gamma_1}{\beta_1}\right)\left(M - \alpha - \frac{\varepsilon}{\tilde{\gamma}_2}\right) < 0.$$

When $\gamma_1/\beta_1 < \gamma_2/\beta_2$, then $M < \alpha + \varepsilon/\tilde{\gamma}_2 < \varepsilon_1/\beta_1 + \varepsilon/\tilde{\gamma}_2 < \varepsilon_2/\beta_2 + \varepsilon/\tilde{\gamma}_2$ which contradicts the requirement that $\bar{M}_1'$ be positive. For $\gamma_1/\beta_1 > \gamma_2/\beta_2$ the stability condition of the stationary state $N_2 = \bar{N}_2$, $M_1 = \bar{M}_1'$, $N_1 = 0$ has the form $M > \alpha + \frac{\varepsilon}{\tilde{\gamma}_2}$.

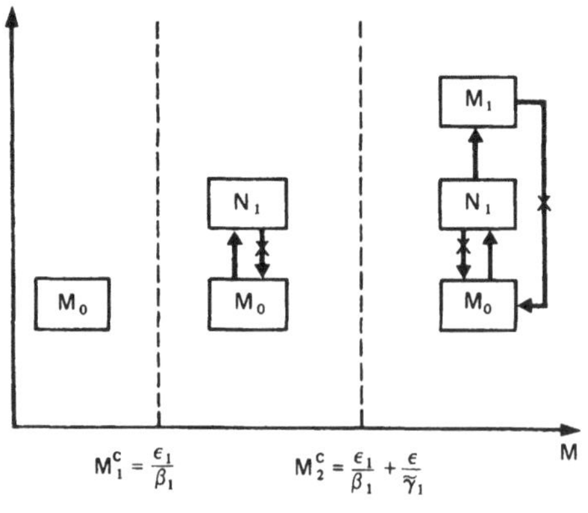

Fig. 23 Example of a simple succession row, when with increased concentration of limiting biogene special diversity monotonously increases.

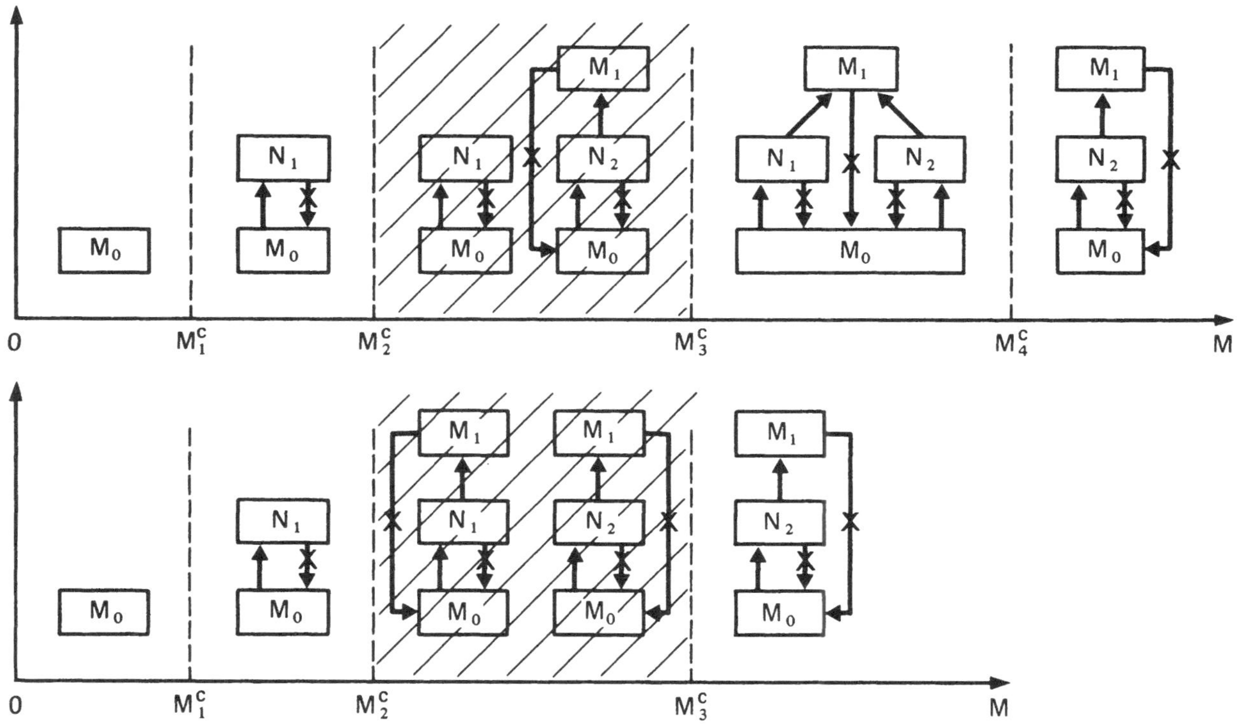

Fig. 24 Examples of complex succession rows.

Figures 23 and 24 represent the succession series obtained from the stability conditions of stationary states. The dashed areas correspond to triggers.

We see that even in this very simple case there is a variety of succession series. It is clear, that there is no direct connection between stability and complexity. Real systems are immeasurably more complicated than the one considered above, but one can still find a number of the important features of successions.

First of all, there is an inambiguity of succession; their dependence on the initial conditions is readily seen. This fact is of practical importance for the study of processes of species introduction. The qualitative character of a succession series is determined by the relations of the ecological coefficients, which gives rise to the necessity to investigate, experimentally, and to classify succession series.

In conclusion we should note that the stability condition following from (14) has a wider application than the condition of local stability since it allows us to judge the global behaviour of the trajectories. For example, at

$$\left(\frac{\gamma_1}{\beta_1} - \frac{\gamma_2}{\beta_2}\right)\left(M - \alpha - \frac{\varepsilon}{\tilde{\gamma}_2}\right) < 0 \;, \quad \frac{\gamma_2}{\beta_2} > \frac{\gamma_1}{\beta_1}$$

we arrive at a global stability condition.

The stationary state corresponding to the presence of three species in the system is curious. The species N_1 and N_2 belong in this case to one and the same ecological group but do not exclude each other.

Mathematical Modelling of Biocenosis on the Basis of Haldane-Semevsky Principle

G. E. Insarov, S. M. Semenov

The general approach to mathematical modelling in biology and particularly to modelling of biocenosis, has always been to use extremal principles. In mathematical modelling of biosystems of different levels of organization, be it a cell, an organism, a population, or biocenosis, researchers have often tended to construct some functional - an optimal criterion, which in the course of natural functioning of the modelled system reaches (or tends to) the extreme value. These methodological tendencies, apparently, have roots in the fundamental tranditions of natural sciences. It is sufficient to note that a very important role has played in physics by the minimum action, or the entropy increase princple. So, the use of extremal principles in mathematical analysis technique of biological problems has always appeared to be very wideranging and attractive to research workers.

At present there are, from these optimal principles many different approaches to ecological problems. Many of them can be found in monographs by Rosen (1967) and Khanin et al. (1978). These reviews, therefore, enable us to discuss only the basic directions of this area of research and not a detailed account of all available procedures.

Classical biology recognises only one quantity which claims the role of *the optimal criterion*. This quantity is the fitness of the individual. It was Darwin who postulated an increase of fitness of individuals over the course of time. The problem of a qualitative measure of fitness remained unclear for a long time. Evidently, beginning with Haldane (1956) one can see in literature an obvious tendency to measuring fitness of individuals by a reproduction coefficient (Cohen, 1966; Levontin, Cohen, 1969; Semevskaya, Semevsky, 1977). In this approach the state, to which the system of inter-

Thermodynamics and Kinetics of Biological Processes

acting species (biocenosis) tends, is characterized by a maximal reproduction coefficient for the individuals of each of the species represented in this state. If the process of variation of numbers in time is not constant, then in the latter formulation one should replace the reproduction coefficient by its geometrical mean with respect to time. We should note at once that the criterion introduced characterizes the individual through interaction with it's environment. This includes not only factors external with respect to a given biocenosis, but also with internal ones, such as predators, diseases, parasites, etc. The principle formulated above is called the Haldane-Semevsky principle (Insarov, Semenov, 1977; Semevskaya, Semevsky, 1977). It should be noted that within this approach, the striving of biocenosis for equilibrium, climax, is by no means the property of biocenosis as a whole. This motion is a consequence of co-evolution and co-adaptation of species mediated through a natural selection which holds at the level of individuals.

A number of scientists have formulated the principle of optimalisation in an analogous form, not for individuals, but for wider subsystems of biocenosis, i. e. populations. A classical effort in this direction was the work of Wynne-Edwards (1962), whithin which a relative rate of population increase is chosen as the criterion for optimalisation. An analogous point of view was expressed by Khanin et al. (1978), who did not only postulate optimalisation (maximalisation) of the population reproduction rate in a stationary state of biocenosis, but also extended these considerations to the transition process of biocenosis evolution to this state. When reasoning this approach theoretically one assumes, either explicitly or implicitly, that the unit of selection is the population as a whole, i. e. natural selection proceeds towards an increase in the (collective) fitness of the population.

Many researchers have also referred to the notion of optimalisation in biological systems of higher organization, in particular, to biocenosis as a whole. This point of view was logically concluded by Margalef (1968). He formulated a statement which proposed that biocenosis evolves towards an increase of relative photosynthetic production per unit time, per unit volume. Recently Matsuno has made an attempt to provide a theoretical basis for Margalef's principle within the framework of the theory of dissipative systems (Matsuno, 1978).

The complexity of natural biocenoses, the high numbers of populations in them, and also a comparatively high degree of knowledge of the processes of matter and energy transformation in biological systems gave rise to hopes for a successful application of thermodynamic descriptions to evolution in biocenosis. In this approach, analogies with chemical processes of matter transformation and accompanying energetic processes are widely used. As an example, we note the paper by Conrad (1973), in which the extremal thermodynamic principle associated with the behaviour of the dissipative function, is used for the description of evolution in biocenosis.

We also note the radical difference in Haldane-Semevsky and Wynne-Edwards approaches from Margalef's and from thermodynamic approaches to the description of evolution in biocenosis. In the first two cases evolution is understood as variation of the structural or functional characteristics of individuals or populations on the strength of the natural selection process. This operates, at the level of individuals or populations, which proceeds as the evolution of interacting species. The other approaches imply the attempt in biocenosis, taken as a whole, to achieve some equilibrium under the action of thermodynamic or some other forces.

The variety of approaches to the modelling of biocenosis from an optimalisation standpoint shows how attractive this method has been for the researcher. The progress made also shows how fruitful and wideranging it has been. Further progress in this field is, in our opinion, impossible without some specification of the initial methodological position and more rigorous verification of the preconditions of the theory as to adequacy to reality. Any hypothesis of optimalisation in biological systems must be confirmed by some natural selection model. We believe that an optimal structure of a biological self-reproducing system may be realized and be stable in time provided, only, that it can be fixed in heredity. Only in this situation will biosystems of optimal structure be realized in the course of evolution, while non-optimal variants will be eliminated. It is commonly accepted that the mechanism of inheritance acts at the level of the individual. It is just the genotype of the individual that possesses a certain stability in reproduction. Neither population nor biocenosis possess such a mechanism of structure reproduction. Neither population nor biocenosis are reproduced as a whole, there is no structure that can be formulated to fix the required information.

By virtue of what has been said we believe that the notion of optimalisation in biosystems cannot be extended to units larger than the individual. In particular, these concepts cannot be extended to populations (Wynne-Edwards) or to biocenosis (Margalef) without especially strict mathematical restrictions.

In our opinion, the most adequate form of the optimalisation principle is that of Haldane-Semevsky. One may construct strict theoretical grounds for this principle on the basis of the widely accepted Darwinian principle of natural selection; survival of the fittest against a background of non-directed variability. The equilibrium state is characterized by the maximal fitness of individuals represented in the biocenosis of the species. It should be particularly emphasized that by its nature this optimalisation is only realized after biosynthesis has reached a final equilibrium climax. Thus, the notion of optimalisation cannot be extended to a transition process.

In the present paper we specify the mathematical meaning of the concepts used in the formulation of the Haldane-Semevsky principle (in particular, the notion of the reproduction coefficient), present the theoretical grounds of this principle, and consider its application to succession modelling.

Reproduction Coefficients of Individuals

One of the main peculiarities of the living is the capacity for reproduction. Using the resources of the environment, the individual grows and reproduces i. e. produces individuals of the same biological species. The most important quantitative characteristic of reproduction is the individual reproduction coefficient. An individual reproduction coefficient, in a particular case, is measured for an asexual reproduction by the number of offspring who have reached the same stage of individual development, and, in the case of sexual reproduction, by a half of this number (for clarity we only consider the case of a population with a 1:1 ratio of sexes, the individuals being equal). Of course, this quantity may be different for individuals within the population. For example, an individual may die before beginning reproduction, and, in this case, the reproduction coefficient is equal to zero. The survival of another individual may appear in exclusively favourable conditions and reach the corresponding stage of individual development without loss. In this case the reproduction coefficient is equal to the overall number of offspring (i. e. fruitfulness of the individual) for an asexual reproduction and to half of this amount for a sexual reproduction. Fertility will be designated by q. In the general case the corresponding stage of individual development will be reached only by some of these individuals, which is usually called survival rate. This quantity will be designated as f, $0 \leq f \leq 1$. Thus, the individual reproduction coefficient may be calculated by the formula $k_n = qf$.

Natural populations usually consist of a rather large number of individuals, which makes the use of the individual reproduction coefficient inconvenient (and unnecessary) for the characteristic of reproduction of the population as a whole. To characterize this process one usually takes a statistical quantity - the mean reproduction coefficient k. If random reproduction processes are similar for all the individuals in a population and do not correlate, and if the development of individuals proceeds synchronously, and different generations do not overlap in time (i. e. by the time of occurence of the next generation the individuals of the previous generation have died), the quantity k coincides with the mean expectation value of the individual reproduction coefficient k_n.

From the definition of the reproduction coefficient it follows directly that, with the assumptions described above the numbers of individuals in the population, measured in two successive generations at identical stages of individual development, obey the relation

$$y(t + t_g) = k(t)\, y(t) \qquad (1)$$

where, t, is the time of enumeration of the previous generation, $t + t_g$ is the time of enumeration of the succeeding generation and $k(t)$ is the reproduction coefficient.

The quantity, t_g, characterizing the time interval between the times required to reach the same stage t of individual development in two successive generations, is usually t-independent, or varies only slightly. Quantity, t_g, is referred to as the lifetime of the generation.

Relation (1) is the basic to the theory of the dynamics of variation in the numbers of biological populations for the case of a simple age structure. If one chooses a different time scale, namely, the lifetime of the generation as the unit of measurement, then relation (1) takes the form

$$y(t+1) = k(t)\,y(t) \tag{2}$$

What is the range of variation of the density, y? For real, stable populations the range of numerical variation in time should always be restricted from below, and separated from zero from above.

The restriction from above is due to the limited capacity of the environment, namely, to a limited number of dwelling-places, for age reserves etc. From below the numbers of natural populations must always be restricted to unity, otherwise the population would die out. From the foregoing it follows that

$$\lim_{T\to\infty} \left(\frac{y(t_0+T)}{y(t_0)} \right)^{1/T} = 1\,,$$

and, consequently, that,

$$\tilde{k} = \lim_{T\to\infty} \left(k(t_0)\ldots k(t_0+T-1) \right)^{1/T} = 1\,.$$

The latter relation reflects the main property of the reproduction coefficient - its geometric mean with respect for a stable population is equal to unity.

The Quantitative Measure of Fitness and Substantiation of the Optimalisation Principle

Fitness of the individual is not a once and for all quantity determined only by the property of the individual. Fitness reflects the process of interaction of the individual with the environment as a whole, and, for one individual it is, generally speaking, different for different states of the environment.

The most convenient and universal measure of fitness is the value of the reproduction coefficient.

Let the state on an abiotic medium be characterized by the vector c, whose components are physical, chemical, hydrometeorological etc. characteristics of the medium. Then, let the state of a biotic medium be characterized by the presence of N populations of individuals, carriers of vector ecological parameters $a_1, \dots a_N$, the numbers (densities) of these populations being $y_1, \dots, y_N$. The components of the vector ecological parameters a_i may characterize properties of individuals such as fertility, resistence to the factors of mortality, agressiveness etc. We arrange the ecological parameters (chosen in a certain way) for individuals of various biological species in a special set A (such sets - the results of reflection of all the variety of the existing species in the space of the parameters - are called Matthew-Kermack sets) and introduce the quantity

$$k(a; y_1, \dots, y_N; a_1, \dots, a_N; c),$$

which represents the reproduction coefficient of an individual-carrier of the ecological parameter, a, under the condition that the state of abiotic medium is c and the individual-carriers of the parameters $a_1, \dots, a_N$, are present in the medium, their numbers of population being $y_1, \dots, y_N$, respectively. The quantity introduced is a quantitative measure of the adaptation of the individual-carrier of the parameter, a, to a given state of the environment, characterized by the quantities $a_1, \dots a_N, y_1, \dots, y_{N,}c$.

Suppose that the biocenosis being modelled is in a state of stable equilibrium, i. e. there exist certain stationary levels of numbers (density), $y_1^*, \dots, y_N^*$, of the constituent species-carriers of the parameters $a_1, \dots, a_N$. In addition, let this biocenosis be in a climacteric state. This means that during the process of formation of biocenosis specific composition has come to an end, i. e. the succession is over. The formation of climacteric biocenosis, and the process of succession may be imagined if we perform the following mental experiment. We place in a given environment several individuals of each species existing in nature. There, starts the process of growth and reproduction of organisms, interaction among individuals etc. Some of the species introduced die, others survive. In the course of time the whole system reaches equilibrium - the constituent species do not die out, their mean reproduction coefficient is equal to unity. Thus, a climacteric biocenosis has been formed, each species of which is optimally adapted to its environment, i. e. to abiotic and biotic (the rest of the species) constituents of the ecological system in question. It is clear that if we introduce into this biocenosis some individuals of a certain species from the initial manifold which did not survive in the selection, their offspring will die out since these species are worse adapted to the surrounding conditions than are each of those survived.

In real biocenoses selection does not, of course, embrace all the contemporary variety of species. In this process only those forms participate which can for this or that reason penetrate inside a given ecological system. What conditions are imposed on the function K, the index of fitness, by the assumption of the absence of succession in a given biocenosis? It can be formally shown that for a climacteric biocenosis the fitness function

$$\varphi(a) = k(a; y_1^*, \ldots, y_N^*; a_1, \ldots, a_N; c)$$

at all $a \in A$ does not exceed unity, and at $a = a_1, \ldots a_N$ is equal to unity.

Thus, the ecological parameters of individuals of the biological species, constituting a climacteric biocenosis, are the maxima of the fitness function in the Matthew-Kermack set.

The latter statement is just the Haldane-Semevsky principle. In the case when biocenosis is not in a state of stable equilibrium, but, on the contrary, some oscillation process in time of variation in the numbers of constituent populations takes place, then the formulation of the optimalisation principle remains unchanged, except for the function φ which now is to be understood as the geometric mean with respect to time of the reproduction coefficient, k.

Modelling of Exogeneous Succession

At each state c of the abiotic environment the succession process ends with transformation of a biocenosis into a climacteric biocenosis. This climacteric state of biocenosis is determined by the state of the abiotic environment c. If we replace the state of the abiotic environment c by c', then the former climacteric state of biocenosis will change. There then arises a transition process in moving to a new climacteric state corresponding to c'. This process is an exogenous succession - a succession in response to a change of the exogenous parameter c.

The Haldane-Semevsky principle formulated above makes it possible to predict the result of this succession (a climacteric state of the biocenosis) within the framework of the following scheme. In models, the set A is usually represented as an open submanifold of points in multidimensional space, $R^l, a = (a^1, \ldots, a^l)$. Proceeding from this, the ecological parameters $a_1, \ldots, a_N$ and the equilibrium numbers $y_1^*, \ldots, a_N^*$ of the species constituting a stable climacteric biocenosis in the case when the state of abiotic environment is c, satisfy the set of equations

$$\begin{aligned} &k(a_i; y_1, \ldots, y_N; a_1, \ldots, a_N; c) = 1 \qquad i = 1, \ldots, N \\ &k(a; y_1, \ldots, y_N; a_1, \ldots, a_N; c)\Big|_{a=a_i} = \max \qquad i = 1, \ldots, N \end{aligned} \tag{3}$$

If $a_1, \ldots, a_N$ are internal points of the set A, the latter set of equations may be replaced by the following set of equations

$$\begin{aligned} &k(a_i; y_1, \ldots, y_N; a_1, \ldots, a_N; c) = 1 \qquad i = 1, \ldots, N \\ &\frac{\partial}{\partial a^j} k(a; y_1, \ldots, y_N; a_1, \ldots, a_N; c)\Big|_{a=a_i} = 0 \qquad \begin{matrix} i = 1, \ldots, N \\ j = 1, \ldots, l \, . \end{matrix} \end{aligned} \tag{4}$$

This is a set of N(l + 1) equations with respect to N(l + 1) variables $y_1, \ldots, y_N; a_1^1, \ldots, a_1^l; a_2^1, \ldots, a_2^l; \ldots, a_N^1, \ldots a_N^l$. The solutions of this set of equations depend on the exogenous parameter c, the state of the abiotic environment. Thus, the model suggested (4) enables us to find changes in the ecological parameters of individuals of the populations constituing a climacteric biocenosis and changes in stationary numbers (densities) of these populations occurring as a response to the changes in the abiotic environment.

In conclusion we should note that the problem of the description of all possible stable climacteric biocenoses, in the assumption that the state c of abiotic environment is constant, is reduced to the solution of the set of equations (3) at different N, and, to a subsequent analysis of the stability of the levels of the numbers $y_1^*, \ldots, y_N^*$. This analysis can be performed by means of the usual methods of stability theory.

Natural Selection and Optimality Principles

M. A. Khanin, N. L. Dorfman

The problem of optimizing the cost functions in biology has been considered by many investigators (Hess, 1914; Lotka, 1925; Murray, 1926; Cohn, 1955; Horsfield and Cumming, 1967; Rosen, 1967; Khanin and Bukharov, 1970; Rashevsky, 1973; Khanin and Dorfman, 1973; Cody, 1974; Kamiya, 1974; Yamakoshi et al., 1976; Mylings, 1977; Dorfman and Ovsyannikov, 1978). Rashevsky (1973) has put forward the principle of adequate design. According to this principle, "the design of an organism is such that the organism performs its necessary functions adequately and with a minimum expenditure of energy and material both in the performance of the functions and in the construction of the organism". A comprehensive survey of the above problem has been made by Rosen (1967). The principle of adequate design is regarded as a hypothesis. The validity of this hypothesis can be tested by comparing the calculated structural and functional parameters with the actual ones. The cost function are formulated on the basis of analysis of the biological, physical and chemical processes involved. A heuristic approach to the problem is also possible. A major difficulty in applying optimalisation principles in biology is the formulation of the cost function. In most of the above works this difficulty has been successfully overcome. The cost functions determining the structural and functional parameters have been formulated and good agreement between theory and experiment has been established. The optimalisation criteria used are listed in Table 1.

Table 1, although incomplete, illustrates a variety of optimalisation criteria applied in biology. Their applicability is determined by natural selection. A formalized description of natural selection and a formulation of special optimalisation criteria based on it are therefore of interest. The field of application for these criteria must also be determined. A formalization of the principle of persistence in ecological considerations will be dealt with in

Thermodynamics and Kinetics of Biological Processes

Table 1

Optimalisation criterion	Parameter	Reference
Minimum vessel resistance	Angle of branching	Murray, 1926
Minimum energy expenditure in structuration and maintenance	Tree structures	Yamakoshi et al., 1976
Minimum power loss in dissipation and maintenance	Cardiac output, erythrocyte count, blood oxygen tension	Khanin, Bukharov, 1970
Maximum relative population growth rate	Growth curve	Khanin, Dorfman, 1973; Leon, 1976
Maximum steady state population size	Reproduction power	Insarov, 1975
General optimalisation principle	Birth weight in mammals	Dorfman, Ovsyannikov, 1978

the following section. By such a formalization we mean a phenomenological approach to determining the final state reached in the course of natural selection. The genetic mechanisms of variability and selection and hence their kinetics are not considered.

Formalization of the Principle of Persistence

The problem is to find structural and functional parameters for a species in an ecosystem under given ecological conditions. It is assumed that the system of equations describing the ecosystem dynamics can be obtained. In the general case, the system of equations is as follows:

$$\dot{x}_{ik} = L_{ik}(X, A) \quad (1)$$

where x_{ik} are the dynamic variables characterizing the state of the ecosystem; A is a set of structural and functional parameters for the species in the ecosystem. The latter vary with time much more slowly than the former. x_{ik} is the ith dynamic variable characterizing the state of the kth population. α_{sr} is the s th structural and functional parameter of the r th population. The influence of structural and functional parameters are taken into account in equation (1). If N is the number of elements in the ecosystem and S the number of dynamic variables characterizing the state of the kth element (kth population then (1) includes $\sum_{k=1}^{N} S_k$ equations. The dynamic variables x_{ik} can be divided into two groups: extensive and i intensive. The extensive dynamic variables are the population size, some age group size etc. The intensive dynamic variables characterize the properties of the species (mean body weight, body composition etc.). Most of the equations in (1) are balance equations. Some of these represent the laws of energy and mass conservation. In the general case, the mutations, interbreeding and genetic mechanisms can be taken into account in (1). Unfortunately, our knowledge of these mechanisms is still scanty. A complete description of evolutional dynamics is therefore impossible at present. A quantitative description of ecosystem dynamics, however, is possible. The system of equations describing ecosystem dynamics permits us to determine, at least, some structural and functional parameters.

To solve this problem we represent the system of equations (1) in another form. Let us consider the simpler problem of finding one evolutional optimal parameter value α_{jm}, other parameter values α_{sr} being fixed ($s \neq j$, $r \neq m$). The mth population can be divided into two sub-populations differing only in the parameter value α_{jm}. Consequently in (1) each equation related to the m th population must be replaced by two analogous equations related to sub-populations with certain parameter values $\alpha_{jm}^{(1)}$ and $\alpha_{jm}^{(2)}$

$$\begin{aligned} \dot{x}_{im}^{(1)} &= L_{im}^{(1)}(\bar{X}, x_{im}^{(1)}, x_{im}^{(2)}, \bar{A}, \alpha_{jm}^{(1)}, \alpha_{jm}^{(2)}) \\ \dot{x}_{im}^{(2)} &= L_{im}^{(2)}(\bar{X}, x_{im}^{(1)}, x_{im}^{(2)}, \bar{A}, \alpha_{jm}^{(1)}, \alpha_{jm}^{(2)}) \end{aligned} \tag{2}$$

where $\bar{X}$ is a set of all the dynamic variables but for the m th population; $x_{im}^{(1)}$ and $x_{im}^{(2)}$, a set of dynamic variables characterizing the 1st and 2nd sub-populations of the m th population, respectively; $\bar{A}$ is a set of all the parameter values but for α_{jm}; $\alpha_{jm}^{(1)}$ and $\alpha_{jm}^{(2)}$, the parameter values characterizing the 1st and 2nd sub-populations of the m th population, respectively.

Thus, the system of equations (1) takes the form

$$\dot{x}_{11} = L_{11}(\bar{X}, x_{im}^{(1)}, x_{im}^{(2)}, \bar{A}, \alpha_{jm}^{(1)}, \alpha_{jm}^{(2)})$$

$$\cdots\cdots\cdots\cdots\cdots\cdots\cdots\cdots$$

$$\begin{aligned} \dot{x}_{im}^{(1)} &= L_{im}^{(1)}(\bar{X}, x_{im}^{(1)}, x_{im}^{(2)}, \bar{A}, \alpha_{jm}^{(1)}, \alpha_{jm}^{(2)}) \\ \dot{x}_{im}^{(2)} &= L_{im}^{(2)}(\bar{X}, x_{im}^{(1)}, x_{im}^{(2)}, \bar{A}, \alpha_{jm}^{(1)}, \alpha_{jm}^{(2)}) \end{aligned} \tag{3}$$

$$\cdots\cdots\cdots\cdots\cdots\cdots\cdots\cdots$$

$$\dot{x}_{nN} = L_{nN}(\bar{X}, x_{im}^{(1)}, x_{im}^{(2)}, \bar{A}, \alpha_{jm}^{(1)}, \alpha_{jm}^{(2)}) .$$

If

$$\alpha_{jm}^{(1)} \rightarrow \alpha_{jm}^{(2)} = \alpha_{jm}$$

the system of equations (3) takes the previous form (1)

$$L_{im}^{(1)} + L_{im}^{(2)} \rightarrow L_{im}$$

$$x_{im}^{(1)} + x_{im}^{(2)} \longrightarrow x_{im} .$$

The solution of the system of equations (3) can be used to describe, quantitatively, ecosystem dynamics and, in addition, the struggle for existence between the two sub-populations. Thus, the problem is to analyze the asymptotic properties of the system of equations (3). This problem is considered below. The type of dynamic variables (extensive or intensive) is important in interpreting the asymptotic properties of the system of equations (3). The present paper is concerned with extensive dynamic variables. Let us consider possible variants.

1.
$$\begin{aligned} \lim_{t\to\infty} x_{pm}^{(1)} &= 0 \\ \lim_{t\to\infty} x_{pm}^{(2)} &= 0 \end{aligned} \tag{4}$$

where the subscript, p, denotes some extensive dynamic variable. As the extensive dynamic variable we can choose the population size or density. The 1st variant is of no interest. This indicates the elimination of the m th species

2.
$$\begin{aligned} \lim_{t\to\infty} x_{pm}^{(1)} &> 0 \\ \lim_{t\to\infty} x_{pm}^{(2)} &> 0 . \end{aligned} \tag{5}$$

This indicates the coexistence of two sub-populations differing in the parameter value α_{jm}. In other words, there exists polymorphism of the character specified by the structural or functional parameter α_{jm}. There arise two possibilities: 1) true polymorphism, i. e., a coexistence of two sub-populations with different parameter values is advantageous from the evolutionary viewpoint. 2) the chosen parameter values are equivalent in persistence. Nevertheless, in this respect there exist better parameter values. In case 1, if the parameter values are significantly different a subspeciation may appear.

3. $$\lim_{t\to\infty} x_{pm}^{(1)} > 0, \quad \lim_{t\to\infty} x_{pm}^{(2)} = 0 \tag{6 a}$$

This indicates the persistence of the individuals with the parameter value $\alpha_{jm} = \alpha_{jm}^{(1)}$. The individuals with the parameter value $\alpha_{jm}^{(2)}$ will be eliminated in the course of evolution. Thus, the parameter value $\alpha_{jm}^{(1)}$ is advantageous to $\alpha_{jm}^{(2)}$ from the evolutionary point of view.

4. $$\lim_{t\to\infty} x_{pm}^{(1)} = 0, \quad \lim_{t\to\infty} x_{pm}^{(2)} > 0 \tag{6 b}$$

The 4th variant is essentially the same as the 3rd.

Let us now define the evolutionary optimal parameter value. According to the law of natural selection, in the course of evolution certain values of structural and functional parameters will be established (individual distributions by parameters). These parameter values will be called the evolutionary optimal ones. The sub-population with the evolutionary optimal parameter value persist in the struggle for existence while a sub-population with any parameter value non-equal to the optimal value will be eliminated. The problem of finding the optimal parameter value is then to analyse the asymptotic properties of the system of equations (3) for various $\alpha_{jm}^{(1)}$ and $\alpha_{jm}^{(2)}$. *By definition,* α_{jm}^{*} *is the evolutionary optimal value of the parameter* α_{jm} *if there exists a value* $\alpha_{jm}^{(1)} = \alpha_{jm}^{*}$ *such that the conditions*

$$\lim_{t\to\infty} x_{pm}^{(1)} > 0, \quad \lim_{t\to\infty} x_{pm}^{(2)} = 0. \tag{7}$$

are fulfilled for any $\alpha_{jm}^{(2)} \neq \alpha_{jm}^{*}$.

Definition (7) can be regarded as an optimalisation criterion formalizing the principle of persistence. It should be noted that we are concerned with persistence rather than fitness optimization. By analyzing the asymptotic properties of the system of equations (3) and using definition (7) the optimal parameter values can be found. A direct method of determining the optimal parameter values is to solve, numerically, (3) and to use (7). This method, however, is very time-consuming. If the system of equations (3) is deterministic the calculated optimal parameter values are not affected by the initial conditons. If (3) is stochastic some restraints must be placed on the initial conditions. When the initial sub-population size is small enough its elimination is not necessarily due to the fact that the parameter value in question is non-optimal. As noted above, only one parameter value is optimized, other being fixed. In fact, the optimal parameter values are interrelated. The optimalisation criterion defined in parameter space is, in principle, the same as (7). In this case the procedure of finding the optimal parameter values, however, is quite laborious.

The Principle of Persistence and Special Optimalisation Criteria in Biology

The general optimalisation criterion (7) formalizing the principle of persistence can be applied directly to determine the optimal biological parameters. Namely, we can now proceed 1) to solve the system of equations (3) for vaious parameter values, 2) to analyse the asymptotic properties of these solutions, 3) to determine the evolutionary optimal parameter values by the search method. The method of determining the optimal parameter values without solving the system of equations (3) is more effective. The method used is based on the special optimalisation criteria following from (7). The field of application for these criteria must also be determined. Before formulating the special optimalisation criteria we shall discuss the possibility of deriving the system of equations (1) taking into account the influence of various parameters. This problem is very difficult and far from being solved. Based on the ideas of Lotka (1925) and Volterra (1931) population mathematical models permit us to elucidate the laws governing population dynamics. These models, however, cannot be used to determine the optimal biological parameters since in such models the influence of various parameters is not taken into account. Fortunately, to formulate the special optimalisation criteria we only need 1) to represent (3) in a simplified form and 2) to establish the main limiting factors affecting the population dynamics. A more complete description of the ecosystem properties is fiven at the next step when formulating the cost functional of the problem. As a simple mathematical model describing the ecosystem dynamics we consider the model of an isolated homogeneous population. Let the extensive parameters be the population size x and the intensive parameter be the mean body weight p . In this case the system of euqations (3) takes the form

$$\dot{x} = (\lambda - \mu)x = f(x, p, \alpha)x \tag{8}$$

$$\dot{p} = \Delta W / c_{b} = h(x, p, \alpha), \tag{9}$$

where λ and μ are the birth- and death-rate, respectively; ΔW is the energy inbalance (the difference between the assimilated and expended power); α is the structural or functional parameter; c_{b} is the body specific caloricit caloricity. Equation (8) is the balance equation for the population size, (9) represents the law of energy conservation. Therefore, in such a model energy consumption is the principal limiting factor. There may be situations with other limiting factors (chemical elements and compounds etc.). In equations (8) and (9) environmental conditions are assumed to be accounted for.

We now formulate the special optimalisation criteria. According to the general optimalisation criterion (7), the difference in parameter values between two sub-populations must be accounted for. Then, equations (8) and (9) take the form

$$\begin{aligned} \dot{x}_1 &= f(x_1 + x_2, p_1, \alpha_1)x_1 & \dot{x}_2 &= f(x_1 + x_2, p_2, \alpha_2)x_2 \\ \dot{p}_1 &= h(x_1 + x_2, p_1, \alpha_1) & \dot{p}_2 &= h(x_1 + x_2, p_2, \alpha_2) \end{aligned} \tag{10}$$

where the subscripts 1 and 2 denote the groups differing in the parameter value α (α_1 and α_2). Assuming that the parameter values are not significantly different we have:

$$\alpha_1 = \alpha, \quad \alpha_2 = \alpha + \varepsilon, \quad \varepsilon \ll \alpha . \tag{11}$$

We take ths solution of equation (10) in the form

$$x_1(t) = a_1(t)\tilde{x}(t) \qquad p_1(t) = \tilde{p}(t)$$
$$x_2(t) = a_2(t)\tilde{x}(t) \qquad p_2(t) = \tilde{p}(t) + \Delta p(t), \tag{12}$$

where $\tilde{x}$ and $\tilde{p}$ are the solutions of (8) and (9). By expanding the right-hand side of (10) in powers of ε and Δp using (11) and (12) and retaining linear terms we obtain

$$\dot{a}_2 = a_2(b_{11}\Delta p + b_{12}\varepsilon)$$
$$\Delta\dot{p} = b_{21}\Delta p + b_{22}\varepsilon \tag{13}$$

where

$$b_{11} = \frac{\partial f}{\partial p}; \qquad b_{12} = \frac{\partial f}{\partial \alpha}; \tag{*}$$
$$b_{21} = \frac{\partial h}{\partial p}; \qquad b_{22} = \frac{\partial h}{\partial \alpha}.$$

The derivatives in (*) are functions of $\tilde{x}(t), \tilde{p}(t)$ and α. Note that $\tilde{x}(t)$ and $\tilde{p}(t)$ are rapidly varying periodic functions of time whereas $a_2(t)$ and $\Delta p(t)$ vary only slowly. Therefore, in considering the asymptotes of $a_2(t)$ and $\Delta p(t)$, $\tilde{x}(t)$ and $\tilde{p}(t)$ can be substituted by their mean values $\bar{x}$ and $\bar{p}$. Biologically $b_{21} < 0$. Thus, the asymptotic solution of the system of equations (13) is given by

$$\Delta p = -\varepsilon b_{22}/b_{21}; \quad a_2 = C\exp\{[\varepsilon(b_{12} - b_{11}b_{22}/b_{21}) + \varepsilon^2 F]t\} \tag{14}$$

where C is integration constant; $\varepsilon^2 F$ are the second order terms in a power series of f and h.

It follows from (7) and (11) that

$$\lim_{t\to\infty} a_2(t) = 0 \tag{**}$$

for any ε if $\alpha = \alpha^*$ where α^* is the optimal value of the structural or functional parameter in question.

The necessary condition for (**) is

$$b_{12} = b_{11} b_{22} / b_{21} . \qquad (***)$$

So, α^* is the root of the equation

$$\frac{\partial f}{\partial \alpha} - \frac{\partial h}{\partial \alpha} \left(\frac{\frac{\partial f}{\partial p}}{\frac{\partial h}{\partial p}} \right) = 0. \qquad (15)$$

The cost functions used to determine the biological parameters have been discussed in detail by Rosen (1967). These functions are usually formulated by analysing the processes involved. The above mentioned special criteria (cost functions) and more general criteria can be obtained from condition (15).

Consider the special cases.

1) The parameters characterizing the reproduction are implicitly dependent on the optimized parameter, i. e. $\partial f / \partial \alpha = 0$. In this case (15) becomes

$$\frac{\partial \Delta W}{\partial \alpha} = 0 \qquad (16)$$

i. e., we are faced with the problem of maximizing the energy imbalance

$$\Delta W = max. \qquad (17)$$

By applying criterion (17) the optimal parameter values can be found that affect fertility (mortality) through energy imbalance alone. The energy imbalance is the power consumption minus the power expenditure. Condition (17) can then be used to formulate the special optimalisation criteria (maximum power consumption, minimum power expenditure, minimum basal metabolism, maximum body efficiency etc). The complentary conditions determining the functional activity of various organs and systems must also be given when using criterion (17). For instance, in considering the oxygen transport system body oxygen consumption is given.

Let us now consider the case in which the parameters do not affect the energy imbalance, i. e. $\partial h / \partial \alpha = 0$. However, reproductive power is explicitly dependent on theses parameters. In this case (15) becomes

$$\frac{\partial f}{\partial \alpha} = 0 \qquad (18)$$

i. e., we are faced with the problem of maximizing the relative population growth rate

$$f = max. \tag{19}$$

This cost function (19) has been suggested by the investigators cited above. In (19), f is assumed to be a function of parameters characterizing the phenotype. The condition (19) is commonly known as Fisher's criterion. Note that Fisher's criterion is a special case of criterion (19) if $\partial h/\partial \alpha = 0$ or $\partial f/\partial p = 0$. The latter condition can be interpreted as follows. The reproductive power is implicitly dependent on body weight, i. e., there are no factors limiting the population size. Thus, Fisher's criterion is applicable in the following cases: 1) the energy imbalance is implicitly dependent on the parameter in question. 2) the factors limiting the population size are lacking. Condition (15) can be used to determine the optimal biological parameters in a more general case.

$\varphi = -\left(\frac{\partial f}{\partial p}\right)/\left(\frac{\partial h}{\partial p}\right)$ is a weight function determining the relative importance of the energetics and reproduction with respect to persistence.

In formulating criterion (15) the energy is assumed to be the main factor limiting population size. If the limiting factor is not the energy but any substance then the derivation of the necessary condition for optimalisation is the same as (15). In this case (15) becomes

$$\frac{\partial f}{\partial \alpha} - \frac{\partial g}{\partial \alpha}\left(\frac{\frac{\partial f}{\partial S}}{\frac{\partial g}{\partial S}}\right) = 0 \tag{20}$$

where g is the substance imbalance ($g = g(x, S, \alpha)$); S is the limiting substance concentration in the body.

We can also consider other special cases, for instance, the optimalisation criterion can be formulated as

$$g = max, \tag{21}$$

when $\partial f(x, S, \alpha)/\partial \alpha = 0$.

A number of special optimalisation criteria can be formulated from the condition for maximum substance imbalance (minimum substance consumption (organ functions given), maximum substance assimilation coefficient etc.).

Let the state of a population be determined by two limiting factors (energy and substance). In this case the system of equations analogous to (8) and (9) can be written as

$$\begin{aligned} \dot{x} &= f(x, p, S, \alpha)\, x \\ \dot{p} &= h(x, p, S, \alpha) \\ \dot{S} &= g(x, p, S, \alpha). \end{aligned} \qquad (22)$$

By analogy with (15) we obtain the equation for the optimal parameter value

$$\frac{\partial f}{\partial \alpha} + \frac{\partial h}{\partial \alpha} \left| \frac{\frac{\partial f}{\partial S}\frac{\partial g}{\partial p} - \frac{\partial f}{\partial p}\frac{\partial g}{\partial S}}{\frac{\partial h}{\partial p}\frac{\partial g}{\partial S} - \frac{\partial h}{\partial S}\frac{\partial g}{\partial P}} \right| + \frac{\partial g}{\partial \alpha} \left| \frac{\frac{\partial f}{\partial p}\frac{\partial h}{\partial S} - \frac{\partial h}{\partial p}\frac{\partial f}{\partial S}}{\frac{\partial h}{\partial p}\frac{\partial g}{\partial S} - \frac{\partial h}{\partial S}\frac{\partial g}{\partial p}} \right| \; . \qquad (23)$$

By using (23) the special criteria can be formulated. If the energy imbalance is the only limiting factor then, $\partial f/\partial S = 0$ and $\partial h/\partial S = 0$. Hence (23) will be (15) since $\partial g/\partial S \neq 0$. If substance imbalance is the only limiting factor then $\partial f/\partial p = 0$ and $\partial g/\partial P = 0$. In this case (23) will become (20). A number of special criteria can be formulated in the following cases:

1) one or two of the functions f, h, g are implicitly dependent on α,
2) one of these functions is implicitly dependent on α, the other on some limiting factors.

Concluding Remarks

An optimalisation criterion formalizing Darwin's law of natural selection has been considered. According to this criterion, the evolutionary optimal biological parameters are determined by the asymptotic properties of the system of equations describing ecosystem dynamics. The criterion can be applied to calculate, in particular, the structural and functional parameters. The proposed criterion can be reduced to simpler criteria (minimum energy expenditure, maximum relative population growth rate etc.). The field of application for these criteria has also been determined.

Statistical Thermodynamic Model of Stationary Population

A. K. Prits

We shall consider a population as a single system, the structural elements of which are individuals. Like any living system, the population must be an open, self-regulating and self-reproducing system. The feature which enables us to distinguish one individual from another, will be the age of the individuals belonging to this population.

Let the total number in the whole population be N and N_t be the number of the individuals of the t -th age group of this population, $t = 0,1,2,\ldots,n$, where n is the limiting age of the population.

By analogy with statistical thermodynamics we shall introduce, as macroscopic parameters characterizing the population, the following quantities: the number in the population N, the mean age, $\langle t \rangle$, of the population, the population entropy S.

We shall take as ecological parameters of the population the following features: birth and death rates, biotic potential, type of the population increase etc.

The present paper is aimed at establishing a connection between the statistical and ecological parameters of the population and determining its optimal age.

Thermodynamics and Kinetics of Biological Processes

Initial Equation and Distribution of Numbers with Respect to Age

Let us consider the population to be in a stationary state. This suggests that the population parameters do not change in the course of time in spite of the presence of substance and energy flows through the population as a system. If the fixed number in the population is N, then the initial equation will be

$$N = \sum_{t=0}^{n} N_t . \tag{1}$$

The second statistical parameter will be the age of the population determined by the expression

$$\langle t \rangle = \frac{\sum^{n} t N_t}{N} . \tag{2}$$

The age structure of the population as a system is established by the expression

$$S = - \sum_{t=0}^{n} \frac{N_t}{N} \ln \frac{N_t}{N} . \tag{3}$$

Let us find out the conditions under which the stationary satisfying formulae (1), (2) and (3) is realized.

In these equations, N, $\langle t \rangle$ and S are constant, time-independent values, and therefore the variations of these parameters with respect to the variable N_t are equal to zero. Using the method of Lagrange multipliers we find, after trivial transformations, that the constant value of these parameters corresponds to the following age-dependent distribution of numbers (Prits, 1974).

$$N_t = \frac{N \cdot e^{-\beta t}}{\sum_{t=0}^{n} e^{-\beta t}} , \tag{4}$$

where β is some constant of dimensionality T^{-1} which characterizes the age-dependent distribution.

Connection between Statistical Parameters and Ecological Characteristics

Formula (4), characterizing the age-dependent distribution of numbers, in cludes ecological characteristics of a stable (stationary) population. In fact, the birth rate R is understood as the number of individuals of zero age per unit population. Assuming that t = 0 in the numerator of formula (4), we obtain

$$R = \frac{N_0}{N} = \frac{1}{\sum_{t=0}^{n} e^{-\beta t}} = \frac{1}{Z}, \tag{5}$$

where Z is the statistical sum of the age distribution.

The survival rate of the population is the ratio of the numbers of the subsequent age group to the numbers of the previous (born in the same year) group. For a stable population

$$N_{t+1} / N_t = e^{-\beta} \tag{6}$$

and

$$\beta = -\ln (N_{t+1} / N_t). \tag{7}$$

Formulae (6) and (7) suggest that β is the death rate coefficient. Introducing the quantity $\Theta = \beta^{-1}$, where Θ is the modulus of the statistical age-dependent distribution of numbers, one can rewrite formula (4) as follows

$$N_t = N e^{-t/\Theta} / \sum_{t=0}^{n} e^{-t/\Theta} \tag{8}$$

where Θ has the dimension of time.

Introducing the designation

$$\Psi = -\Theta \ln Z = -\Theta \ln \sum_{t=0}^{n} e^{-t/\Theta} \tag{9}$$

we obtain

$$N_t = N e^{(\Psi - t)/\Theta}, \tag{10}$$

where

$$\Psi = \Theta \ln R, \tag{11}$$

characterizes the ability of the population to persist according to the state of the environment and the age structure of the population. Therefore, the quantity Ψ is called a biotic potential.

Connection of Biotic Potential with the Age Structure of Population and the Equation of the State of the Population

Substituting expression (10) in (3) and taking account of expression (2) one can easily show that

$$\Psi = \langle t \rangle - \Theta S \tag{12}$$

Expression (12) suggests that the biotic potential is determined by the mean age, $\langle t \rangle$, of the population, the age structure, S, and the death rate coefficient, β, since $\Theta = 1/\beta$.

To establish the connection between the death rate coefficient β and the mean age $\langle t \rangle$ of the population, we replace, in formula (2), the quantity, N_t, by its value from formula (4) and we obtain

$$\langle t \rangle = \frac{\sum_{t=0}^{n} t e^{-\beta t}}{\sum_{t=0}^{n} e^{-\beta t}} = -\frac{\partial}{\partial \beta} \ln \sum_{t=0}^{n} e^{-\beta t}. \tag{13}$$

Accounting for the value of β one can extend the sum of the terms to infinity, then

$$\sum_{t=0}^{\infty} e^{-\beta t} = \frac{1}{1 - e^{-\beta}}, \tag{14}$$

and $\langle t \rangle$, the mean age of the population, will be determined by the expression

$$\langle t \rangle = \frac{1}{e^{\beta} - 1} \quad . \tag{15}$$

Consequently, the death rate coefficient is a function of the macroscopic parameter

$$\beta = \ln\left(1 + \frac{1}{\langle t \rangle}\right). \tag{16}$$

The number of individuals of age, t, dying within the interval between two succeeding moments of reproduction will be

$$\Delta N_t = N_t - N_{t+1} = N_t(1 - e^{-\beta}) \; .$$

Replacing β by its value, expressed in terms of the mean age, and summing ΔN_t over all the age groups, we get the connection between the number of individuals dying between the two succeeding instants of reproduction, ΔN , ("waste") (Romanovskyi et al., 1975), net number and mean age of the population

$$N = \Delta N\,(1 + \langle t \rangle) \tag{17}$$

$$N = N_0\,(1 + \langle t \rangle). \tag{18}$$

Equations (17) and (18) are state equations of any stable population in which the numbers of age groups are sufficiently large or, which is the same, the limiting age of individuals is much higher than the time interval between two succeeding instants of reproduction. N_0 is the number in the zero-age group (replenishment).

Optimal State of a Stable Population under Exploitation

The state providing the maximal increase dM/dt of the population biomass per unit time, the population biomass itself being conserved, will be taken as the criterion of the optimal state of a stable population.

An increase is known to be determined by the age structure of a population under exploitation. The optimum in the age structure providing the maximum rate of biomass increase, is given by the expression

$$dM/dt = max \tag{19}$$

or

$$d^2M/dt^2 = 0, \tag{20}$$

where M is the population biomass.

By applying different equations to the increase of population elements, one can establish the mean age of the population at which the biomass increase is greatest.

As calculations have shown, using equations for biomass increase (Zotina, Zotin, 1967; Zotin, 1972) we arrive at

$$m = m_\infty e^{-k(t_m - t)^2} \tag{21}$$

where m_∞, k, t_m are constants determined by the experiment, and t is the age. This suggests that the optimal age of a herring population in the North Sea must be equal to 3,2 years (at present it is equal to 1,56). From biological data is is known that an age of 3,2 years does not only provide the maximum biomass increase, but also gives the maximum reproduction (replenishment).

Taking into consideration the fact that the death rate coefficient β is formed of the natural death coefficient M and that of being fished, F, one can easily estimate the coefficients which do not lead to biomass decrease in populations under exploitation (see formula (16)).

By evaluating β for each population under exploitation, one can easily find the number of equipment necessary to exploit this population.

VI. Reaction of Organisms to External Factors

Effects of External Influences on the Types of Transition Processes Observed in the Energy Metabolism of Organisms

V. A. Grudnitzky †

As is known, the necessary condition for the existence of living organisms is a constant exchange of substance and energy with the environment. This property which is the basis of living processes is also the characteristic which determines the concept of an open system in thermodynamics. Notwithstanding an evident conformity between the thermodynamic concepts of an open system and the properties of biological subjects under investigation, it seems expedient to emphasize the following. At present a concept of an open system in thermodynamics implies a requirement for the exchange of substance and energy with the environment. The boundary conditions for such a system are not determined. In practic for applied analysis such bounds have to be found by intuition, although it is not the best possible way. However among the biological systems studied, open systems can be found, which are naturally precisely limited in space and have well expressed, essential properties. It is an entire living organism. While some biological phenomena, specific metabolic chains for example, require a special decision, as to whether some conjugated link should be included in the open system under examination, no such questions arise when the entire organism is concerned. Consequently particular evidence of the open system's properties in biology may be obtained from experimental investigations of the entire organism properties.

Taking into account the above, it is expedient to examine the degree of conformity between existing notions about the means of transition in an open system, from one stationary state to another, and the known types of the response of the organism's energetic metabolism to external influence. Concepts of the type of transition process in an open system were developed when analyzing the properties and living conditions of stationary states. The most demonstrative example of a system in a stationary state is the simple hydrodynamic model analyzed by Burton (1939). It consists (Fig. 25) of two vessels A and B connected with the environment by tubes. The water flow in the latter is regulated by taps K_o and K_z which set the rate of inflow and outflow from the system. Tap K regulates the flow rate from A to B. Vessels S and Z represent the external medium.

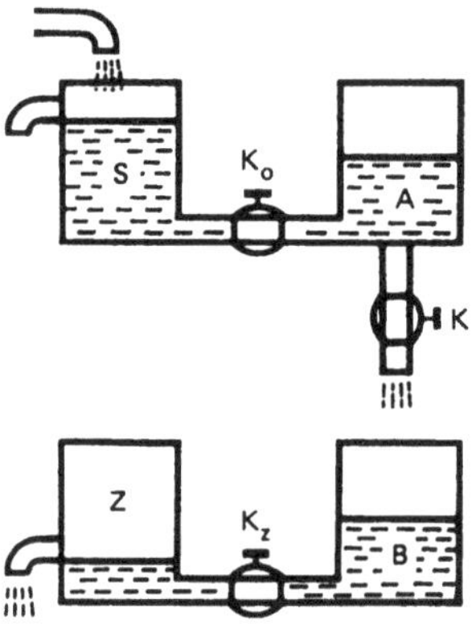

Fig. 25 Hydrodynamic model of Burton's open system (explained in the text).

Under an established regime the water level will be constant. Any change in the liquid flow by taps K_o, K and K_z results in the establishment of new levels in vessels A and B. Because the rate of water outflow is not constant and depends on the hydrostatic pressure of the liquid column the transient process will have some maximum or minimum (with increase in K_o or K_z, respectively).

A model of a chemical reaction similar in structure to Burton's hydrodynamic model was analyzed by Pasynsky (1957, 1968). The scheme is as follows:

$$S \xrightarrow{k_o} \; \vdots \; A \underset{k_2}{\overset{k_1}{\rightleftharpoons}} B \; \vdots \; \xrightarrow{k_z} Z$$

where k_o and k_z are the constants of free diffusion or permeability of the membrane; k_1 and k_2 are constants of chemical reaction or transformation A $\rightleftharpoons$ B; S and Z reservoirs of the external medium; the dotted line marks the system boundaries.

For each of the substances in the system the rate of concentration change will be equal to the sum of the rates of the elementary transition processes through the system boundary and inner chemical transformations. For the simplest conditions, when a chemical system of the first order is taken, and the transition is determined by a diffusion process obeying Fick's law (in proportions to the concentration gradient) we have

$$\frac{da}{dt} = k_0(a_0 - a) - k_1 a + k_2 b$$
$$\frac{db}{dt} = k_2(b_0 - b) + k_1 a - k_2 b, \quad (1)$$

where a_0 and b_0 are initial concentrations; a and b are concentrations at a given moment. The general solution for this particular system is a sum of exponential terms:

$$a = \bar{a} + C_{11} e^{-\lambda_1 t} + C_{12} e^{-\lambda_2 t}$$
$$b = \bar{b} + C_{12} e^{-\lambda_1 t} + C_{22} e^{-\lambda_2 t} \quad (2)$$

where C_{11}, C_{12} etc are the integration constants; λ_1 and λ_2 the roots of the equation; $\bar{a}$ and $\bar{b}$ concentrations of substances in a stationary state. Equation (2) describes a change of the system in time. It follows thereby that with real and positive values λ of the equation roots the system moves over time to the stationary state, i. e. $a=\bar{a}$; $b=\bar{b}$. At different values of C and λ various changes in the concentrations a and b over time are obtained, i. e. the transition kinetics between the stationary states is rather complex. For instance with arbitrary parameters $\lambda_1 = 1$; $\lambda_2 = 3$; $C_{11} + C_{12}$ $C_{11} + C_{12} = 10$ the value $(a - \bar{a})$ changes along one of the curves given in Fig. 26 versus the accepted C_{11} value. In all the cases presented the transition from a higher stationary level to the lower one takes place. In curve A the concentration passes through a lower intermediate level, in curve D it increases at first and then falls below the initial level. Curve C represents a simple exponential transition from one state to another. A curve like type A is designated an *overshoot curve*; and a curve of type D a *false start curve*.

The above models are the simplest models for open systems. Even a small complication in their structure requires a much more complex mathematical description. Therefore, it seems unreal to expect that a system as complex as an organism could be described by kinetic equations. Nevertheless, general conclusions concerning possible types of transition processes from one stationary state to another can be probably transposed to any open system, including the living organism.

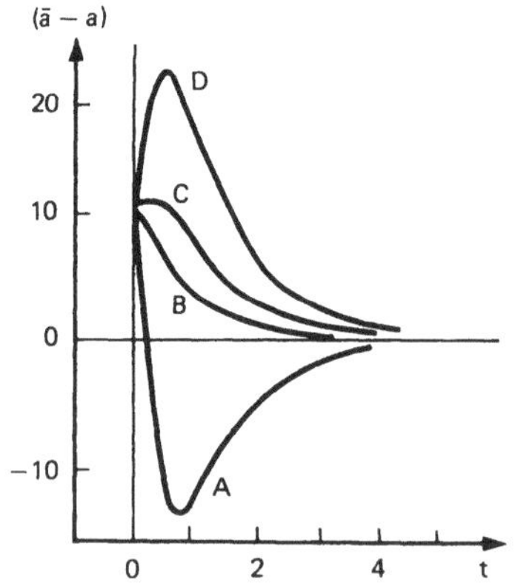

Fig. 26 Transition processes in the Pasynsky model of open system.

To be convincing, transition processes obtained experimentally on biological material ought to be considered. Changes of energetic metabolism in animals, caused by different stimuli, have been studied. However, it is unnecessary to examine the effect of numerous factors on the energetics of the organism. It is well known that the effect of any factor capable of changing the energetic metabolism resolves itself only as a rise or fall in this level, i. e. the responses of the organism's energetics to external influences are non-specific. Taking the above into account it is believed that specific features of transition processes are also common properties of a biological, open system. Particular examples of the transition processes presented below are obtained when the effect of external temperature on the organism's energetics is analyzed. Within these themes problems of temperature adaptation of animals were considered which are of interest in transition processes. It should be noted that the final objective of these studies is, as a rule, the examination of the adaptation mechanism for the energetics of animals. Therefore, metabolic reactions are usually neglected in experiments where observations are performed (from time = initial time) on animals placed in a medium with an unusual temperature. Some researchers explain the initial abrupt changes in energy metabolism by an increase in the animal's mobility, which, in their opinion, has no bearing upon the mechanisms of temperature acclimation. These initial stages of energy metabolism are of special interest in the analysis of types of transition processes, since the mechanism of the process itself would be less interesting in terms of a thermodynamic approach. The transition processes exhibited at this stage show a great variety. Nevertheless the main features on the transition processes mentioned above also manifest themselves in profound reconstructions of the organism's energetics.

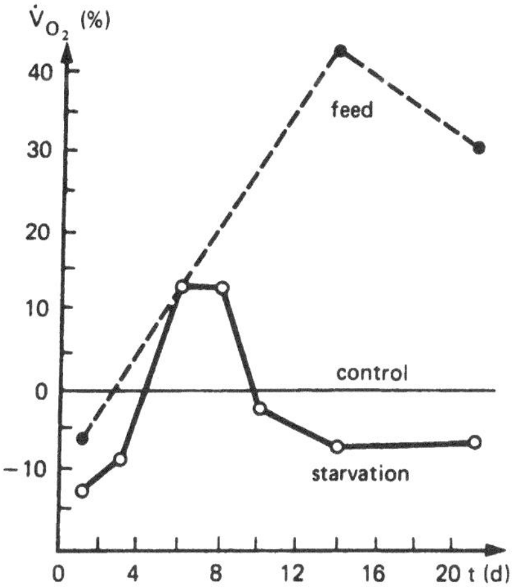

Fig. 27 Change in respiration at temperature acclimation of *Uca pugnax* (Vernberg, 1959).

Fig. 27 shows a transition process in the acclimation of the fiddler crab *Uca pugnax* adapted initially to 27 °C (Vernberg, 1959). The respiratory velocity of the animals was measured during acclimation to 15 °C. Observations were conducted for 21 days. Changes in metabolism of crabs which were starved and then regularly fed during the experiment were analyzed. In both cases the same peculiarity of the transition process is observed, namely overshoot.

Fig. 28 presents graphs of transition processes during temperature acclimation in the milliped *Ortomorpha gracilis* (Gromysz-Kalkowska, 1976). The scheme of the experiment was as follows. *Myriapods* initially adapted to 15 °C were placed into a 30 °C temperature environment. A parallel experiment was performed on acclimation to 15 °C of myriapods initially adapted to 30 °C. In both transition processes overshoot was observed.

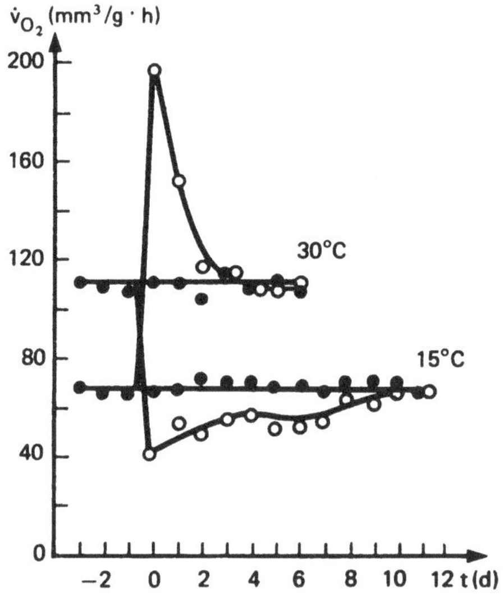

Fig. 28 Change in respiration intensity at temperature acclimation of *Orthomorpha gracilis* (Gromysz-Kalkowska, 1976).

Subsequent figures show transition processes for temperature acclimation in some invertebrates which were obtained under similar conditions. Measurements of respiration were taken 30 minutes after placing the animals into the new temperature environment.

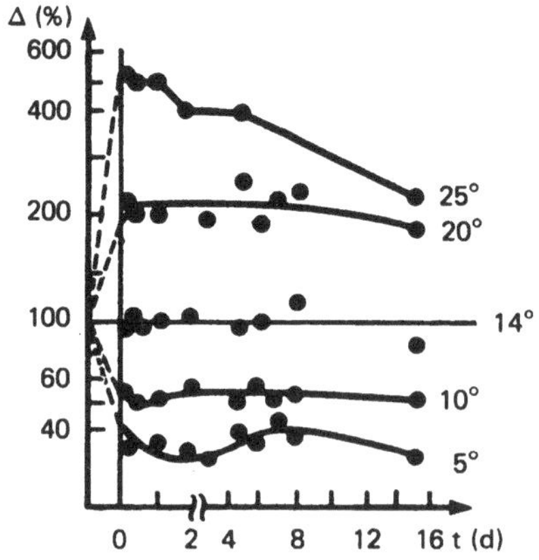

Fig. 29 Change in the level of energetic metabolism at temperature acclimation of *Modiola adriatica* (Ivleva, 1972).

Fig. 29 gives curves for changes in respiration rate during acclimation of the sea muscle *Modiola adriatica* to different temperatures (Ivleva, 1972). Although the dynamics of the initial stages of the process are unknown, the author believes that, in this case, exponential transition to a new stationary level were observed under the experimental conditions.

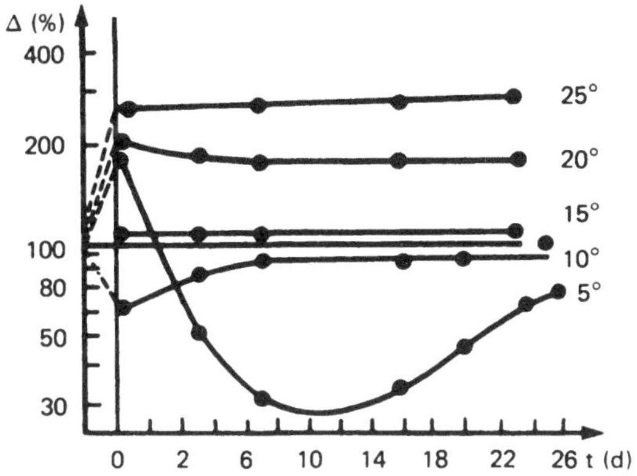

Fig. 30 Change in the level of energetic metabolism at temperature acclimation of *Branchiostoma lanceolatum* (Ivleva, 1972).

Fig. 30 presents experimental data on the acclimation of the lancelet *Branchiostoma lanceolatum* (Ivleva, 1972). Special attention should be paid here to the graph of the transition process obtained during the acclimation to 5 °C. The respiration rate increases before reduction to a new stable state. This means that the transition process has the form of a false start. During acclimation to 10 °C an overshoot is observed.

Fig. 31 shows graphs for respiration changes in the prawn *Palaemon adspersus* (Ivleva, 1972). Overshoot is observed on two of them (28 °C and 10 °C). During acclimation to 5 °C the overshoot is weakly expressed, therefore this transition may be considered as approximating to an exponential process.

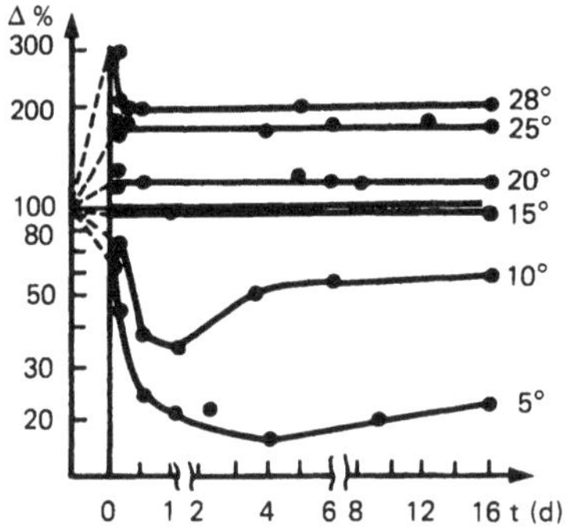

Fig. 31 Change in the level of energetic metabolism at temperature acclimation of *Palaemon adspersus* (Ivleva, 1972).

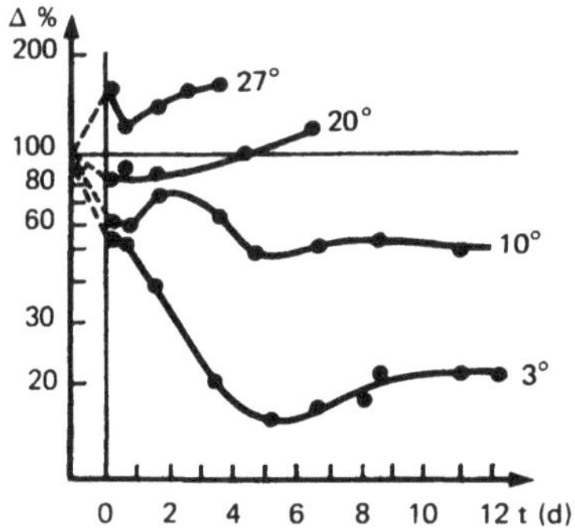

Fig. 32 Change in the level of energetic metabolism at temperature acclimation of *Nereis diversicolor* (Ivleva, 1972).

Fig. 32 shows transition processes in energy metabolism in the clam worm *Nereis diversicolor* (Ivleva, 1972). Attention should be paid to the fact that some curves show oscillatory behaviour which approximates to the level of the respiration rate in the stationary state.

Thus analysis of the patterns of transition processes in biological systems shows that the first stages of temperature acclimation are of great interest. Fig. 33 presents transition processes in the energy metabolism of the Crustacea *Hemimysis lamorna* during transition from 20 °C to 10 °C and vice versa. These data were obtained whilst recording the very first stages of the acclimation process (Grainger, 1956). The transition process proceeds for about two hours and is represented by oscillations approximating to the stationary level.

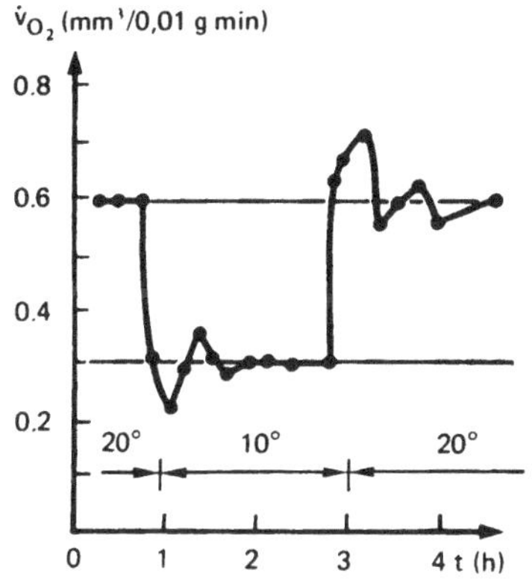

Fig. 33 Change in respiration intentsity at temperature acclimation of *Hemimysis lamorna* (Grainger, 1956).

It seems that the above experimental examples of transition processes in the level of energy metabolism in animals are sufficient to show that the peculiarities of transition processes found for simple models of open systems are also valid for biological open systems. It does not mean, however, that the mathematical modelling of the kinetics of these model systems can be directly applied to the analysis of such a complex system as the entire organism.

In fact, the model of energetic regularities in an organism, even on the basis of general concepts, has a more complex structure than Burton's and Pasynsky's models. Using the analogy with a hydrodynamic model, a pattern of energy transformation in a general form can be represented by the scheme shown in fig. 34. Unbroken arrows show energy flow paths, the dotted lines controlling couplings. Unit 1 is a reservoir which contains the energy source which periodically enters as food. Unit 3 shows a transformation mechanism of the supplied energy (respiration, glycolysis). Unit 4 represents the energy involved in the realization of vital activity and realised externally. Unit 2 shows the energy stored by the organism whose expenditure maintains some degree of independence in the state of units 3 and 4. It also controls the periodic oscillations in the state of unit 1 resulting from irregular meals.

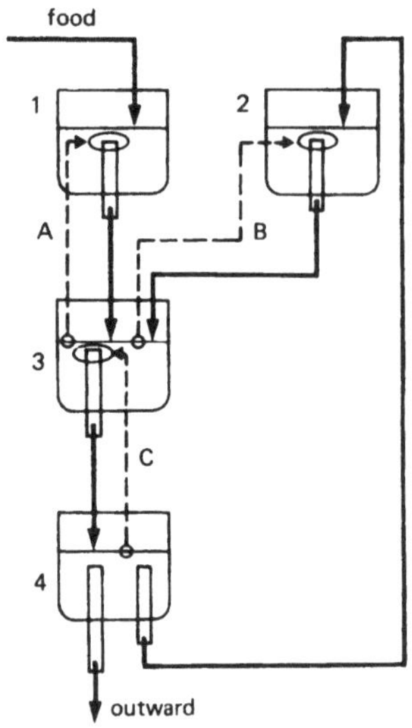

Fig. 34 Model of energy transformation in organisms (explained in the text).

The analogy between the units and reservoires with liquid is convenient because in this case the extensive and intensive characteristics of the units state (volume and level of liquid) are obvious. Their values determine the state of each unit. The rate of energy out-flow from the unit depends on the liquid level within it, and the rate of the level change is related to liquid exchange. The state of the whole system depends on the rate of filling of unit 1, however, particular states of the units are established as a result of mutual regulations which in a general form are shown in the figure. In addition a feedback mechanism exists from unit 4 which, by means of the work produced, controls the filling of unit 1. This is realized via the external medium, however, and is excluded from this model. It is a difficult at present to establish strict laws of governing the controlling couplings. Therefore there are no grounds for the analysis of kinetics of this model. Nevertheless, some qualitative conclusions about transition processes can be drawn from this model for an organism's energetic state.

The first conclusion that follows is that the mathematical description in Burton's and Pasynsky's models is not applicable to the entire organism. Nevertheless, as shown above, the transition processes in the models and in the organism's energetics have common features. Proceeding from the model of energy transformation in the organism we shall consider possible instances of initiation in typical transition processes of energetic metabolism at the change of the external temperature.

The level of the animal's metabolism in relation to the temperature of the medium can be described, approximately, by Krogh's normal curve (Krogh, 1916; Winberg, 1956 a), according to which the rate of metabolism increases with rise in temperature. In conformity with this regularity stationary levels of metabolism are established. A concrete type of transition process will depend on the state of each unit of the model. In particular, under small temperature drops or under sufficiently slow temperature changes the changes in energy loss are small, and an exponential transition of metabolism to a new level should be expected. That is, this transition can be continued by such rates of changes in the level of unit 4, within which is felt the influence of A and B regulations damping the change in the level of unit 3. If the rate of change in the level of unit 4 is so high that A and B cannot change rapidly enough to regulate the energy currents from units 1 and 2 to unit 3 in accordance with the change in the level of unit 3, than an overshoot is initiated in the transition process.

Two possible mechanisms can be suggested for the appearance of a false start in the transition process. In one variant the initial energy expenditures for the increased motor reaction of the animal can be sufficiently large for a temporal shift of the level of energetic metabolism in a direction opposite to the final value determined by Krogh's normal curve. Another possibility is associated with the interaction between the change in the level of unit 3, caused by temperature change, and the degree of filling of units 1 and 2. Thus when a just fed, non-starved animal is placed into a cold medium under the effect of a known metabolic increase after a meal (the so-called specific dynamic action of food) a false start may appear. With minimal filling of units 1 and 2, and movement of the animal to a higher temperature, after a small rise in the metabolic level due to exhausted supplies of units 1 and 2, the level of unit 3 will be lower than that expected from the Krogh's curve and, possibly lower than the initial level. In the latter case the transition process will have a region of the false start. Something of this kind can be seen in fig. 27 by comparing the transition processes in non-starving and starving crabs.

The generalized model of energy transformation in the entire organism presented shows that biological, open systems are of great complexity. First of all because of mutual regulations between single processes in energy transformation. Quantitative regularities in these processes seem to show species-specificity and are changed in the course of ontogenesis and in adaptation to different living conditions. Nevertheless evaluation of possible initiation of specific features of transitions processes in energy metabolism, and particular experimental examples, indicates that the types of transition processes described are peculiar to open systems, notwithstanding the degree of complexity.

Rate of Temperature Adaptation of Respiratory Metabolism in Soil Invertebrates

Yu. B. Byzova

The new studies that have appeared on regularities and mechanisms of temperature adaptation of respiratory metabolism in animals has meant that results obtained at the beginning of this century have had to be revised and the experiments repeated. New theoretical bases and approaches to the problem have made this necessary. It has been clearly shown that changes in respiration metabolism rate with temperature variation involves three stages: there exists a comparatively short period of primary response to the change of temperature (acute, primary reaction) and a subsequent longer period of acclimation resulting in the establishment of a stable level of metabolism (Ivleva, 1972; Galkovskaya, Sushchenya, 1978). Some progress has been made in a theoretical analysis and practical description of the mechanisms of the primary metabolic reactions to temperature changes and metabolic temperature adaptations (Hochachka, Somero, 1977). In the light of these new ideas the existing theoretical concepts and the accumulated facts are being reconsidered. It was found that many of those studies which formed the basis of theoretical generalizations were not founded on reliable evidence. Analysis of the data in the literature on the relationship between respiration intensity and temperature in poikilotherms (mainly aquatic animals) showed (Ivleva, 1972) that the experimental material presented was generally formed around the conclusions drawn. This was due to the fact that the investigators neglected the stages mentioned above for the course of temperature adaptation of respiratory metabolism. Some indications of the complex nature of acclimation were, however, known long ago. They were, for instance, mentioned in the work of Blazka (1955) when dealing with gas exchange in the edible snail *Helix pomatia (Pulmonata, Helicidae)*. Neglect of this complexity resulted in the attempts to compare non-comparable characteristics.

Thermodynamics and Kinetics of Biological Processes

These comments apply equally to studies of gas exchange in soil invertebrates. For these the temperature factor seems to have even greater significance than it does for aquatic animals (Byzova, 1972, 1977). There is a shortage of data, based on modern concepts and obtained by modern methods in this field, too. This is true both for studies of the values for primary reactions to change in temperature, for levels of stable respiratory metabolism resulting from acclimation and for investigations of the rate of metabolic adaptation. On the strength of the data available for water invertebrates we conclude that there is a great variety of such characteristics, even in systematically close forms (Ivleva, 1972). On the other hand, theoretical consideration of possible genetic mechanisms of adaptation suggests that the time of establishment of metabolic adaptation varies greatly in higher and lower eucariots; 2 to 4 and 20 to 40 days, respectively (Labas, Khlebovich, 1976).

Very little is known about the reaction of respiratory metabolism to temperature changes in soil invertebrates. Few experimental studies have dealt with the acclimation rate of soil in vertebrates when a stable heat regime is substituted by another regime. These data were presented in a recent review (Byzova, 1977). Without going into details only some general principles will be described and these must form the basis for investigation into respiratory metabolism-temperature relationships. These principles are as follows.

When studying the temperature dependence of the intensity of oxygen uptake, two groups of closely related characteristics are obtained: a temperature curve of stable metabolism and a value of the acclimation rates with temperature changes over all possible intervals and in different directions. Data on the pattern of the primary reaction of respiratory metabolism to the substitution of a stable heat regime by another regime are of special interest. It is necessary, as a minimum, to obtain evidence about the nature of gas exchange at various temperatures resulting from natural living conditions of the animals. From all this evidence on ecological-physiological description of the species temperature-respiration relationship can be obtained and used in differing comparative investigations.

The following observations relate to procedures to be used in studying soil invertebrates. These species, living on relatively solid substrates, have as a rule, thigmotaxis. In respiration vessels covered with a damp folded paper filter these invertebrates use the paper as a cover and do not move during recordings of oxygen uptake. Standard metabolism in this case seems to be close to basic metabolism.

The second observation concerns the feeding regime of the invertebrates under study. It remains unclear, up to the present, how the intensity of oxygen uptake is connected with the process of food consumption and digestion. The effect of starvation on the rate of oxygen uptake in most soil saprophages and predators is not well understood. Therefore it is impossible to standardize either the feeding regimes, or the time of starvation before measuring gas exchange. These characteristics should be, evidently, estimated in each particular case. This will require much investigation. We believe that it is possible to obtain reliable data, if, before observation, the animals are kept in their natural substrate with sufficient food. Thus individuals at different stages in digestion and between meals (a state more natural than deprivation of food) will be introduced into the experiment. Average values of numerous measurements are in this case adequate for specification of a standard metabolism.

There are few studies to be found in literature which deal with experimental estimation of the rate of transition from one stable level of respiratory metabolism to another with change in constant temperature regimes.

Blazka (1955) studying *Helix pomatia* has found that, with a temperature change from 15 °C to 30 °C, complete acclimation of gas exchange is reached after two weeks.

Edney (1964 a, b) studied the effect on the wood louse *Parcellio laevis* and *Armadillum vulgare* of acclimation to constant temperatures on temperature/life thresholds. He has found that with temperature increase by 10 °C, within the range of 10 °C to 30 °C, adaptation of gas exchange is reached after two weeks, the most intensive changes in oxygen uptake being observed during the first week.

It was shown on nematodes *Panagrellus redivivus* and the vinegar eel *Turbatrix aceti* that with a temperature increase from 10 °C and 20 °C to 30 °C a stable level of oxygen uptake intensity is established within a week, corresponding to that for temperature change from 20 °C to 30 °C (Cooper, Ferguson, 1973).

Data are available to support the view that acclimation rate can be dependent on the direction of temperature change. Thus, in the milliped *Orto-morpha-Oxydus gracilis* (Diplopoda, Paradoxosomatidal) acclimated to 15 °C and 30 °C and transferred to 30 °C and 15 °C respectively respiratory metabolic rearrangement with temperature increase continued for three days, whereas with temperature decrease for ten days (Gromysz-Kalkowska, 1976).

Data on the acclimation rates of respiration in soil invertebrates when one stable temperature regime is substituted by another one are scarce. The evidence is fragmentary and incomplete. It is not usually related to the general characteristics of the relationship between stable respiratory metabolism and temperatures found for natural temperature ranges. Such a relationship seems, however, absolutely necessary for the following reason.

Examination of data in the literature (Byzova, 1972) has shown that the relationship between stable respiration intensity and temperature can be either strictly linear or complex. In the first case the temperature coefficient of oxygen uptake, Q_{10} used for comparing the experimental data is, as a rule, reduced with temperature increase. In a more complex case the relationship is expressed graphically by a broken line having a plateau which reflects the stable level of oxygen uptake intensity within some definite temperature range. Our experimental data have shown that such differences in the temperature curves of stable intensity of oxygen uptake are observed in close related species of one family and also in the individuals of one species taken from different geographic points (Byzova, 1970; Byzova, Zalesskaya, 1980). It appears that acclimation rates may differ with temperature changes, within the plateau, at its boundaries, and in the case of a linear dependence of metabolism on temperature in different temperature ranges. In addition, these examples differ as to the direction of temperature changes and these differences correlate with the ecological peculiarities of the forms.

Several examples can be presented which illustrate different relationships between the level of stable intensity of oxygen uptake and temperature, as well as various acclimation reates resulting from temperature changes.

Sarmatiulus kessleri (Diplopoda, Julidae) shows a complex curve of respiration intensity-temperature relationship (Fig. 35). A plateau is observed on the curve in the temperature range 10 °C to 20 °C. This shows an almost stable metablic level. With 10 °C to 20 °C successive temperature changes in this range a stable level of oxygen uptake is reached on the second day. In addition, the moderate slope of the curve is of note. The *S. kessleri* investigated were collected in their most populous habitat, namely in the Ukrainian steppe regions and forest plantations. Sharp temperature changes in the soil of this region are observed in summer. The ability to preserve respiratory metabolism at a stable level and to summer hibernation under extremely unfavourable conditions (high temperature, dryness) testifies to their high vital ability. Their number reaches 400 individuals per m^2. They disintegrate and reprocess deciduous plants with such activity that *S. kessleri* is, here, a basic factor in the biological cycle (Ghilarov, 1957).

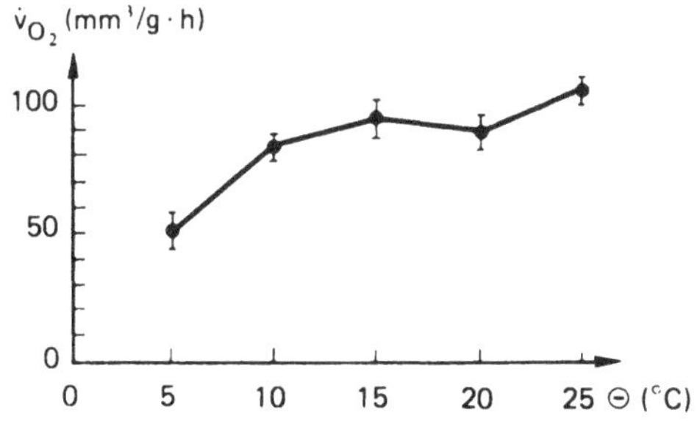

Fig. 35 Relationship between respiration intensity and temperature in *Sarmatiulus kessleri* C. Koch

Fig. 36 presents graphs of oxygen uptake-temperature relationships in the myriapos *Monotarsobius curtipes* (Chilopoda, Lithobiidae). The range of this species is greatly extended along the meridians. Three geographic points where *M. curtipes* were collected reflect this range; they differ substantially in the soil temperature regime, i. e. in total heat content in summer,

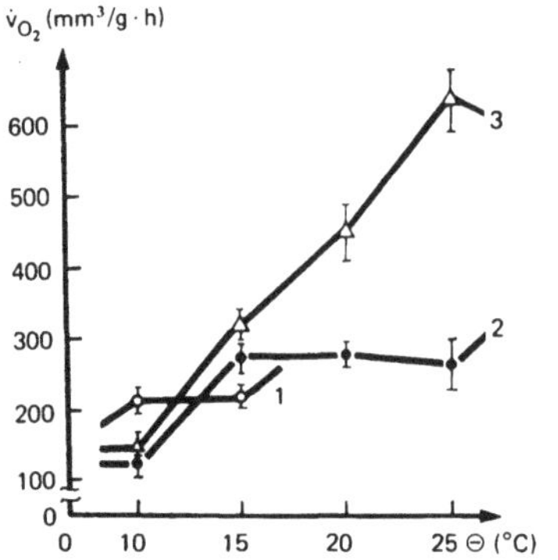

Fig. 36 Relationship between respiration intensity and temperature in *Monotarsobius curtipes* C. Koch from different areal parts:
1 - Murmansk region 2 - Moscow region 3 - Rostov region.

in prevailing summer temperatures etc. (Byzova, Zalesskaya, 1980). Accordingly *M. curtipes* of three populations have different upper heat thresholds of life; they differ from one another in the level of stable respiration at similar temperatures, and, which is more interesting, in the pattern of their stable respiration-temperature relationship. The shape of the curves varies. For *M. curtipes* from an Arctic population (Murmansk region) the short summer gives rise to average soil temperatures in the range 10 °C to 15 °C and the intensity of oxygen uptake is essentially stable. A broken temperature curve of stable respiration is typical of the individuals derived from the Moscow area, the plateau being shifted to the 15 °C to 20 °C (average summer temperature in the Moscow district). The stable oxygen uptake level in *M. curtipes* from a more southerly population, where temperatures fall in the range 10 °C to 25 °C, depends on temperature (Rostov region). Some data are available on the acclimation rates of these insects during such temperature changes. Individuals from the Moscow population adapted to 15 °C and transferred to 25 °C reach the initial level of oxygen uptake in three days. When the individuals adapted at 25 °C are transferred to 15 °C acclimation is completed in a week. Such is the rate of reconstruction of respiratory metabolism described by the plateau. It is a reconstruction because there is always a primary reaction to the change of temperature. When placing *M. curtipes* (Moscow region) acclimated to 15 °C at 10 °C ((temperature decrease at the plateau boundary) the stable intensity level of oxygen uptake is reached at the end of two weeks. In southern individuals acclimation following a temperature change (increase) takes no less than two weeks.

It is necessary also to touch upon the pattern of the primary reaction of respiratory metabolism to temperature changes, although there are almost no such data for soil invertebrates. It is but natural that there is no appreciation of the problem as to what should be taken as the primary reaction: the state of metabolism in the first hour, first few hours or first day after change in living conditions. It is apparent that a most detailed picture, if only for the first day, should be obtained.

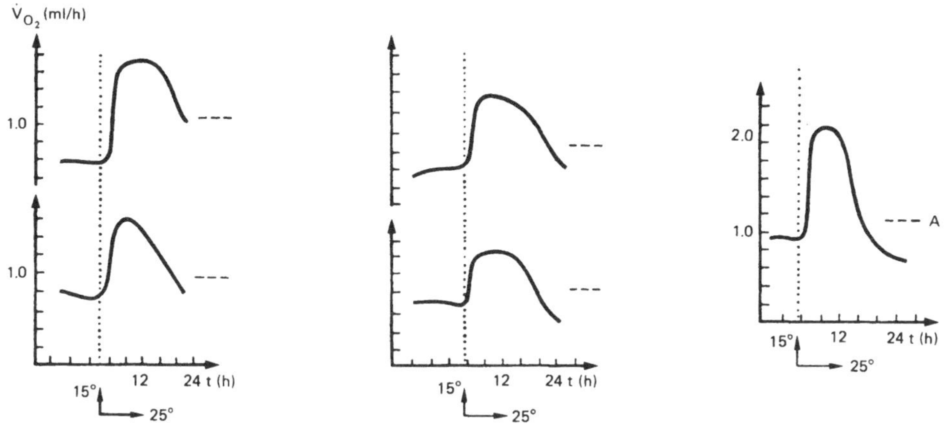

Fig. 37 Primary reaction to a temperature change in some individuals of *Helix pomatia L.* (Blazka, 1955). Notations: vertical broken lines show the time of temperature change, horizontal ones the mean level of metabolism on the 2nd and 3rd day after the temperature change.

Up to now only one study is known in the literature concerning the reconstruction of respiratory metabolism on the first day after temperature change. This study by Blazka (1955) presented individual changes in oxygen uptake by *Helix pomatia* individuals in the initial period of acclimation to a temperature change from 15 °C to 25 °C (Fig. 37). In the first three hours sharp changes of oxygen uptake were recorded; then, for 12 hours a high and stable level of metabolism was observed. After the period of high respiratory intensity a decrease up to some stable level, which is preserved during the following days, was noted. Then, the intensity of oxygen uptake gradually increase, and complete adaptation was achieved, as was already noted, two weeks later.

Our studies, conducted on wood lice, myriapods and rain worms show that the primary reaction is well-defined in all situations; for a temperature change within the plateau region at its boundaries, as well as within different ranges of the temperature of linear dependence of stable intensity of oxygen uptake on temperature.

In the pill bug *Armadillidium pallasii* (Isopoda, Armadillidae) adapted at 10 °C and transferred to 20 °C, in the first hour after temperature change the intensity of oxygen uptake rises to half the value of the metabolic level reached after two-weeks acclimation (Fig. 38). A linear relationship between stable intensity of oxygen uptake and temperature was recorded in this range. Then for 24 hours the respiratory intensity varies but does not exceed the maximum level achieved in the first hour and then gradually decreases; 24 hours later the level of metabolism at 20 °C exceeds the initial level (at 10 °C), by only 23 %. Further acclimation proceeds smoothly during the first week, then, 10 days after the beginning of acclimation the second sharp rise in oxygen uptake intensity is observed.

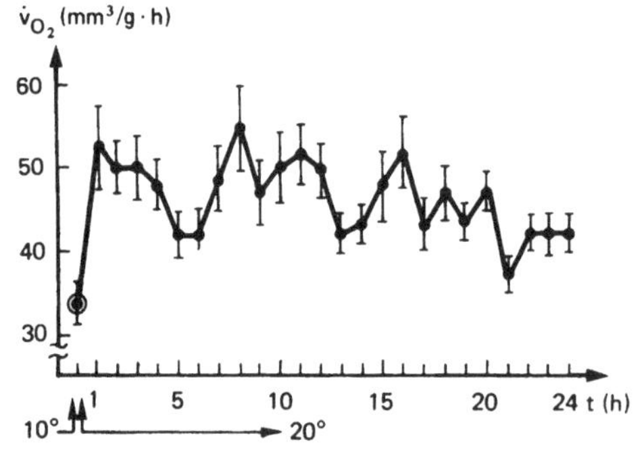

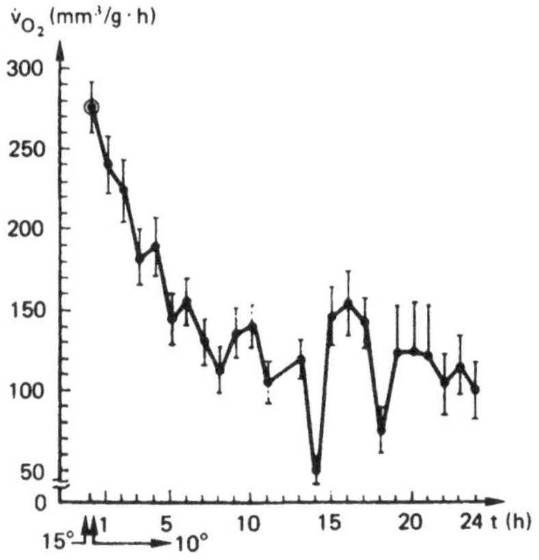

Fig. 38 Respiration intensity of *Armadillidium pallasii* Brand on the first day after temperature change (n = 48).

Fig. 39 Respiration intensity of *Monotarsobius curtipes* C. Koch from Moscow population on the first day after temperature change (n = 48).

We present here some examples of primary reactions to temperature changes in *Monotarsobius curtipes* (for individuals of Moscow population in particular; pattern of temperature curve of stable oxygen uptake Fig. 36). Fig. 39 shows a graph of the change in the intensity of oxygen uptake on the first day after temperature change, a decrease in the plateau boundaries from 15 °C go 10 °C. From the first hour the oxygen uptake intensity smoothly decreases and, by the 4th hour it has reached 33 % of the initial value. For 12 hours the respiratory intensity decreases and then it oscillates around a level which is less than half the initial value (close to the level of metabolism at the end of two weeks acclimation). After the first day appreciable variations of respiration intensity, particlarly in the first week, are observed, they then become gradually smaller.

Earthworms *Eisenia nordenskiöldi* (Oligochaeta, Lumbricidae) collected in the northern most part of their habitat, in Taymir, have an invariant level of stable oxygen uptake in the temperature range 4 °C to 8 °C (usual summer temperature for this habitat). Temperature decrease in this range is accompanied by a smooth decrease in oxygen uptake intensity (Fig. 40). In the first hour the level of respiration intensity is maintained at a constant level, but, during the following 10 hours the respiratory intensity becomes only half as high. During the rest of the first 24 hours oxygen uptake intensity varies insignificantly.

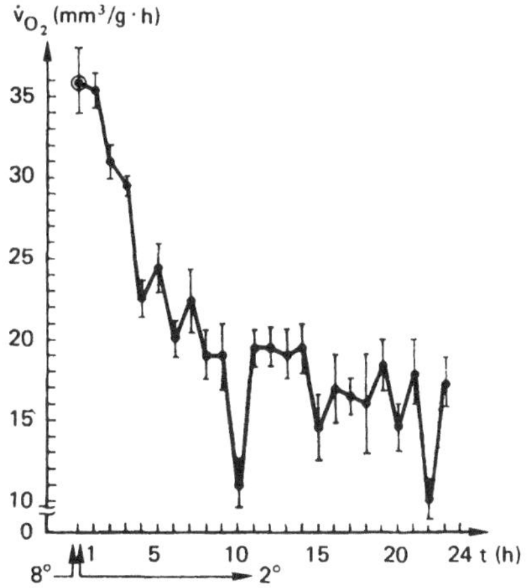

Fig. 40 Respiration intensity of *Eisenia nordenskiöldi*(Eisen) from arctic population on the first day after temperature change (n = 25).

The experimental data presented here are insufficient to allow generalizations. However, it is possible to note some features of similarity and difference, between the primary reactions of respiratory metabolism to change of temperature (during the first 24 hours with hourly readings). With an increase, or decrease, of temperature within the temperature range characterized by a linear relationship between oxygen uptake intensity and temperature, the value of the first response (first hour) depends on the abruptness (5 °C, 10 °C or more), as well as on the direction of temperature change. When the temperature is decreased within the plateau region, then in the first hour the level of oxygen consumption may remain the same. It should be noted that according to Blazka's evidence, as well as to our own observations, an increase in respiratory intensity in the first hour after temperature change is not associated with intensification of motor activity.

Consequently, the material presented here suggests certain differences in acclimation rates of soil invertebrates when one constant temperature regime is substituted by another. These differences are conditioned by several causes. We believe that the acclimation rate depends, first of all, on the pattern of stable levels of oxygen uptake in the particular temperature range (linear relationship, plateau), in the direction of temperature change (increase, decrease) and on its abruptness. The same phenomena seem to determined the differences in the primary reactions to the substitution of a different temperature regime.

Relationship Between Maximal Metabolism, Body Weight and Standard Metabolism of Animals

G. V. Dontsova, A. I. Zotin

The value of the maximal, possible increase of respiration intensity under the effect of external or internal factors is of great importance for vital activity in animals. The value determines, amongst other things, the maximal power developed by organisms, rates of responses to changed environment etc. It appears that evolution in animals proceeds towards increasing maximum levels of metabolism in organisms with similar weight, since this provides substantial advantages in the struggle for life.

What properties of the physical world are used in the course of evolution to select organisms for maximal metabolism? It is difficult to give an comprehensive answer to this question, however, some assumptions can be advanced as to the properties of open systems to which all living organisms pertain. One of the assumptions is that the value of response of an open system, in a stationary state, to external influence depends on the degree of remoteness of the open system from the equilibrium state. Since the stationary state of the open system is determined by the value of the system's dissipation function, this assumption is in accord with the statement that, maximal deviation of the open system from the stationary state is a function of the dissipation function:

$$\Delta\Psi = \Psi_m - \Psi_{st} = f\Psi, \qquad (1)$$

where $\Psi_{st} = \frac{T}{V} \cdot \frac{d_i S}{dt}$ is a specific dissipation function of the system in a stationary state; Ψ_m the maximal possible value of this function for deviation from the stationary state; T the absolute temperature; V the system volume; d_iS/dt the rate of entropy production in the system.

Thermodynamics and Kinetics of Biological Processes

Unfortunately, the meaning of function (1) has not yet been stated in thermodynamic terms, therefore an empirical method has to be used in its determination. If it is believed (see Zotin, 1972) that the dissipation function for living systems is proportional to the respiratory intensity of the organism, then an attempt can be made to find the nature of function (1) for the living organism from experimental data on the relationship between maximal metabolism and standard (basal) metabolism. Thus this work has two aims; on the one hand to reveal the relationship between maximal and standard metabolism in animals, and, on the other - to show with this example, that it is possible to establish the concrete nature of function (1). To do this it is necessary to consider such concepts as basal and standard metabolism, active and maximal metabolism, as well as the factors upon which they depend.

Usually basal, or resting, metabolism is estimated by the rate of oxygen uptake in the state of complete rest of an animal. This correlates with minimal respiration intensity, i. e. in terms of the thermodynamics of irreversible processes with the stationary state of the living system (Zotin, 1972). As a rule, it is impossible to achieve complete rest in animals (man exluded) under experimental conditions. Therefore, usually standard metabolism and not basal metabolism is measured. Standard metabolism corresponds to the respiratory intensity of the animal in an inactive state. Inaccuracy in, and the conventional character of, such a definition of standard metabolism is evident. Many small animals, invertebrates in particular, cannot be in an inactive state, therefore in the literature the term *routine* or *common* metabolism is used in addition to standard. All this produces certain difficulties when comparing energy metabolism in different animals; and to obtain real relationships a great quantity of data has to be collected.

In spite of this quite exact relationship for standard metabolism are obtained now; the most important relationship is that between the value of standard metabolism and the body weight of animals. This relationship, found as a result of numerous investigations, for all animal species (see Zeuthen, 1953, 1970; Winberg, 1956 a, 1976; Ivlev, 1959, 1963; Hemmingsen, 1960; Lasiewsky, Dawson, 1967; Dolnik, 1968; Zar, 1968; Schmidt-Nielsen, 1970, 1975; Zar, 1970; Byzova, 1972; Sushchenya, 1972; Zotin, 1972; Bennett,

Dawson, 1976; Prosser, 1977 and others) is one of the few securely established, quantitative relationships in biology. In a general form it may be written as

$$\dot{Q}_{O_2} = aW^{k} \tag{2}$$

or

$$\dot{q}_{O_2} = aW^{-b} \tag{3}$$

where $\dot{Q}_{O_2}$ is the rate of oxygen uptake; W the body weight; $\dot{q}_{O_2} = \frac{\dot{Q}_{O_2}}{W}$ the respiratory intensity; a, k and $b = 1 - k$ are constants.

Firstly, the significance of (2) became clear when it was found that (2) holds for all mammals "from mouse to elephant" (Benedict, 1938; Brody, 1945; Kleiber, 1947, 1961; Slonim, 1952; Schmidt-Nielsen, 1975; Prosser, 1977). It is usually assumed that in this relationship values of the constants for mammals are, respectively, a = 18.0* mW and k = 0.75 (Zotin, Konoplev, 1978) - it is the so-called "Kleiber curve". Some mammals deviate significantly from the Kleiber curve, the *Soricidae* (Insectivora), or representatives of *Edentata* for instance. It should be noted that although for the group of mammals the relationship $\dot{Q}_{O_2} = 18.0\ W^{0.75}$ is true, for each order, however, the values of the constants vary (Konoplev et al., 1978).

Much evidence showing that standard metabolism depended on body weight has been obtained for birds (see Brody, 1945; King, Farner, 1961; Lasiewsky, Dawson, 1967; Dawson, Hudson, 1970; Schmidt-Nielsen, 1975; Pimm, 1976; Prosser, 1977 and others). In this case a total King-Farner curve is used as standard (Fig. 41). This is characterized by the constants a = 21.0 and k = 0.76 (Zotin, Konoplev, 1978). As for mammals, the data obtained for small birds do not correspond to the King-Farner curve (Zotin, Konoplev, Grudnitzky, 1978). Essential differences were found between *Passeriformis* and all other birds (Lasiewsky, Dawson, 1967; Aschoff, Pohl, 1970; Berger et al., 1970; Schmidt-Nielsen, 1975), as well as variation in the measurement conditions of standard metabolism (Fig. 41).

---

* Here and below the constant 'a' given by the authors in cal/hour or other similar units has been converted into mW.

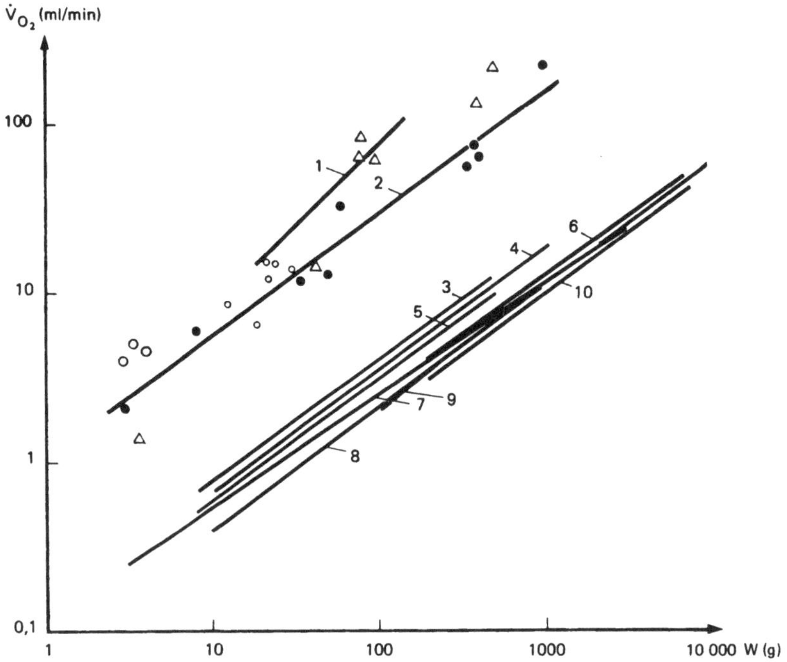

Fig. 41 Relationship between standard and maximal metabolism and body weight in birds (from Hart, Berger, 1972)
Resting metabolism: data for all birds (7) and those with the weight above 100 g (9) (King, Farner, 1961); for passering (4) and nonpasserine (8) birds (Lasiewski, Dawson, 1967); for passerine and nonpasserine in the dark, at day (3), (6) resp. and at night (5), (10) (Aschoff, Pohl, 1970). Metabolism at flight: formation of CO_2 in 13 species of birds (1) (Teal, 1969); total data on respiration intensity in birds (2) (Hart, Berger, 1972).

In addition to mammals and birds, many studies have been conducted on *Crustacea* (Winberg, 1950, 1956, b; Reichle, 1968; Sushchenya, 1972; Marshall, 1973; Ivleva, 1974), *Mollusca* (Winberg, Belyatskaya, 1959; Turpaeva et al., 1970; Alimov, 1975); *Insecta* (Edwards, 1953; Wigglesworth, 1974); *Teleostei* (Ivlev, 1954; Winberg, 1956 a; Pritchard et al., 1958; Belokopitin, 1968; Bashamchideen, Parvatheswararao, 1976); *Reptilia* (Bartholomew, Tucker, 1964; Calvao et al., 1965; Bennett, Dawson, 1976) and other groups of animals. All these studies confirm the validity of equations (2) and (3) and their wide applicability, not only to homoiothermic, but to poikilothermic animals as well.

Different environmental factors (oxygen content, salinity, pH and others), but especially temperature, exert certain influences on basal and standard metabolism (see Winberg, 1956 a; Fry, 1957; Wolvekamp, Waterman, 1960; Farmanfarmaian, 1966; Ivleva, 1972; Sushchenya, 1972; Schmidt-Nielsen, 1975; Bennet, Dawson, 1976; Prosser, 1977 and many others). The effect of these factors may be of purely physical character, for instance acceleration of chemical reactions with temperature increase, necessity for osmotic work with change of environment salinity and so on. The effect may, however, also depend on specific biological reactions related to, for example, the damaging influence of environmental factors on the organism, to differences in the rate of adaptation, to changes in the environment, to an increase of motor activity etc. We shall not give a detailed account of these phenomena, since this can lead us far from the main subject of this contribution. It should be noted that, as is clear from the above, we shall consider not only standard metabolism, but active metabolism as well.

The notion of active metabolism is as indefinite as that of standard metabolism, since its definition does not include a value for the animals' motor activity. Recently some progress has been made in studies of active metabolism in animals because of availability of various methods for determination of the respiration of animals under conditions of accurately dosed loads (see Durnin, Passmore, 1967; Matyukhin, 1973; Schmidt-Nielsen, 1975). The most reliable data were obtained when investigations into the relationship between respiration and the velocity of running, flying or swimming in animals and men were performed. In all these cases active metabolism increased

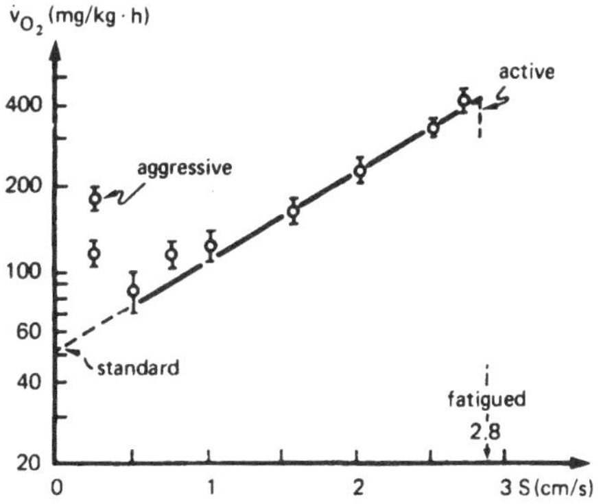

Fig. 42 Dependence of active metabolism $\dot{v}_{O_2}$ of pumpkinseed on swimming speeds at 20 °C (Brett, Sutherland, 1965).

abruptly of running, flying or swimming in animals and men were performed. In all these cases active metabolism increased abruptly with an increase in physical load. Fig. 42 exemplifies the data on the enhancement of respiratory intensity with change in the swimming velocity of pumpkinseed *Lepomis gibbosus* (Brett, Sutherland, 1965). From the data on the relationship between active metabolism and velocity of movement both values can be obtained, that of standard metabolism by extrapolating the curve to zero velocity of movement or of maximal metabolism, if the limiting possible velocity of the organism movement is known. Estimation methods for maximal volume would be essentially simplified if total analytical expressions could be obtained for the relationship between respiration intensity and motor activity in animals. Unfortunately, this is difficult at the present time and, from Fig. 42, it seems that this relationship is of an exponential character. However, according to other observations of fishes this relationship can be expressed, by a parabolic equation (Ivlev, 1962; Alekseeva, 1967; Matyukhin, Stolbov, 1970), and in mammals by a linear equation (Taylor et al., 1970; Raab et al., 1976 etc.). Much work will be required to express this relationship quantitatively for different animal species, and there is no guarantee that universal constants similar to the constants a and k in (2) will be obtained. Therefore at present the value of maximal respiration of animals can only be obtained experimentally. Thus possible errors due to incorrect estimation of maximal load and to the methods used for obtaining data (short-term, long-term and super-long-term loads and others) should be taken into account. In spite of the above it is possible to obtain interesting relationships for maximal metabolism by using the data of various authors and collecting adequate statistical material.

As with standard metabolism the maximal level of animal respiration depends on body weight: the higher the body weight, the higher the value of maximal metabolism (see Brody, 1945; Wilkie, 1959; Brett, 1965; Brett, Sutherland, 1965; Jansky, 1966; Tucker, 1969, 1970; Lomov, 1970 a, b, 1975; Berger et al., 1970; Pasquis et al., 1970; Hart, Berger, 1972; Thomas, 1975; Bennett, Dawson, 1976; Casey, 1976 and others). This relationship is similar to that found for standard metabolism and body weight (2) (see also Wilkie, 1959; Brett, 1965; Farmer, Beamish, 1969; Berger et al., 1970; Thomas, 1975; Casey, 1976)

$$\dot{Q}_{O_{2m}} = a_m W^{k_m} . \tag{4}$$

For fishes it is expressed by the formula $\dot{Q}_{O_{2m}} = 4.9\, W^{0.75}$ (Winberg, 1956 a); for reptiles $\dot{Q}_{O_{2m}} = 16.0\, W^{0.82}$ (Bennett, Dawson, 1976); for birds $\dot{Q}_{O_{2m}} = 356\, W^{0.79}$ (Hart, Berger, 1972); for mammals $\dot{Q}_{O_{2m}} = 151\, W^{0.73}$ and $\dot{Q}_{O_{2m}} = 228\, W^{0.79}$ (Pasquis et al., 1970).

Brody (1945) (see Wilkie, 1959) seems to be the first to have suggested that maximal metabolism in animals is determined by the value of basal metabolism. Lomov (1970 b) wen still further believing that relative maximal power (vigour) of animals (which was determined from maximal metabolism) is strictly associated with basal metabolism and that there exists some constant ratio between them. In other words, according to these authors, maximal metabolism in animals is proportional to the value of basal or standard metabolism. Thus the similarity between equation (2) and (4) is understandable. In addition, this statement allows the form of equation (1) to be found for living systems.

To support this statement we used to prove that the constant in the equation (2) does not differ from constant K in equation (4). In fact if $\dot{Q}_{O_2} = aW^{k}$ and $\dot{Q}_{O_{2m}} = a_m W^{k_m}$ then $\dot{Q}_{O_{2m}} = a_m W^{k_m}/aW^{k} \cdot \dot{Q}_{O_2}$ and at $k = k_m$

$$\dot{Q}_{O_{2m}} = \frac{a_m}{a} \dot{Q}_{O_2} = \alpha \dot{Q}_{O_2} . \tag{5}$$

Proceeding from the data given in Figs. 41 and 43 (similar observations were also made in some other studies, such as, for example, those of Wilkie, 1959; Berger et al., 1970; Lomov, 1970 b; Pasquis et al., 1970) for birds and mammals, it can be seen that $k = k_m$. There are some observations, however, which do not support this conclusion (Brett, 1965; Rao, 1968). According to Brett (1965), for instance, for one species of Salmonidae (the sockey salmon *Oncorhynchus nerka)*, $k = 0.77$, and $k_m = 0.97$, i. e. $k \neq k_m$. Possibly, these observations are erroneous, because earlier, Job (1955), obtained other values of the constants k and k_m for *Salvelinus fontinalis,* the eastern brook trout. However, in this study different values of these constants were again obtained; at 5 °C $k = 0.85$ and $k_m = 0.94$;

at 10 °C k = 0.84 and k_m = 0.86; at 15 °C k = 0.84 and k_m = 0.85; at 20 °C k = 0.80 and k_m = 0.75 (Job, 1955). Judging from Fig. 43, where Job's data are presented, these differences do not seem reliable. Nevertheless, much work is required to get the final proof for the equality between the constants k and k_m for different animal groups.

The second principal problem to be solved is whether constant α in equation (5) is a universal constant or if it varies in different animal groups and has specific peculiarities connected with evolution or ecology in certain animal groups. If the constant α is proved to be universal, it would mean that it reflects the properties of the organism behaving as an open system.

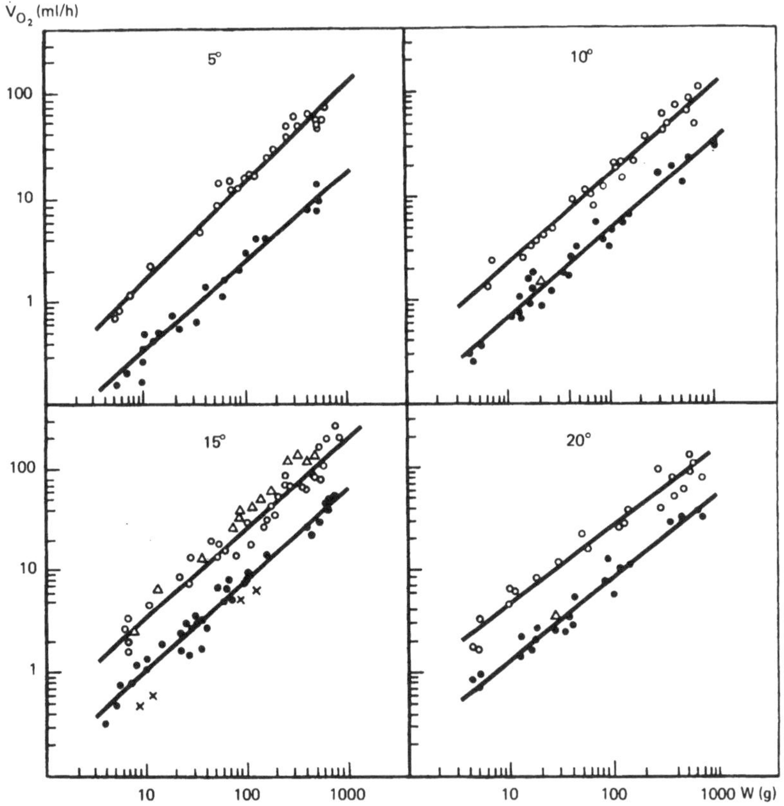

Fig. 43 Relationship between standard and active metabolism $\dot{V}_{O_2}$ of *Salvelinus fontinalis* and body weight W and environmental temperature (Job, 1955).

To obtain a preliminary solution of the problem the literature data relating to standard and maximal metabolism for different animal groups are arrised in Table 2. As is seen from Table 2, the constant α varies significantly for different species. In most cases, however, its values were similar for various classes and even types of animals. Against this background, the data obtained for some species of insects and birds stand out (Table 2). They require therefore special consideration. It is well known (see for instance, Schmidt-Nielsen, 1975; Prosser, 1977) that some animals (and twilight animals in the day-time) become torpid at night-time with a very low level of metabolism. If the respiratory intensity in the torpid state is taken as standard metabolism, high values of α can be obtained. For instance, in the mouse *Peroganthus californius* in the torpid state body temperature drops to 10 °C, and respiration to 0.2 ml/g/hour (Tucker, 1965). If the level of standard metabolism is counted from this value then α = 58.1. However, during active movement the body temperature increases (especially for insects), and the maximum metabolism which strongly depends on temperature is measured practically, under temperature conditions different from those under which standard metabolism was obtained. For instance, body temperature during free flight of *Sphingida* or the bee reaches 40 °C at environmental temperatures of 16 to 33 °C (Heinrich, Casey, 1973; Heinrich, 1975). Therefore, the maximal metabolism of these animals cannot be compared with a standard metabolism recorded at environmental temperatures much lower than 40 °C. It is interesting to note that, in contrast to other poikilothermic animals, e. g. reptilia (see Bennett, Dawson, 1976), respiration during insect flight is independent of the environmental temperature (Heinrich, 1971, 1975). Consequently, not all the data presented in Table 2 should be taken into account when comparing standard and maximal metabolism. On the basis of the above considerations it was concluded that only those observations of maximal and standard metabolism where $\alpha < 20$ can be used for this purpose. It goes without saying, that the minmal values of the constant α ought also to be limited, since many studies have yielded reduced values for maximal metabolism. Here, we have excluded results if α , calculated from data in the literature, was below 4.

Table 2
Standard ($\dot{q}_{O_2}$) and maximal ($\dot{q}_{O_{2m}}$) metabolism in various animal species.

Species	W	t	$\dot{q}_{O_2}$	$\dot{q}_{O_{2m}}$	α	Author
	g	°C	mW	mW		
1	2	3	4	5	6	7
			Insects			
Drosophila repleta	0.0033	25	9.70	122	12.6	Chadwick, Gilmour, 1940
Drosophila americana	0.0011	20	9.07	127	14.0	Chadwick, 1947
Schistocerca gregaria	1.6	28	3.66	75	20.5	Krogh, Weis-Fogh, 1951
Sarcophaga bullata			10.0		25.2	McCann, Boettiger, 1952
Drosophila melanogaster	0.0009	42	28.9	191	6.6	Hocking, 1953
Simulium vittatum	0.0021		28.0	287	10.2	"
Simulium venustum	0.0025	38	25.6	159	6.2	"
Aedes nearcticus	0.0031		32.3	119	3.7	"
Aedes flavescens	0.0035		22.3	127	5.7	"
Tabanus septentrionalis	0.062		6.57	323	49.2	"
Tabanus affinis	0.16		4.81	125	26.0	"
Apis mellifica	0.10		18.3	351	19.2	"
Various *brachycere*			8.7	122	14.0	Ludwig, 1963
Melolontha melolontha			2.15	230	107	"
Amphimallus solstitialis			2.21	221	100	"
Aphodius prodomus			0.81	84.3	104	"
Necrophorus vespillo			5.58	93.6	16.8	"
Coccinella septempunctata			3.14	56.4	18.0	"

1	2	3	4	5	6	7
Chrysopa vulgaris			6.57	97.7	14.9	Ludwig, 1963
Schistocerca gregaria	2.0		14.3	128	8.9	Weis-Fogh, 1964
Celerio lineata	0.39	26	34.9	453	13.0	Heath, Adams, 1967
Hyalophora cecropia	1.4	25	10.1	93.0	9.2	Henegan, Heath, 1970
	2.7	25	5.46	49.6	9.1	
Euconocephalus nasutus	0.65		13.0	107	8.2	Don Stevens, Josephson, 1977
Neoconcephalus rubustus	0.87		11.2	91.8	8.2	"
		Arachnoids				
Lycosa insignita	0.04	15	0.81	5.35	6.6	Miyashita, 1969
		Fishes				
Oncorhynchus nerka	746	15	0.29	2.98	10.3	Brett, 1965
Lepomis gibbosus	45	20	0.19	1.66	8.7	Brett, Sutherland, 1965
Carassius auratus	61	15	0.08	0.87	10.8	Smit et al., 1971
Salmo gairdneri	264	15	0.30	2.67	8.9	Webb, 1971
Thymallus arcticus	292	4	0.25	1.39	5.5	Stolbov, 1972
Coregonus autumnalis	321	4	0.24	0.94	3.9	"
Oncorhynchus nerka	1792		0.22	2.86	13.0	Brett, 1973
Thymallus arcticus	388	4	0.25	2.72	10.9	Matykhin, 1973
Spicara smaris	50	12	0.01	0.044	4.4	Muravskaya, Belokopytin, 1975
Tiliapia milotica	19		0.24	3.35	13.9	Yarzhomlek et al., 1977

1	2	3	4	5	6	7
			Amphibia			
Hyla regilla	2.8	20	0.35	1.98	5.6	Bennett, Licht, 1973
Batrachoseps attenua	1.5	20	0.35	1.06	3.0	"
Scaphiopus hammondii	11	20	0.54	4.73	8.9	Slymour, 1973
Bufo cognatus	40	20	0.65	6.80	10.4	"
Rana pipiens	38	20	0.56	2.14	3.8	"
Rana catesbeiana	44	20	0.22	0.92	4.2	"
			Reptilia			
Varanus sp.	714	20	0.16	1.03	6.4	Bartholomew, Tucker, 1964
Iguana iguana	795	24	0.46	3.20	6.9	Moberly, 1968
Cnemidophorus tigris	18	20	0.30	2.38	7.9	Asplund, 1970
Sphenodon punctatum	430	10	0.05	0.37	7.4	Wilson, Lee, 1970
Pituophis catenifer affinis	548	20	0.11	1.57	14.2	Greenwald, 1971
Sauromalus hispidus	574	20	0.10	1.51	15.1	Bennett, 1972
Varanus gouldii	674	20	0.14	1.86	13.3	"
Pseudemus scripta	305	40	0.52	5.98	11.5	Gatten, 1974
Terrapene ornata	354	20	0.13	1.21	9.3	"
Physignathus lesueuri	585	20	0.18	1.35	7.5	Wilson, 1974
Amphibolurus barbatus	306	20	0.23	1.74	7.5	"
Egernia cunninghami	260	20	0.20	1.43	7.1	"
Tiliqua rugosa	424	20	0.18	1.36	7.5	"
Amblyrhynchus cristatus	489	20	0.11	0.68	6.2	Bennett et al., 1975
Dipososaurus dorsalis	51	30	0.52	5.17	9.9	Bennett, Dawson, 1976

1	2	3	4	5	6	7
Tiliqua scincoides	493	20	0.15	0.78	5.2	Bennett, Dawson, 1976
Trachydosaurus rugosa	461	30	0.53	2.61	4.9	"
Sceloporus occidentalis	13	25	0.52	5.23	10.0	Bennett, Gleeson, 1976
Uromastix aegyptius	1.1	42	0.41	2.15	5.2	Dmi'el, Rappoport, 1976
Chelonia mydas	735		0.40	1.45	3.6	Prange, 1976
Crotalus viridis	301	35	0.46	3.20	6.9	Ruben, 1976
Masticophis flagellum	262	35	0.46	5.93	12.9	"
Coluber constrictor	262	35	0.46	5.93	12.9	"
Lichanura roseofusca	314	32	0.43	1.45	3.4	"
		Birds				
Calypte anna	3.8	24	62.2	494	7.9	Pearson, 1950
	3.8	24	4.9	494	101	"
Selasphorus sasin	3.2		22.0	692	31.4	Lasiewski, 1963
Selasphorus rufus	3.9		19.8	442	22.3	"
Stellula calliope	2.7		23.2	385	16.6	"
Calypte anna	5.0		22.7	500	22.0	"
Calypte costae	3.1		17.4	246	14.1	"
Melopsittacus undulatus	35		18.4	122	6.6	Tucker, 1968
Zonotrichia albicollis	28		22.9	266	11.6	Teal, 1969
Ramphocelus flammin	44		26.6	299	11.2	"
Icterus galbula	38		15.0	263	17.5	"
Scardafella squamata	51		22.9	263	11.5	"
Cyanocitta cristata	89		13.2	207	15.7	"
Agelaius phoeniceus	69		28.7	234	8.1	"

1	2	3	4	5	6	7
Euplectes arix	19		29.5	216	7.3	Teal, 1969
Cuiraca caerulea	22		26.6	213	8.0	"
Heterophasis capistrata	38		15.0	187	12.5	"
Spreo superbug	54		21.3	166	7.8	"
Passer domesticus	29		34.5	225	6.5	"
Colaptes auratus	146		15.0	171	11.4	"
Hesperiphona vespertina	59		13.9	198	14.2	Berger et al., 1970
Anas rubripes	1026		6.34	80	12.6	"
Anas platyrhynchos	1081	23	5.41	22.2	4.1	Prange, Schmidt-Nielsen, 1970
Rhea americana	22000	25	2.21	25.6	11.1	Taylor et al., 1971
Larus atricilla	322		8.49	66.3	7.8	Tucker, 1972
Corvus ossifragus	275		2.39	24.5	10.2	Tucker, 1973
Columbia livia	442		7.08	69.8	9.8	Butler et al., 1977
		Mammals				
Canis familiaris	10000		2.93	32.1	10.9	Young et al., 1959
Perognathus californicus	22	30	7.96	67.4	8.5	Tucker, 1965
Rattus norvegicus	400	20	6.22	32.1	5.2	Jansky, 1966
Peromyscus sp.	24	20	9.24	62.2	6.7	Segrem, Hart, 1967
Mus musculus	33		8.35	60.1	7.2	Pasquis et al., 1970
Ochrotomys nuttalli	100		6.31	41.2	6.5	"
Rattus sp.	376		4.53	31.5	6.9	"
Cavia sp.	841		3.71	23.4	6.3	"

1	2	3	4	5	6	7
Canis familiaris	6300		2.24	24.9	11.1	"
Homo sapiens	68100		1.24	17.7	14.3	"
Equus caballus	677000		0.70	13.1	18.7	"
Dipodomys merriami	36	25	8.72	64.0	7.3	Yousef et al., 1970
Phyllostomus hastatus	101	25	38.5	160	4.1	Thomas, Suthers, 1972
	90	25	4.42	109	24.7	"
Mus musculus	21	24	13.1	50.0	3.8	Yousef et al., 1973
Dipodomys merriami	31	25	13.4	75.0	5.6	"
Dipodomys spectabilis	104	24	6.63	22.3	3.4	"
Ammospermophilus leucurus	86	25	8.72	34.0	3.9	"
Pteropus poliocephalus	750		2.91	58.1	20.0	Carpenter, 1974
Canis familiaris	12600		3.08	20.3	6.6	Raab et al., 1976

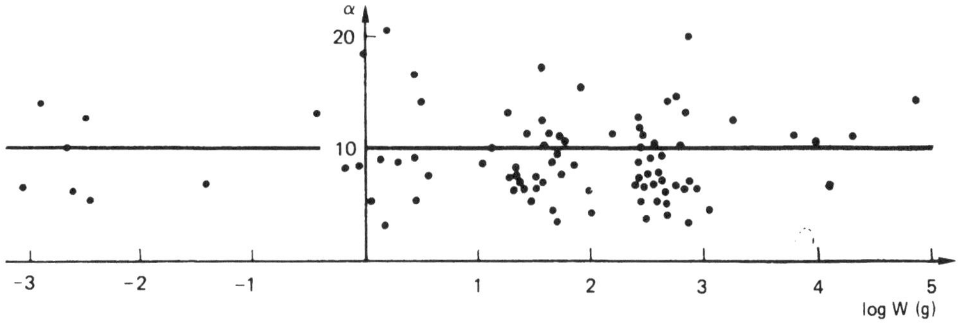

Fig. 44 Dependence of constant α (relationship between maximal and standard metabolism) on body weight W of animals.

The relationship between maximal and standard metabolism, within the indicated limits $4 < \alpha < 20$ is presented in Fig. 44. It is seen that maximal and standard metabolism are in a direct proportion, i. e. relationship (5) is valid. It follows that α is a universal constant equal for animal groups differing in ecology and systematics to,

$$\alpha = 9.8 \pm 0.4 \, . \tag{6}$$

Certainly, additional special investigations are required to prove this conclusion, but the exceptions noted above are insignificant.

Thus on the strength of the data presented in Table 2 and in Fig. 44 it is concluded that the value for maximal metabolism is determined by the level of standard (basal) metabolism and for living systems relationship (1) is expressed by the formula:

$$\Delta\Psi = \Psi_m - \Psi = \beta\Psi , \tag{7}$$

where β is a constant equal to $\alpha - 1 = 8.8$.

It should also be noted that maximal possible intensification of respiration strongly depends on the size of the animals, as follows from formulae (2), (4), (6) and (8) and is expressed by the relationship:

$$\Delta \dot{q}_{O_2} = \dot{q}_{O_{2m}} - \dot{q}_{O_2} = a \beta W^{-b} \quad (8)$$

The smaller the animal the higher specific power it can produce with similar constants a and b. However conclusion and the formula analogous to (8) were known earlier (Lomov, 1970 b, 1975).

Kinetics of Transition Processes in Living Systems - Load, Adaptation, Stability

E. Z. Rabinovich

One of the poorly studied and urgent aspects of physiology is the energetics of organisms reacting to various environmental effects, i. e. the energetic aspects of transition processes in living systems. It seems to us that the search for criteria providing optimization of energetic expenditures during adaptation, i. e. under prolonged unidirectional effects in the environment, is of great interest.

To elucidate this problem a general consideration of specific features of transition processes in living systems must be carried out, in particular adequate models ought to be described. The modern literature on transition processes in the organism includes many studies where concepts in common use in the theory of linear processes, such as transmission function, control grade, feedback, stability etc. are used. These studies describe some models of control systems providing, in a general form, a certain pattern of transition processes for living systems. At the same time such an approach gives no explanation for the selection of these characteristics of the transition process which are defined at each instant and are valuable for the integral, living system. Thus it has proved impossible, in terms of this theory, to consider such important characteristics of living matter as adaptation and evolution. This means essentially that there is no biological advantage in using this theory. Rosen (1967), Mesarovich (1971), Grodins (1963) point to such incosistencies.

In a general transition process patterns are observed for step-like or impulse effects on the organism, or for some system changes in the organism or its systems. An attempt can be made to analyse such transition processes by solving a differential equation of the general form:

Thermodynamics and Kinetics of Biological Processes

$$a_n \varphi^{(n)} + a_{n-1} \varphi^{(n-1)} + \ldots + a_0 \varphi = R(t) \quad (1)$$

where φ is the studied characteristics; a_i $(i = 1, \ldots, n)$ are constants; R(t) the inlet effect.

For this purpose equation (1) should be interpreted in terms of the problem considered. Unfortunately, one cannot know what order limits the relationship. In some cases this order may be found by extrapolation of dynamic curves. Thus if the curves are of an aperiodic character, the solution of (1) is the sum of exponents:

$$\varphi(t) = C_1 e^{\pm r_1 t} + \ldots + C_n e^{\pm r_n t} + B. \quad (2)$$

This proves that a large class of transition processes in living systems is described by a second order equation.

To derive an equation for a closed control system we introduce a negative feedback for the controlled parameter and get:

$$a_2 \ddot{\varphi} + a_1 \dot{\varphi} + a_0 \varphi = k[R(\varphi_i) - \varphi], \quad (3)$$

where R(φ_i) is the predetermining controlling effect.

In the general case R(φ_i) is some time function, in particular a harmonic function.

With such an approach variations of φ in a transition process and in an established state are determined not only by the properties of the system itself, but by a harmonic pattern in the controlling effect.

The basic problems remain unsolved. In spite of the fact that the method proposed for analyzing transition processes in the organism makes it possible to elucidate some important details of the regulation of homeostatic characteristics under extremal exposures and, that it explains, to a considerable extent, the differences in the established parameter values during the course of the transition processes. It is absolutely unclear why, under equal stimulation, essentially different controlling effects are formed in different subjects, as well as in one individual (depending on its state). If the transition process dynamics are reliably described in terms of equation (3), the little can be said about the mechanisms of the process itself. For this purpose then the coefficients ought to be interpreted in physiological terms. This is, practically, impossible, since the above approach presents the inverse problem in which equation (3) was obtained not from biological prerequisites but was assumed a priori.

The above considerations draw our attention to the following fact: it seems impossible to obtain a useful (in a biological sense) kinetic equation to describe a transition process for some physiological parameter without taking into account the dynamics of its interaction with other parameters of another functional system in terms of which this parameter is regulated.

Any functional system of the organism can be described by means of phenomenological equations of interaction between its dynamic parameters:

$$P_i(\varphi_j, \ldots, \varphi_n) = 0 \qquad (4)$$

where, in the general case providing a quasistationary change of dynamic parameters for any φ_k an equation of the dynamic system is obtained from equation (4):

$$\dot{\varphi}_k + A(t)\cdot\varphi_k = B(t)\cdot U(t), \qquad (5)$$

where $A(t) = A(\varphi_j)$; $B(t) = B(\varphi_j)$; $U(t) = U(\varphi_j)$; $j \neq k$.

Let us consider as an example impulse or step-like effects in human blood circulation: physical loads, changes in the body posture in space, emotional reactions, stresses and so on. For the analysis of pulse frequency we obtain an equation of the dynamic system from basic phenomenological ratios for the blood circulation system (Shumakov et al., 1971):

$$P_a = R\cdot\varphi\cdot V_{st}; \quad Q = \varphi\cdot V_{st} \qquad (6)$$

$$V_{st} = N/P_a; \quad N = b\,V_{f.d.},$$

where P_a is the arterial pressure, R the peripheral resistance; S the pulse frequency; V_{st} the stroke volume; Q the minute volume; $V_{f.d.}$ the final diastolic volume; N the contraction energy.

These ratios (6) are the unknown phenomenological equations of interaction (4) for the blood circulation system.

After differentiation, substitution and transformation from (6) we obtain an equation for the dynamic system:

$$\begin{aligned} &\dot{\varphi} + A(t)\cdot\varphi = B(t)\cdot U(t), \\ &A(t) = \frac{1}{V_{f.d.}}\left(\dot{V}_{f.d.} + \frac{\dot{R}}{R} + \frac{\dot{P}_a}{P_a}\right), \\ &B(t) = 1/N; \quad U(t) = Q . \end{aligned} \tag{7}$$

The main advantage of the approach suggested consists of the fact that analysis of the kinetics of transition processes in the organism according to equations (5), (6) and (7) provides for a physiological interpretation of the coefficients of the dynamic system. However, an important question remains unsolved: why does only one transition characteristic out of the whole family of curves, exist for the physiological parameter studied for some organism system in a definite, initial state. To answer this question let us turn again to equation (5), the equation for a dynamic system. With such an approach U(t) is a governing function, which minimizes some function determined for a set of permissible relationships of [U(t)] . According to Rozen (1969) such a function, i. e. an estimating function for the living systems, is the metabolic cost of the processes proceeding in them. Minimization of this function determines the rigid form of the controlling function and, consequently, it gives single-valued transitional characteristics of the system.

If by metabolic cost total metabolism of the organism, i. e. specific uptake of oxygen, is meant, it is necessary to minimize the enthalpy. For the biological system the latter is the sum of biological work and heat dissipated from the system, and, according to the First Law of thermodynamics, is equivalent to total metabolism. But this approach is inadequate to the real phenomenology of transition processes in the living systems, which often tend to the maximization of heat transfer, even at the expense of additional metabolic consumption, as for instance, during intensive work of muscles.

One of the basic metabolic functions in the organism is the accumulation of free energy, and this state function has no component of external heat dissipation. In view of this it is useful to interpret metabolic cost as the intensity of free energy expenditure. A change of free energy value in time for an open thermodynamic system, such as an entire organism, or of its sub-system can be written (Haase, 1963) as:

$$\frac{dF}{dt} = -\left(S\frac{dT}{dt} + \frac{dW}{dt}\right) + \sum_{r} A_r \frac{d\varepsilon_r}{dt} + \sum_{k} \mu_k \frac{d_e n_k}{dt}, \quad (8)$$

where F is the free energy; S the entropy; T the absolute temperature; W the work; A_r the affinity and ε_r the degree of completeness of the chemical reaction; μ_k the chemical potential; and n_k the number of substance particles in transfer processes.

It is seen that F grows, due to an increase of chemical energy inside the system, and, due to the supply of free energy accompanying the substance. Reduction of F is caused by working expenditure and by temperature increase. It is clear that minimization of the velocity of F expenditures falls into two parts, i. e.

$$\min\left(\frac{dF}{dt}\right) = \min\left(-S\frac{dT}{dt} - \frac{dW}{dt}\right) + \max\left(\sum_{r} A_r \frac{d\varepsilon_r}{dt} + \sum_{k} \mu_k \frac{d_e n_k}{dt}\right). \quad (9)$$

Then according to the theory of dynamic systems (Rozen, 1969), the function $\mathfrak{F}_B = dF_B/dt$ is written in the form

$$\mathfrak{F}_B = \int_{t_1}^{t_2} L[t;\ \varphi(t),\ U_B(t)] \quad (10)$$

where the sufix "B" points to a change under forcing effects. For the example under consideration the nucleus of function L associates the studied characteristics of the blood circulation system with the intensity of expenditures of the free energy F in a transition process. The solution of the optimization problem will now be reduced to a search for those transition characteristics for which $\mathfrak{F}_B$ is minimized.

In the theory of dynamic systems, adaptation is understood as that type of behaviour in which the system optimizes some processes or states in accordance with the optimum criterion, i. e. in conformity with the minimization of the estimating function-functional. Then, in accordance with our supposition for the living system the reaction to the stimulus will be an adaptive one, if it is characterized by a decrease of expenditure of free energy in a transition process. To understand the mechanism of prolonged adaptation under repeated impulses let us analyze the experimental material about the effect of stimulus amplitude on the character of transition process (Arshavsky, 1967, 1971, 1976 a). It has been shown that the monophasic pattern of transition processes corresponds to small values of extremal load stimulus. This type of reaction is exhibited, in adult animals, by a number of characteristics, for example under short tetanus contraction of skeletal muscle under the effect of a weak stimulating current. In early post-natal ontogenesis a vast numbers of the transition characteristics of various physiological parameters are of the same nature, e. g. the well-known action potential. When the stimulus amplitude increases up to a certain value in adults, a multiphase pattern of the transition process is exhibited by some characteristics. Initially a negative phase of the parameter change appears; this is named the anabolic phase (Arshavsky, 1967, 1971). Then the normal reaction to the induced effect takes place, and, after the end of the effect there is an oscillatory regime during the course of restitution. For some parameters there are no variations, and the transition regime is of an exponential character.

Attention ought to be paid to an important feature of the restitution course of the transition process in living systems. After certain effects the dynamic parameters of the system are not restored to the initial state for a long time. Thus according to the data of sports physiologists (Temkin, 1974) the pulse remains reduced for 24 to 28 hours after training with considerable loads. After swimming to exhaustion the glycogen concentration in rat skeletal muscles is elevated for two weeks. After contraction the muscle remains in a hyper-relaxation state for a long time. The ratio between the time of forced parameters change under the effect of load and the restitution time is of the order of 1 : 170. Naturally, for these examples it is nonsense to discuss some regulatory quality of the stability of the initial state. However, instead of the realization of the transition process in the normal sense, i. e. return to the initial state, there is another approach which considers that after a sufficiently intensive stimulus the system transfers to a new state, near to the initial one and the pradox expressed above can be excluded. Such an approach implies relative instability or metastability of the initial state as a basic property of the system.

It should be noted that Glansdorff and Prigogine (1971) have shown that open chemical systems, far from equilibrium in an essentially non-linear region, have a metastable limit cycle in the zone of thermodynamic instablity. This means that in such systems the concentration varies around some mean value, which determines the state of the system that is stable towards small, and unstable towards limiting, excitations under which the system transfers to a new state of the limit cycle. Under certain conditions such transition is accompanied by a change in the spatial order of the system structure, praticularly in its complexity, occurrence of heterogeneity and asymmetry, i. e. by the decrease of entropy and the respective increase in free energy. The initiation of such systems in the zone of instability was shown earlier by Turing (1952). He pointed out that a heterogenous spatial structure appears in a homogenous system, as a result of instability in the system and the presence of two interacting processes - diffusion and chemical reaction.

Referring to the examples of transition processes under consideration it appears that transition to a new state during restoration, which is possible in the case of metastability in the initial state, under certain effects proceeds in such a way that the system in its new state acquires a higher absolute value along with a decrease in the intensity of free energy expenditures. Thus a decrease in pulse beat in sportsmen at rest is accompanied by a decrease in the level of daily expenditures from 10 to 14.10^3 kgm to 5 to 8.10^3 kgm, i. e. by a factor of two. Muscle in a state of extreme relaxation has a higher free energy since it is capable of accomplishing greater work. An increase in the content of high-energy compounds in muscle after work also results in a free energy increase. According to an energetic rule for motor activity (Arshavsky, 1967), optimal activity is always accompanied by excess anabolism during restoration. At an early age this excess mainly provides accumulation of the active mass, in the form of excess anabolism, in old age, after the achievement of the standing position in mammals, excess anabolism ensures accumulation of potentials and therefore an increase in free energy of the organism. According to this rule excess anabolism is carried out only within the limits of adaptive reaction to the load, i. e. when the load does not exceed some limit after which excess anabolism of the system cannot be realized. This points to some principle in the living system's reaction to an effect, which may be termed the principle of determining effects. It also means that, besides other factors, the amplitude of the effect necessarily influences the kinetics of any physiological parameter. But for dynamic systems such a relationship is equivalent to a change of coefficients in the dynamic system equation (5). Therefore formally this principle may be written in the form

$$A = A(\Delta h, t) \quad \text{and}$$
$$B = B(\Delta h, t), \qquad (11)$$

where Δh is the effect amplitude.

Referring again to the dynamic system equation (5) and as shown above, the selection of the controlling function U (t), which optimizes the transition curve of the parameter under consideration following induction effects, is reduced to minimization of the function J_B. The established parameter level is determined by the value of the effect, the initial state of the system, i. e. by the form of A(t) and B(t) and by the character of $U_B(t)$. After the cessation of the effect the selection of new $U_r(t)$, optimizing the transition characteristics of the parameter in the course of restitution, will be reduced to maximization of the intensity of free energy accumulation. This follows from the energy rules fro motor activity, i. e. function $J_P = \frac{dF_P}{dt}$ should be maximized.

$J_P \neq J_B$ since, first, $J_B < 0$ and $J_P > 0$, second, according to (8) $\frac{dF_P}{dt}$ does not contain the parameters of external work, and other factors in (8) will also vary during the course of restitution. In particular, the rate of chemical energy accumulation, $\sum_r A_r \, d\varepsilon_r/dt$ increases significantly due to the intensification of anabolism. The new level of the parameter under study which is established after the effect depends not only on the form of U(t), but also on the change of coefficients A(t) and B(t); these are determined by the principle of determining disturbance. Consequently the conclusion reached is that in order to model adaptive behaviour of the living system the very dynamic system which describes the complete transition process ought to undergo structural changes.

Such a change in the dynamic system during the course of the process is usually reduced to setting the so-called parametric feedback (Rozen, 1961), i. e. to setting the relationships $A = A(\varphi)$ and $B = B(\varphi)$ and the final equation of the dynamic system will then nonlinear and have the form:

$$\dot{\varphi} + A(\varphi)\varphi = B(\varphi) \cdot U(t) \qquad (12)$$

In summary this approach leads to the conclusion that the transition process in living systems i. e. stimulus, reaction, adaptation, can, using the variation principles presented, be completely described in terms of nonlinear dynamic systems of the type (12).

It is also worth noting that utilization of the suggested principle of energetic optimalization, i. e. minimization of losses and maximization of free energy accumulation when the living system responds to the environmental effects, allows prediction of adaptive changes in the organism to be suggested. However, this proves only if the living systems are considered as dynamic and nonlinear systems.

Dynamic Characteristics of Heat and Water Exchange in Man

A. A. Glushko

According to modern concepts the thermal state of man is described by a definite level of the thermoregulatory system. The stress ratio of the latter is finally determined by a relationship between the thermal balance components, i. e. the specific heat production ($\dot{q}_t$), heat transfer ($\dot{q}_s$) and heat content of the organism ($\Delta \dot{q}_m$), and is written as an equation:

$$\dot{q}_t = \dot{q}_s + \Delta \dot{q}_m \tag{(1)}$$

Heat content ($\dot{q}_m$) is interpreted for an object under regulation while heat production and heat transfer play the role of a regulating factor. In this sense the general tendency of adaptive functioning in the organism is directed to the preservation of thermal balance, i. e. to a correlation between $\dot{q}_t$ and $\dot{q}_s$ for which change in enthalpy would tend to zero ($\dot{q}_t \cong \dot{q}_s$).

It is also known that regulation of q_t is directed to change in oxidative phosphorylation, and of $\dot{q}_s$ to the optimal mode of heat exchange. The latter is achieved through various responses of the organism, e. g. external respiration, diaphoresis regulation of thermal resistance of the body peripheral region etc.

Studies of the kinetics and dynamics of these processes are important in the solution of a number of problems.

Such investigations are of great importance in the design of isolating clothing (diving, pneumatic and waterproof suits). These are sometimes complex engineering devices which are composed of liquid cooling (heating) systems, full-pressure and heat-isolating covers together with the necessary ventilation. For such designs changes in heat, mass transfer and humidity

Thermodynamics and Kinetics of Biological Processes

exchange are very complicated. They produce a variegated pattern of temperature, humidity and velocity fields in the undersuit space.

In view of the above it is necessary to find new methods and techniques for studying these processes.

Methods and Procedures

On the basis of general relationships of energy, heat exchange and thermoregulation in man a complex approach to the investigation of man's thermal state has been proposed by us. This approach allowed a number of new methodological trends to be developed:

1) Method of contact thermometry (Glushko et al., 1971 a), calorimetry (Glushko et al., 1971 b) and hygrometry (Glushko, 1976 a) based on the construction of special climatic dresses, in which thermoelectric transformations are reduced to the measurements of electric resistance (R) of thermosensitive elements of length l, evenly located on the human body surface and related with temperature through the equation

$$R = \frac{\rho}{S} \int_0^l dl + \frac{\alpha_t \rho}{S} \int_0^l t(l)\, dl \,, \tag{2}$$

where ρ, S, α_t are specific electric resistance parameters and temperature coefficients of the thermosensitive element resistances respectively.

2) Methods of non-contact thermography based on the pyroelectrical effect. For the measurement of this the heat energy currents ($W(\tau)$) are modulated, for instance, according to an exponential law of the form $W(\tau) = W_0(\tau)[1 + \exp i\omega\tau]$ the pyroelectrical tension, U, of the detector is measured and the desired parameter determined from the equation:

$$W(\tau) = \frac{(1 + i\omega\varepsilon_\tau)(1 + i\omega\varepsilon_l)}{(i\omega\varepsilon_0)\exp(i\omega\tau)} \cdot \frac{U}{R S_0 \xi} \,, \tag{3}$$

where ε_T, ε_l are heat and electric constants of pyroelectrical detectors respecitvely; ε_0 the absorbing ability of the surface, S_0 accepting the energy current; $R^{-1} = R_H^{-1} + R_A^{-1}$ the total electric conductivity of the detector load; ξ the pyroelectrical coefficient.

3) Method for investigating moisture exchange (js) in man which is found by accounting for gas flow velocity (u), and by measuring the psychrometric difference v_j from the equation

$$j = k_j \left(v_j + u^{-0.35} \frac{dv_j}{d\tau} \right), \tag{4}$$

where k_j is a calibration coefficient and τ the time.

4) Method for studying skin-lung humidity formation in man (j) enclosed in isolating costumes which is based on measurments of changes in humidity content of gas volume (G_r) ventilating the undercostume space and is determined from the empirical equation

$$j = 0.374\, G_r \{[(\exp 0.0521\, t_2)(\varphi_2 10^{-2} + \varepsilon_j \frac{d\varphi_2}{d\tau})] - \exp(0.0521\, t_1)\varphi_1 10^{-2}\} \tag{5}$$

where t, φ are temperature and relative humidity of the gas at input (1) and output (2) of the isolating costume respectively; ε_j is the thermal constant for humidity exchange.

5) Calorimetric method measured by thermoelectric effect, systems for which are incorporated in the design of a heat- and moisture-measuring suit. This measure is coordinated with a device for recording the velocity of temperature changes (Fesenko, Glushko, 1976). Changes in heat liberation dynamics ($\dot{q}_S$) are calculated from the expression:

$$\dot{q}_s = k_o \left(v_o + \varepsilon_o \frac{dv_o}{d\tau} \right) + k_j \left(v_j + \varepsilon_j \frac{dv_j}{d\tau} \right), \tag{6}$$

where v_o is the temperature difference between the body surface and the "dry" thermometer of the heat- and moisture-measuring dress; ε_c, ε_j are time constants of convective and evaporating heat exchange; and k_o, k_j calibration coefficients.

6) Method of direct calorimetry of man in an isolating costume based on delta-enthalpy effect recorded by the heat expenditure measuring devices

(Glushko, Selivanov, 1974). These are designated for the examination of change dynamics, as conductive ($\dot{q}_{cd}$), convective ($\dot{q}_c$) and evaporating ($\dot{q}_j$) heat exchange in isolating costumes and are found from the expressions:

$$\dot{q}_{cd} = c_l G_l \left(v_l + \varepsilon_{cd} \frac{dv_l}{d\tau}\right) \tag{7}$$

$$\dot{q}_c = c_g G_g \left(v_g + \varepsilon_c \frac{dv_g}{d\tau}\right) \tag{8}$$

$$\dot{q}_j = j \cdot r, \tag{9}$$

where c, G are specific heat capacity and volume expenditures of liquid (l) and gas (g), respectively; ε_{cd}, ε_c are time constants of conductive and convective heat exchange; r the latent heat of vapour production.

7) Method for investigating the heat transfer in bodies with complex geometrical configurations, e. g., in multilayered covers of the isolating costume. This is based on the thermogradient effect realized in heat measuring devices. The latter involves elastic bands (Glushko et al., 1974) evenly adjusted on the surface of the isolating costume.

Equations (2) to (9) include the parameters τ and ε, which directly characterize the kinetics and dynamics of heat and moisture exchange processes in man. Some experimental methods suitable for this approach have been comprehensively described (Gorodinsky et al., 1976), others will be outlined below. It should be noted that all these studies were derived from a long-term program which involved the modelling (mathematical and physical) of heat and water exchange, and thermoregulation in man together with experimental studies of the heat exchange parameters etc.

Experimental Results

Modelling of heat and moisture exchange was carried out on a thermal model of man (Glushko, 1976 b) and has been described comprehensively in a monograph (Gorodinsky et al., 1976).

Investigations of human heat and water exchange whilst at rest were conducted under three thermal regimes: at environmental temperature t_o = 20, 35 and 50 °C, relative humidity φ = 40 ± 20 % and velocity of the air movement 0.5 ± 0.2 m/sec.

The results have shown that under comfortable conditions (t_o = 20 °C) a stable heat balance and corresponding topography of heat and water liberation are maintained; their mean statistical values are presented in tables 3 and 4.

Under restricted heat exchange (t_o = 35 °C) shifts in thermal state, measured by standard technique (temperature, energy expenditures), seem to be not great (0.4 °C and 40 to 50 W, respectively). However, such events are of interest in studies of transition states of the organism, when the differences between the temperature of the body surface and that of the environment is close to zero.

Table 3

Heat balance of man at relative rest

Parameters	Value, W	(%)
Convective-radiation heat exchange from the body surface ($\dot{q}_{cr}$)	46.6	(51.7)
Evaporating heat exchange from the body surface ($\dot{q}_{js}$)	31.0	(34.4)
Heat liberation from the body surface ($\dot{q}_{s}$)	77.6	(86.0)
Convective-radiation heat exchange with external respiration ($\dot{q}_{cw}$)	4.3	(4.6)
Evaporating heat exchange with external respiration ($\dot{q}_{jw}$)	8.6	(9.4)
Heat exchange with external respiration ($\dot{q}_{w}$)	12.9	(14.0)
Total value of heat liberations ($\tilde{q}_{s}$)	90.5	(100.0)

Table 4

Topography of heat and water liberation in man at relative rest

Parameters	units	Body surface regions			
		Body	Head and upper extremities	Lower extremities	Total body surface
Sweat liberation (js)	g/h	12.5	7.9	7.1	27.5
Heat liberation ($\dot{q}_S$)	W	43.2	22.5	23.2	88.9

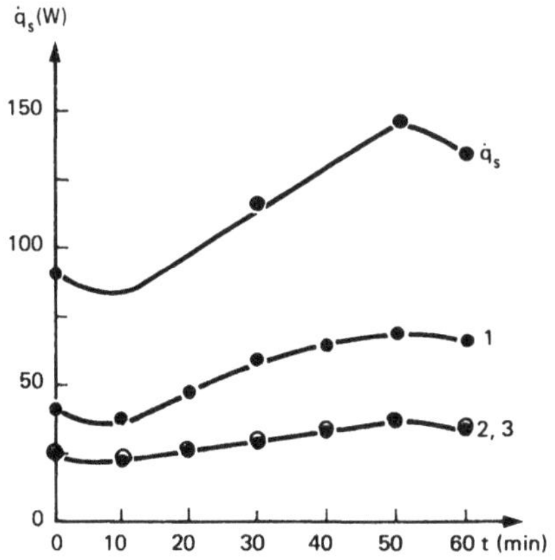

Fig. 45 Kinetics of local heat liberation q_S in man at relative rest with restricted heat exchange: $\dot{q}_S$ - integral data on heat liberation surface: 1 - body; 2 - upper extremities and head; 3 - lower extremities.

Let us examine kinetics of heat liberation in man under these conditions (Figs. 45, 46). In the first phase (up to 12 min) a decrease in heat liberation was reliably recorded; this was accompanied by an concomitant change in heat exchange structure: the intensity of convective-radiation heat exchange rapidly decreased and heat exchange with external respiration increased (2 to 3-fold), the contribution of evaporative heat exchange begins to increase gradually. In the second (transition) phase (12-40 min) the

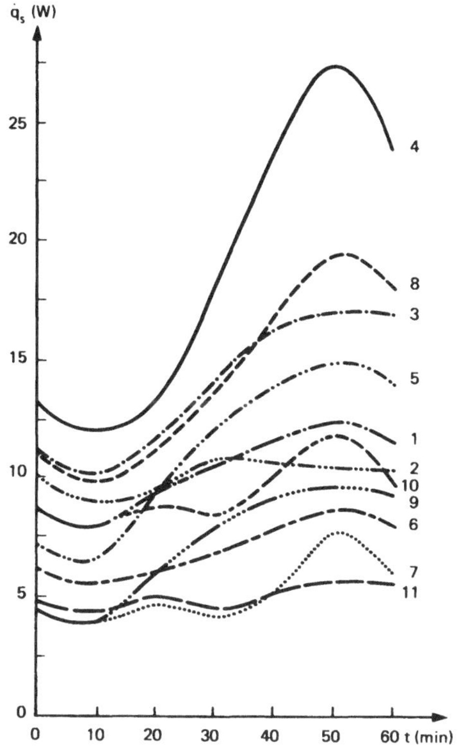

Fig. 46 Kinetics and topography of heat liberation $\dot{q}_s$ in man at relative rest with restricted heat exchange:
1 head, 2 breast, 3 back, 4 abdomen, 5 lumbus, 6 shoulders, 7 forearms, 8 thighs, 9 wrist rear, 10 skin, 11 foot rear.

transformation of metabolic heat retains the same tendency: $\dot{q}_w$ is stabilized at the level 25 to 27 W, $\dot{q}_{cr} = 0$, and $\dot{q}_j$ increases by ~ 2 W/min. By the end of the experiment a new thermal regime has been established, and its parameters have the values presented in the tables.

Table 5

Heat balance in man in conditions of restricted heat exchange

Units	Parameters								
	$\dot{q}_{cr}$	$\dot{q}_{js}$	$\dot{q}_s$	$\dot{q}_{cw}$	$\dot{q}_{jw}$	$\dot{q}_w$	$\dot{q}_s$	$\Delta\dot{q}_m$	$\dot{q}_t$
W	0	105.0	105.0	3.6	27.4	31.0	136.0	85.5	221.5
%	0	77.0	77.0	2.6	20.4	23.0	100.0	38.7	100.0

As is seen, the structure and topographic peculiarities of heat and water liberation under these conditions differ from those under unstressed conditions.

Effects on Heat and Water Exchange of Muscle Activity in Man

These studies were carried out under the similar thermal regimes to those described above and with a range of physical loads (A) from 150 to 1200 kgm/min. Heat and water liberation in man performing dosed physical work show the relationships described by Figs. 47 and 48. Each point on these graphs results from statistical treatment of several kinetic curves, one of which (for A = 400 kgm/min, t_o = 20 °C and different parts of body surface) is illustrated in Figs. 49 and 50.

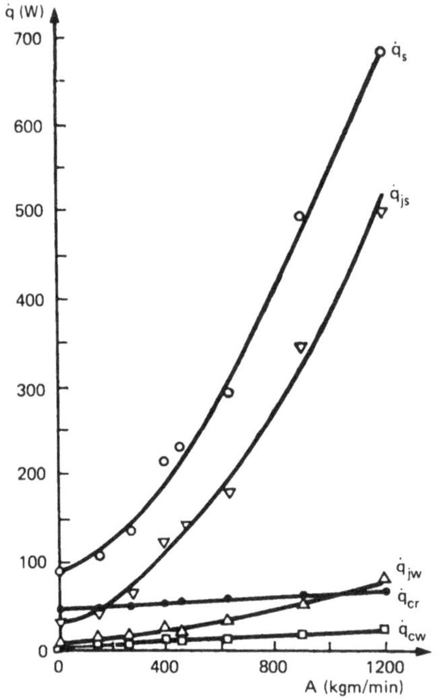

Fig. 47 Relationship between the changes of metabolic heat $\dot{q}$ and human muscle activity:
A power of physical load, $\dot{q}_{js}$, $\dot{q}_{jw}$ intensity of evaporative heat exchange in relation to body surface (S) and external respiration (w) of man, $\dot{q}_{cr}$, $\dot{q}_{cw}$ intensity of convective-radiation heat exchange in relation to body surface and external respiration of man, $\dot{q}_S$ total heat liberation.

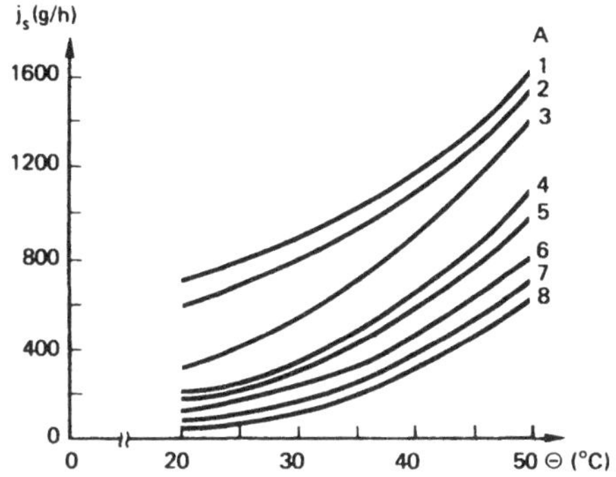

Fig. 48 Dependence of skin sweat liberation on different combinations of physical and thermal loads (φ = 40-60 %, ω = 0.5 $\pm$ 0.2m/s), j_s sweat liberation from body surface of man, Θ_o environmental temperature (according to dry thermometer), A power of physical load (1 - 1200 kgm/min 2 - 900, 3 - 630, 4 - 460, 5 - 400, 6 - 260, 7 - 150, 8 - 0).

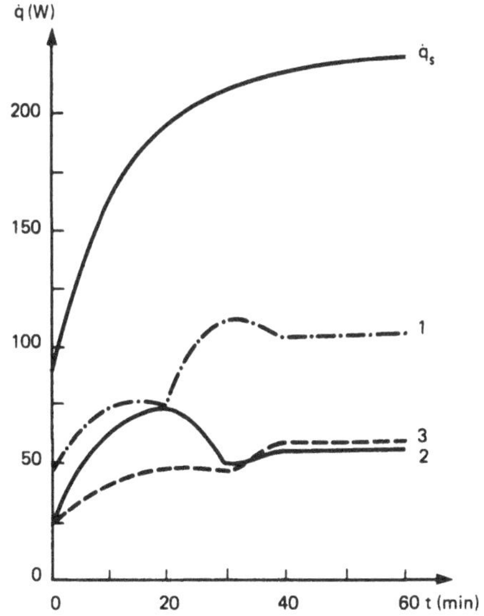

Fig. 49 Kinetics of local heat liberation in man fulfilling averagely hard work (A - 400 kgm/min) under conform microclimatic conditions. Notations same as in Fig. 45.

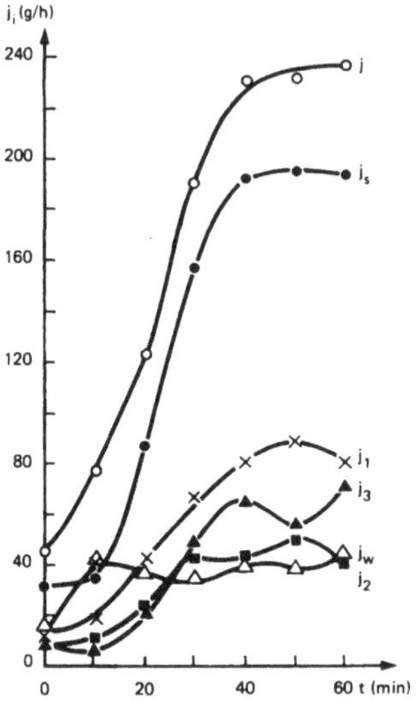

Fig. 50 Kinetics of water balance components in man fulfilling averagely hard work (A = 400 kgm/min) under comfort microclimatic conditions: j_w lung water release, j_1 sweat liberation from body surface, j_2 sweat liberation from head and upper extremities surface, j_3 from lower extremities surface, j_s from overall body surface, j total skin-lung water liberation.

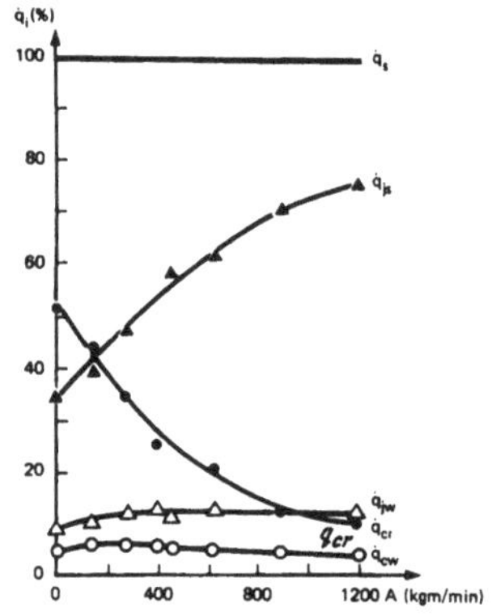

Fig. 51 Heat balance $\dot{q}_i$ of man at muscle work of different intensity A under comfort microclimatic conditions. Notations same as in Fig. 47.

Analysis of the data obtained from 21 experimental series has shown that the dynamics of rearrangement of the thermal state of the organism depends, essentially, on the structure of the energy, thermal and water balance of the organism. Whilst undergoing muscle activity the contribution of each of the balance components is changed within a wide range (Figs. 51 and 52).

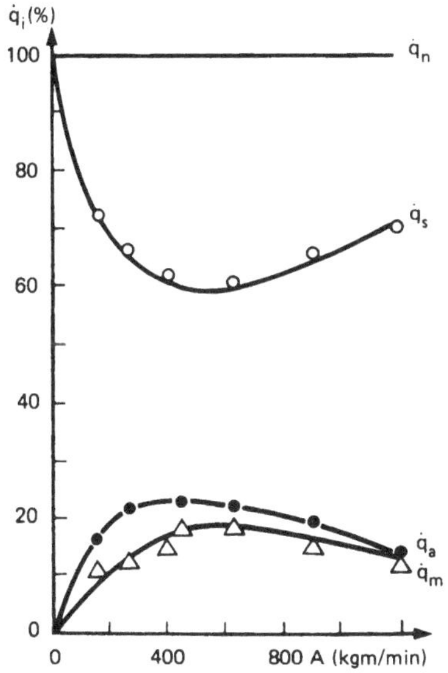

Fig. 52 Energetic balance $\dot{q}_i$ of man at different intensity A of muscle work under comfort microclimatic conditions:
A power of physical load, $\dot{q}_n$ energetic losses, $\dot{q}_s$ total heat liberation, $\dot{q}_m$ delta-entropy of the organism, $\dot{q}_a$ heat equivalent of muscle activity.

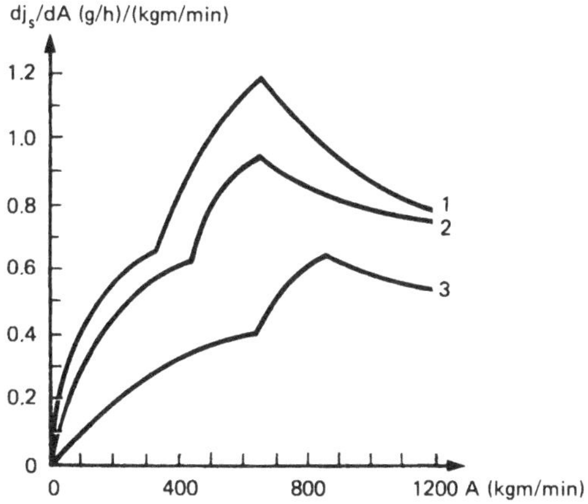

Fig. 53 Dynamics of the change of skin water liberation dj_s/dA under muscle work of man:
A power of physical load, $d\gamma_s/dA$ change rate of skin water liberation at different environmental temperatures (1-50 °C, 2-35 °C, 3-20 °C).

For instance, Fig. 51 shows that at relative rest (A = 0) the convective-radiation heat exchange comprises 51.7 % ($\dot{q}_{cr}$) + 4.6 % ($\dot{q}_{cw}$) of the total heat liberation ($\dot{q}_s$), and with A = 1200 kgm/min 9.7 % and 3.7 %, respectively. And vice versa, under the same conditions the role of evaporative heat exchange increases (with A = 0 : 34.3 % + 9.4 %; with A = 1200 kgm/min: 75.3 % + 11.3 %).

Together with rise of physical power the velocity of water evaporation from the skin surface also increases (Fig. 53). This relationship is described accurately enough by a trinomial exponential equation of the type:

$$\frac{dj_s}{dA} = \exp(-k\Delta A_1) + a\exp(-k\Delta A_2) + b\exp(k\Delta A_3) \qquad (10)$$

where a, b, k are individual constants of the organism; ΔA_i is the load in each of three phases of the process. Thus, for example, ΔA_1 at t_o = 20 °C is within 0 to 660 kgm/min; at t_o = 35 °C to 460 kgm/min and at t_o = 50 °C to 300 kgm/min.

For a wide range of heat states, when the limiting degree of stress in the organism's thermoregulation system has not been fully developed, the dynamic properties of the organism can be estimated from the so-called time constant (ε) of the transition heat regime. These have been determined by a graphical analytical method (from the family of kinetic curves for heat and moisture exchange). The mean statistical values, ε_i of each the i-components of the organism's thermal and energetic balance are given in Figs. 54 and 55. In parallel experiments, having accounted for ε , the general trend in the rearrangement of man's heat state may be determined approximately from the expression:

$$\dot{q}_{es} = \dot{q}\,[1 + \sum_{i=1}^{n} A_\alpha B_s \exp(-\frac{\tau - \tau_o}{\varepsilon_i})] \qquad (11)$$

where $\dot{q}_o$, $\dot{q}_{es}$ are the initial and established values of the organism heat production; A_α is a constant depending on the pattern of heat exchange and thermal physical constants of the organism; B_s a constant depending on topographic peculiarities of heat exchange.

Table 6

Topography of heat and water liberation in man at hindered heat exchange

Parameters	Units	Body surface regions											
		Head	Breast	Back	Stomack	Lumbus	Shoulder	Forearm	Thigh	Wrist	Shin	Foot	Total surface
Skin temperature	°C	36.3	33.7	34.0	34.3	32.2	34.7	34.8	34.0	34.0	35.0	34.4	34.9
Sweat liberation	g/h	24.4	28.0	19.1	26.0	23.6	8.5	8.1	27.1	9.4	8.5	6.8	189.0
Heat liberation	W	12.4	10.5	17.0	27.4	14.9	8.7	8.0	19.4	9.8	11.9	6.1	146.1

The present work has not aimed at discussion of all the complex problems of non-stationary heat and water exchange in man and its regulatory mechanisms. These require a special and more detailed analysis. However, the results obtained support the application of dynamic characteristics to the solution of many applied problems, such as the realization of operative medical control, prediction of the functional state of the organism, control over man's adaptation to specific peculiarities in living activity, selection, the production of the necessary cooling (heating) systems and ventilation of man in an enclosing thermal environment, and the automatic regulation of the parameters of these systems.

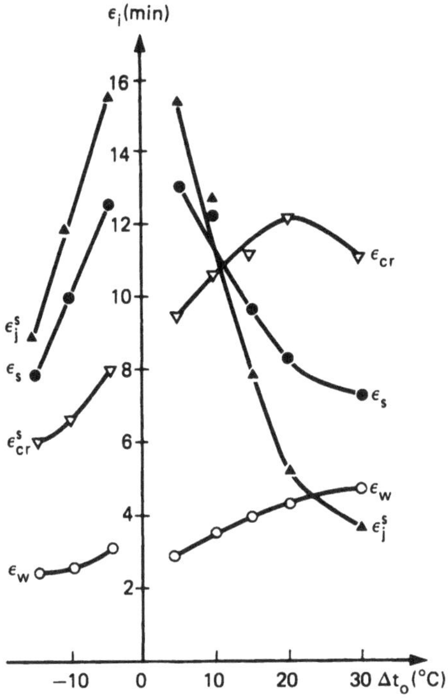

Fig. 54 Change dynamics ε_i of transformation and transfer of metabolic heat by thermic loads ($\Delta t_o = t_o - 20$). ε_s, ε_{cr}^s, ε_j^s, ε_w - time constants of heat metabolism (heat liberation, convective-radiation, evaporative and external respiration, respectively).

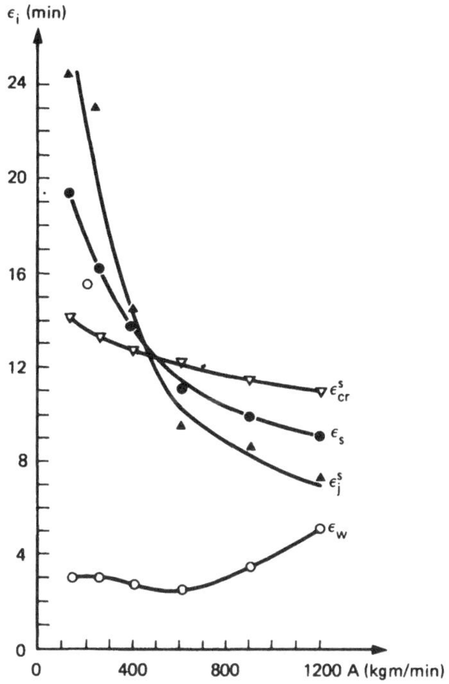

Fig. 55 Change dynamics ε_i of transformation and transfer of metabolic heat at muscle work A of man. ε_s, ε^s_{cr}, ε^s_j, ε_w - time constants of heat metabolism (heat liberation, convective-radiation, evaporative and external respiration, respectively).

Relation of Stimuli Controlling Lung Ventilation During Transient and Stable Periods of Exercise

I. S. Breslav, G. G. Isaev, A. M. Shmeleva

As is known, lung ventilation is regulated via the respiratory centre with stimuli of different modality. Conventionally these fall into two categories: humoral, responsible for the conformity of ventilation to the changing level of metabolic processes in the organism, and neurogenic which produce respiratory reaction to other disturbing factors.

The humoral stimuli act by changing the activity of chemoreceptors. These serve as error detectors responding to deviations in the chemical parameters of the internal medium from physiologically optimal values (settings). Quantitative relationships between ventilation and these deviations have been reliably established (Cunningham, 1974; and others).

The principal role in respiration control is played by the hypercapnic stimulus, determined as the carbon dioxide tension (P_{CO_2}) in blood and cerebrospinal fluid. Within a fixed range, lung ventilation ($\dot{V}$) is linearly related to this stimulus via:

$$\dot{V} = S\,(P_{CO_2} - B) \tag{1}$$

where S is a sensitivity index of the respiration control apparatus to hypercapnic stimulus (measured by ventilation increment against P_{CO_2} increase), B a threshold value of P_{CO_2}, at which the hypercapnic stimulus begins acting upon respiration.

The hypoxic stimulus of respiration reflecting the deficiency of oxygen tension (P_{O_2}) in blood is somewhat less significant. Its effect on ventilation is characterized by the relationship

$$\dot{V} = \dot{V}_0 + A/(P_{O_2} - C)\,; \tag{2}$$

Thermodynamics and Kinetics of Biological Processes

where V_o is the level of ventilation in the absence of hypoxic stimulus (e. g. during oxygen respiration), A the sensitivity index of respiration control apparatus to hypoxic factor, C the asymptote in the $\dot{V}$-P_{O_2} curve; in man taken as P_{O_2} = 32 mm Hg.

Both humoral respiration stimuli, hypercapnic and hypoxic, positively interact. According to modern concepts the hypoxic stimulus provokes a shift in adjustment of the respiratory response to the hypercapnic stimulus, i. e. the threshold value of this stimulus, expressed in equation (1) by parameter B, decreases (Rosenstein et al., 1974).

However all the above is related to the defined state of resting conditions. Respiration control under physical load presents more difficulties for analysis. In this case neurogenic as well as humoral stimuli are involved.

It is believed that these neurogenic stimuli come from motor centres in the brain and the receptors of working muscle (Marshak, 1961; Dejours, 1967). In particular, the steplike rise of ventilation at the beginning of a work regime is accounted for by neurogenic stimuli. At this time the chemistry of the internal medium has not yet changed and consequently the activity of chemoreceptors remains at the rest level. Only after the development of metabolic shifts concerned with prolonged muscle activity is the humoral control or respiration increased.

However, there are few studies on the interaction of these stimuli, and on their quantitative contribution to the regulation of lung ventilation during transient and stable periods of exercise. This work is aimed at filling this gap, at least partially.

Lung ventilation was analyzed in 12 healthy young men engaged in muscular work of different power against a background of different content of oxygen and carbon dioxide in the inhaled air.

P_{CO_2} and P_{O_2} in alveolar gas was recorded with a mass-spectrometer and served as a measure of hypercapnic or hypoxic stimulation of respiration. The persons under examination performed muscular work on a bicycle ergometer for 5 min. Three gradations of load power were used: zero (pedaling an unloaded wheel), 50 Watt and 100 Watt. Lung ventilation was recorded on a modified spirograph device.

Relationship between ventilation and alveolar P_{CO_2} and alveolar P_{O_2} were calculated for three states: at rest; during the transition working period (in the first 20 seconds after load switch-on; and during the stable exercise period (5th min.). During each working period the dependence of ventilation on load power was also estimated. Computations were performed on electronic computer (M-3040).

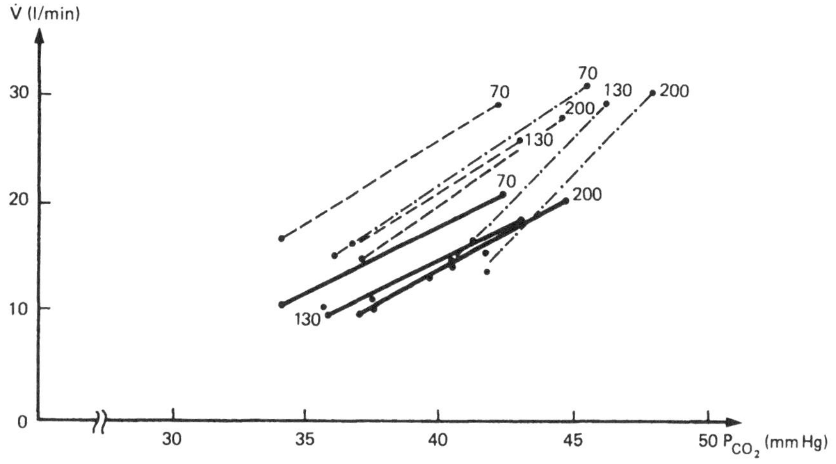

Fig. 56 Dependence of ventilation $\dot{V}$ on P_{CO_2} in transition and stable regimes of muscle activity with zero load under normal conditions, at hypo- and hyperoxia. Notations: full lines - rest, broken ones - transition regime, dot-and-dash ones - stable state (figures designate alveolar P_{O_2} in mm Hg).

Fig. 56 shows the relationship between lung ventilation ($\dot{V}$) and P_{CO_2} found for different values of P_{O_2} under rest and zero load conditions. It can be seen that at rest hypoxia is caused (decrease of P_{O_2} to 50 to 70 mm Hg), as compared to normal and hyperoxic conditions (P_{O_2} 100 to 200 mm Hg), and there is an upward shift of the $\dot{V}$-P_{CO_2} line (here and subsequently only statistically significant (i. e. $p < 0.05$) changes in the studied relationships will be reported). This observation is related to decrease in parameter B in equation (1) and confirms the decrease, mentioned previously in the threshold of sensitivity of the respiratory control apparatus to hypercapnic stimulus under hypoxia, i. e. positive interaction between two humoral stimuli. Such interaction is described by the following equation (compare it with (1) and (2):

$$\dot{V} = S\left[P_{CO_2} - \left(B_0 - \frac{k_1}{P_{O_2} - 32}\right)\right], \qquad (3)$$

where B_0 is the threshold of respiratory response to hypercapnia in the absence of hypoxic stimulus (with P_{O_2} = 200 mm Hg); k_1 the coefficient (its empirical value in the present experimental series was 211), reflecting the effect of hypoxia on the indicated threshold of hypercapnic response.

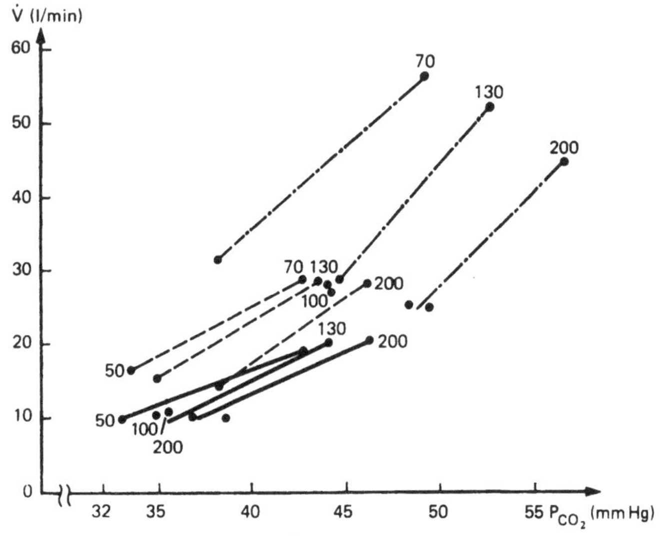

Fig. 57 Dependence of ventilation $\dot{V}$ on P_{CO_2} in transition and stable regimes of muscle activity with 50 Watt load under normal conditions, at hypo- and hyperoxia. Notations same as in Fig. 56.

When the exercise started all the $\dot{V}$ - P_{CO_2} lines were shifted upwards, but the potentiating effect of hypoxia was realized as at rest. In the stable regime of the zero load the relationships tended to return to those of the resting level.

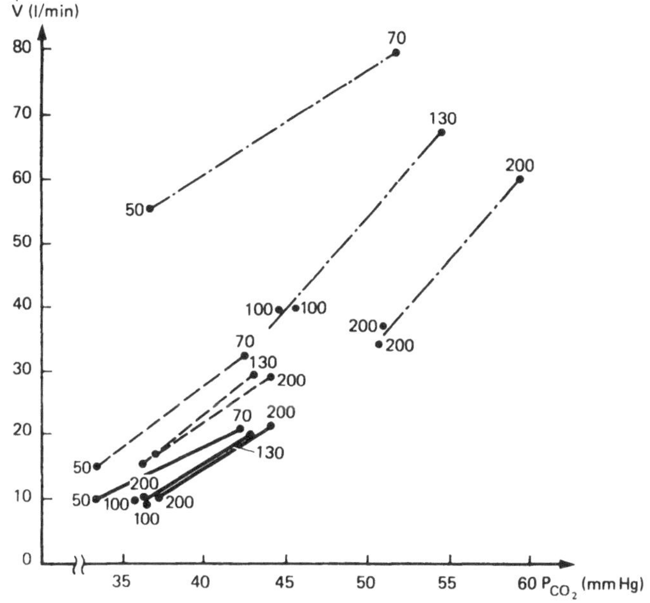

Fig. 58 Dependence of ventilation $\dot{V}$ on P_{CO_2} in transition and stable regimes of muscle activity with 100 Watt under normal conditions at hypo- and hyperoxia. Notations same as in Fig. 56.

The response of lung ventilation under muscular loading - 50 and 100 Watt (Figs. 57 and 58) - during the transition periods did not differ from those described. A shift upwards of the $\dot{V}$-P_{CO_2} lines also took place. This shift seemed not to depend on the power of the work initiated, and its value varied over a wide range. Thus this did not allow any quantitative relationship to be established. The relative contribution of the hypoxic stimulus was not changed.

A different relationship was found for the stable periods of work. Under hyperoxia (P_{O_2} = 200 mm Hg) the $\dot{V}$-P_{CO_2} lines run as if continued the corresponding lines plotted for the rest condition. Some increase in the slope of the lines can be explained by the higher values of P_{CO_2} during exercise. It should be noted that the relationship between ventilation and hypercapnic stimulus is, in fact, not completely linear, and, in its upper range, the parameter S in equation (1) increases somewhat. However, under normoxia (P_{O_2} = 100 to 130 mm Hg) and especially under hypoxia (P_{O_2} = 50 to 70 mm Hg) the $\dot{V}$-P_{CO_2} lines were shifted upwards, which conformed to a decrease in parameter B. This shift was larger the higher was the intensity of the physical load.

In other words, in addition to usual interactions of both humoral stimuli, which was observed at rest and at the initiation of work (equation (3)), in the stable work period an additional potentiating effect of the hypoxic factor on ventilation occurred, this effect being in a direct proportion to the working power. The relationships under examination are described in the following way:

$$\dot{V} = S\left[P_{CO_2} - \left(B_0 - \frac{k_1 - k_2 W}{P_{O_2} - 32}\right)\right] \qquad (4)$$

where k_2 is a coefficient (its empirically found value for this experimental series was 8.4), reflecting the load-dependent effect of hypoxia on the threshold of hypercapnic response, W the working power in Watt.

Naturally, with zero work the second term in the numerator for the hypoxic part of the equation is eliminated. This corresponds to the absence of essential differences in the levels of ventilation at rest and in the steady state of such load.

Consequently in the transition regime of muscular activity the neurogenic stimuli, which are switched on at the instant exercise begins, act additively with the humoral factors on lung ventilation, without changing the ratio between the contributions of hypercapnic and hypoxic stimuli. The neurogenic origin of the stimuli evoking the initial rise of ventilation is confirmed by the fact that it is not related to the load power. Thus a rapid rearrangement of respiration is induced, which only to a first approximation meets the expected increase of gas exchange associated with the forthcoming work.

The stable period of work is described by a necessary increase in the contribution of hypoxic stimulus to respiration control. The hypoxic stimulus potentiates hypercapnic stimulation by direct relationship with the load power. This can be ascribed either to the effect of some metabolic products, liberated in the muscle, or to the action of some neurogenic factors, probably, to muscle receptors discharge.

In any case such augmentation of positive interaction in respiration stimuli during the course of muscle activity must provide more and more exact adaptation of lung ventilation to the level of gas exchange, and thereby to the maintenance of chemical homeostasis in the organism.

VII. Ontogenesis

Kinetics of Constitutive Processes During Development and Growth of Organisms

R. S. Zotina, A. I. Zotin

A constitutive process designates a change in the intensity of heat production or respiration in the course of development, growth and aging of organisms, as well as during wound healing, regeneration and malignant growth (Zotin, 1972, 1974; Zotin, Zotina, 1976). The intensity of heat production or respiration can be with some approximation, equated to the specific rate of entropy production in the system. And consequently thermodynamic criteria of evolution can be used to specify constitutive processes. The matter is, however, more complicated because of two circumstances:

1. the existing evolutionary criteria are of a qualitative pattern and cannot be applied to a quantitative description of constitutive processes;
2. the most common evolutionary criterion based on Prigogine's theorem pertains only to systems near equilibrium.

In this connection we have made an attempt to introduce a new evolutionary criterion based on the concepts of external and bound dissipation functions and on their statistical interpretation. This approach enabled us to obtain an analytical expression for a change in the dissipative functions characterizing the movement of an open system to a stationary state (Zotin, Zotina, 1977).

The present study is aimed at the application of these relationships to the description of the kinetics of constitutive processes during the development and growth of the organism. To do this the concepts of external and bound dissipation and their stochastic interpretation must be analysed.

Thermodynamics and Kinetics of Biological Processes

Functions of External and Bound Dissipation

As follows from the Second Law of thermodynamics any irreversible process is accompnied by the formation of dissipation heat. If the system is far from equilibrium and the rate of formation of dissipation heat high, not all this heat will have time to leave the system, a portion of it will be employed inside the system to perform some work. In other words, not all the heat formed as a result of irreversible processes in an open system will be non-compensated heat. On this basis it has been proposed (Zotin, 1974) that the dissipative function of a system far from equilibrium can be separated into two parts:

$$\Psi = \Psi_d + \Psi_u = \sum_{j=1}^{n} \bar{J}_j X_j \, , \qquad (1)$$

where

$$\Psi = \frac{T}{V} \frac{d_i S}{dt} \, ; \quad \Psi_d = \frac{T}{V} \frac{d_i S_d}{dt} \, ; \quad \Psi_u = \frac{T}{V} \frac{d_i S_u}{dt} \qquad (2)$$

Ψ is the specific dissipation function, Ψ_d the external dissipation function, Ψ_u the bound dissipation function, $\bar{J}_j$ specific thermodynamic flows, X_j the thermodynamic forces, T the absolute temperature, V the volume of the system, d_iS/dt the rate of entropy production.

The subdivision (1) was introduced from considerations of the physical sense of the concept of energy dissipation in open systems, but it can be also obtained in a more standard thermodynamic manner (Zotin, Zotina, 1977; Zotin, Zotina, Konoplev, 1978) for systems to which Prigogine's principle of local equilibrium can be applied. In fact it is implicit in the work of Prigogine (1967) and Rubin (1976) where these authors discuss possible measurements of the rate of entropy production in chemical systems by means of the thermal effects of reactions. It appears also in a paper by Lurie and Wagensberg (1979 a, b). Stochastic substantiation of a possible subdivision of type (1) has been described by others (Landauer, 1973; Wangsness, 1975).

Although, initially, the thermodynamic proof of subdivision (1) was given for purely chemical systems (Zotin, Zotina, 1977; Zotin et al., 1978), it can also be extended to any other system which can be described by the Gibbs' equation. Thus according to Haase (1963) the specific dissipation function can be presented as:

$$\Psi = \frac{1}{V}\,\frac{dW_d}{dt} + \frac{1}{V}\sum_{\varrho=1}^{m} A_\varrho v_\varrho\,, \tag{3}$$

where A_ϱ is the affinity, $v_\varrho = d\xi_\varrho/dt$ the rate of reaction ϱ, ξ_ϱ the degree of reaction completeness, dW_d the dissipation work, which, in a general case, can be equated to:

$$dW_d = \sum_{i=1}^{k} L_i\, dl_i\,, \tag{4}$$

where l_i is the extensive and L_i the intensive parameters for work.

Under the condition T = const. and p = const. the generalized Gibbs' equation takes the form:

$$dG = -\sum_{i=1}^{k} L_i\, dl_i - \sum_{\varrho=1}^{m} A_\varrho\, d\xi_\varrho\,. \tag{5}$$

It follows that

$$L_i = -\left(\frac{\partial G}{\partial l_i}\right)_{p,T} = -\left(\frac{\partial H}{\partial l_i}\right)_{p,T} + T\left(\frac{\partial S}{\partial l_i}\right)_{p,T} \tag{6}$$

and

$$A_\varrho = -\left(\frac{\partial G}{\partial \xi_\varrho}\right)_{p,T} = -\left(\frac{\partial H}{\partial \xi_\varrho}\right)_{p,T} + T\left(\frac{\partial S}{\partial \xi_\varrho}\right)_{p,T}\,, \tag{7}$$

where G is Gibbs' free energy, H the enthalpy. Comparing (4), (3), (6) and (7) we obtain

$$\Psi = \frac{1}{V}\left[\sum_{i=1}^{k}\left(\frac{\partial H}{\partial l_i}\right)_{p,T} v_{l_i} + \sum_{\varrho=1}^{m}\left(\frac{\partial H}{\partial \xi_\varrho}\right)_{p,T} v_\varrho\right] + \frac{T}{V}\left[\sum_{i=1}^{k}\left(\frac{\partial S}{\partial l_i}\right)_{p,T} v_{l_i} + \sum_{\varrho=1}^{m}\left(\frac{\partial S}{\partial \xi_\varrho}\right)_{p,T} v_\varrho\right] \tag{8}$$

where $v_{l_i} = dl_i/dt$.

Now introducing the symbols

$$\Psi_d = -\frac{1}{V}\left[\sum_{i=1}^{k}\left(\frac{\partial H}{\partial l_i}\right)_{p,T} v_{l_i} + \sum_{\varrho=1}^{m}\left(\frac{\partial H}{\partial \xi_\varrho}\right)_{p,T} v_\varrho\right] = \frac{1}{V}\left[\sum_{i=1}^{k} r_{p,T}^{(i)} v_{l_i} + \sum_{\varrho=1}^{m} r_{p,T}^{(\varrho)} v_\varrho\right] = \dot{q} \tag{9}$$

and

$$\Psi_u = \frac{T}{V}\left[\sum_{i=1}^{k}\left(\frac{\partial S}{\partial l_i}\right)_{p,T} v_{l_i} + \sum_{\varrho=1}^{m}\left(\frac{\partial S}{\partial \xi_\varrho}\right)_{p,T} v_\varrho\right] \tag{10}$$

we arrive at subdivision (1) (here $r_{p,T}^{(i)}$ and $r_{p,T}^{(\rho)}$ are thermal effects of some work and chemical reactions respectively; $\dot{q}$ is the intensity of heat production of the system).

The physical meaning of the external dissipation function (9) is clear, it is equal to the heat production intensity of the system. In its turn the physical meaning of the bound dissipation function proceeds from formula (10): in purely chemical systems the Ψ_u-function depends on temperature, the system size, the entropy change related to the degree of reaction completeness, and on the rates of chemical reactions. In more complex systems the first sum on the right-hand side of (10) ought to be taken into account. For instance, when there is an electrical potential difference in the system $dW_d = \sum_{i=1}^{k} \varphi_i de_i$, where φ_i are potentials and e_i charges, then equation (10) should be written in the form

$$\Psi_u = \frac{T}{V}\left[\sum_{i=1}^{k}\left(\frac{\partial S}{\partial e_i}\right)_{p,T} v_{e_i} + \sum_{\rho=1}^{m}\left(\frac{\partial S}{\partial \xi_\rho}\right)_{p,T} v_\rho\right]. \tag{11}$$

In such a system the Ψ_u-function depends on the current strength $v_{l_i} = dl_i/dt$ and on the particle properties carrying the electric charge.

If the system is near equilibrium all processes in it proceed at a small rate. The values $(\partial S/\partial l_i)_{p,T}$ and $(\partial S/\partial \xi_\rho)_{p,T}$ tend to zero, and the Ψ_u-function in such a system may be neglected. In this case (1) is transformed to

$$\Psi = \Psi_d = \sum_{j=1}^{n} J_j X_j . \tag{12}$$

This is valid for an open system near to the stationary state. Equation (12) holds in all cases (see Prigogine, 1967; Rubin, 1976) where the thermal effect of the reactions (if purely chemical systems are involved) is much higher than the entropy change concerned with a change in the degree of reaction completeness. If the Ψ_u-function is not equal to zero, it can be calculated from (1) and (9) by the formula

$$\Psi_u = \sum_{j=1}^{n} \bar{J}_j X_j - \dot{q} . \tag{13}$$

This formula allows an experimental check on the validity of subdivision (1). It is most conveniently exemplified by living systems. In the limit, all dissipative processes in the system are determined by respiration and glycolysis. Therefore the bound dissipation function may be calculated from the formula:

$$\Psi_u = \dot{q}_{O_2} + \dot{q}_{gl} - \dot{q}, \quad (14)$$

where $\dot{q}_{O_2}$ is the specific respiration intensity, $\dot{q}_{gl}$ the specific aerobic intensity of glycolysis. If the aerobic glycolysis can be neglected (which is reasonable for most organisms), then

$$\Psi_u = \dot{q}_{O_2} - \dot{q}. \quad (15)$$

Equation (15) shows that in organisms which are far from their final, steady state* a discrepancy between the data of direct and indirect calorimetry exists. This has been experimentally demonstrated for a number of biological objects (Grudnitzky, 1975; Schaarschmidt et al., 1975; Zotin, 1975; Loehr et al., 1976). Fig. 59 shows the changes in Ψ_u and Ψ_d-functions during oogenesis of the frog *Xenopus laevis.* A large discrepancy between the data of direct and indirect calorimetry is revealed; it is greater when earlier stages of oogenesis are investigated. This is true, also, for embryonic development and growth in animals. Estimates of heat production and respiration in hen embryos (Barott, 1937) has shown that, during embryonic development of these animals, a disparity is observed between the results from direct and indirect calorimetry (Fig. 60). A discrepancy was also observed between results obtained on the course of development and growth in insects (Fig. 71) and for the growth of fish (Fig. 61). These observations, as well as others, substantiate subdivision (1) and the existence of the bound dissipation function.

---

* To designate the stationary state of living systems we use the term steady state.

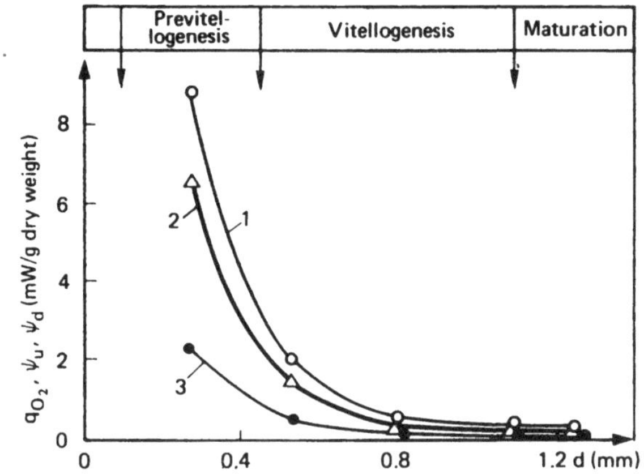

Fig. 59 Change in respiration intensity $\dot{q}_{O_2}$ (1), Ψ_u-function (2) and Ψ_d-function (3) during oogenesis of *Xenopus laevis* as function of the diameter d of the oocytes (Zotin, 1975).

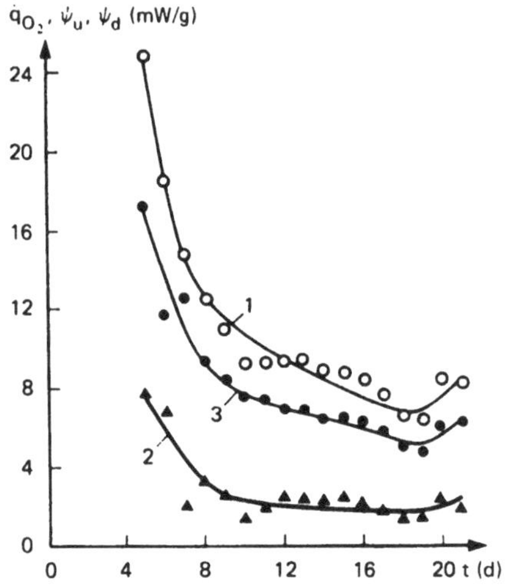

Fig. 60 Change in respiration intensity $\dot{q}_{O_2}$(1), Ψ_u-function (2) and Ψ_d-function (3) with time t during embryonic development of hen (respiration and heat production after Barott, 1937; wet weight of embryos after Romanoff, 1967).

Stochastic Consideration of Dissipative Functions

The approach of a thermodynamic system to an equilibrium or stationary state can be compared to a relaxation in large fluctuations (Gurov, 1978). From this viewpoint the theory of fluctuations and, reduction of fluctuations, makes a great contribution to the understanding of the kinetics of thermodynamic systems approximating to equilibrium or to a stationary state.

As is known (Glansdorff, Prigogine, 1971; Zalevsky, 1973; Gurove, 1978), fluctuation in an isolated system at equilibrium may be described by Einstein's formula:

$$p = C \exp \frac{\Delta S}{k} \tag{16}$$

where $p(x_1, \ldots, x_n)$ is the probability density of deviation in parameters $x_1, \ldots, x_n$ form their equilibrium values $x_1^{(0)}, \ldots, x_n^{(0)}$; $\Delta S = S - S_0$; S_0 the value of entropy at the equilibrium; k the Boltzmann constant. Applying the concept (Nivolis, Babloyantz, 1969; Nicolis, Prigogine, 1971; Nicolis, 1972; Glansdorff, Prigogine, 1971) that Einstein's formula can be used to describe fluctuation initiation near a nonequilibrium stationary state in an open system, and in addition, using some thermodynamic arguments (Zotin, Zotina, 1977; Zotin, Zotina, Konoplev, 1978) we have obtained the relationship between the dissipation function and probability density in the following form:

$$\Psi = \Psi_{st} + \frac{kT}{p}\frac{dp}{dt}, \tag{17}$$

where Ψ_{st} is the value of the specific dissipation function in the stationary state; k a constant depending on Ψ_{st}. In deriving (17) it was proposed that the probability density is a function of time. In Einstein's formula (16) it is a function only of thermodynamic parameters. This is related to the fact that (16) describes a fluctuation which has already been initiated, and the probability density in this case is not a function of time.

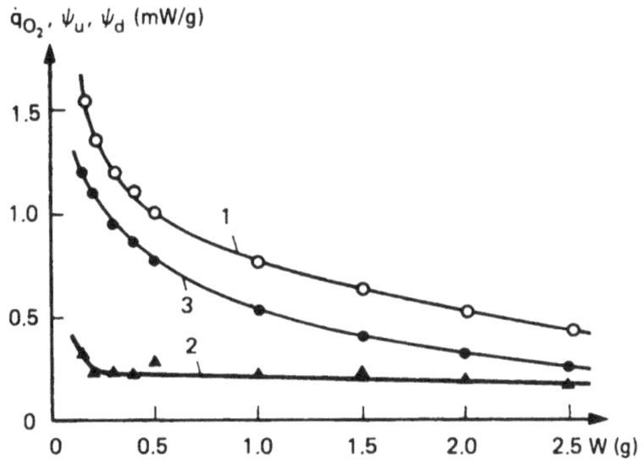

Fig. 61 Change in respiration intensity $\dot{q}_{O_2}$ (1), Ψ_u -function (2) and Ψ_d -function (3) as function of weight W during the growth of *Xiphophorus* (after Grudnitzky, 1975).

In the case studied by us of fluctuation reduction the probability density must be a function not only of thermodynamic parameters, but of time as well. Moreover, since in fluctuation reduction the system parameters themselves are time functions, it is possible that, in this case, the probability density is only dependent on time. Application of Einstein's formula, which forms the basis of (17), to nonequilibrium systems raised some objections (see discussion in: Nicolis, 1972; Glansdorff et al., 1974; Keizer, Fox, 1974; Nicolis et al., 1974; Nitzon, Ross, 1974; Kuramoto, 1975; Landauer, 1975). Without going into detail it may be noted that for many cases and with not too strict limitations then Einstein's formula is also applicable to nonequilibrium systems (Glansdorff, Prigogine, 1971; Nicolis, Prigogine, 1971; Nicolis, 1972; Chakrabarti, 1975; Kuramoto, 1975).

As is seen from equation (17), to obtain the relationship between the dissipation function and the probability of the system state, the rate of change in probability density when the system approaches the stationary state must be known. In our case, it is logical to suppose that,

$$dp/dt = f(p) \tag{18}$$

where the function, f(p), is infinitely differentiated in the vicinity of the point p_{st} (in the vicinity of the system stationary state). Expansion of this function into a Taylor series yields

$$\frac{dp}{dt} = f(p_{st}) + \frac{f'(p_{st})}{1!}(p - p_{st}) + \frac{f''(p_{st})}{2!}(p - p_{st})^2 + \frac{f'''(p_{st})}{3!}(p - p_{st})^3 + \ldots \qquad (19)$$

this requires that when $p = p_{st}$ and $dp/dt = 0$ then we have $f(p_{st}) = 0$.

If the system is not far from the stationary state, then, in expansion (19) we may adapt as limits only the linear member, and then, we obtain

$$dp/dt = \alpha_1 (p_{st} - p) \qquad (20)$$

where $\alpha_1 = -f'(p_{st})$.

Substituting (20) into (17) we obtain

$$\Psi = \beta_1 (p_{st}/p - 1) + \Psi_{st}, \qquad (21)$$

where $\beta_1 = \alpha_1 kT$.

It follows from (21) that the larger is the system state probability, the lower is the value of the dissipation function and vice versa. It is also evident that with $p \rightarrow p_{st}$ the dissipation function tends to its stationary value.

If the system is far from equilibrium, the change in the probability density will be described by equation (19). Confining discussion to two members of the Taylor power series, i. e.

$$dp/dt = \alpha_1 (p_{st} - p) + \alpha_2 (p_{st} - p)^2 \qquad (22)$$

where $\alpha_2 = -f''(p_{st})/2!$ then substituting this equation into (17) we obtain

$$\Psi = (\beta_1 + \beta_2 p_{st}) p_{st}/p + \Psi_{st} - \beta_1 + \beta_2 (p - 2p_{st}), \qquad (23)$$

where $\beta_2 = \alpha_2 kT$.

Now, correlating (23) with subdivision (1) yields:

$$\Psi_d = (\beta_1 + \beta_2 p_{st}) p_{st}/p \qquad (24)$$

and

$$\Psi_u = \Psi_{st} - \beta_1 + \beta_2(p - 2p_{st}). \tag{25}$$

It is evident that if $p = p_{st}$ then $\Psi_d = \Psi_{st}$ must be true, therefore (24) may be written in the form:

$$\Psi_d = \Psi_{st}\frac{p_{st}}{p} \tag{26}$$

where $\Psi_{st} = \beta_1 + \beta_2 p_{st}$. The validity of the correlation between (24) and (25) follows from the following. With $p \longrightarrow p_{st}$, as follows from (26), then $\Psi_d \longrightarrow \Psi_{st}$, which agrees with the general concept of the behaviour of the Ψ_d -function. On the other hand, with $p \longrightarrow p_{st}$; then $\beta_1 - \beta_2(p - 2p_{st}) \rightarrow \beta_1 + \beta_2 p_{st} = \Psi_{st}$ and $\Psi_u \longrightarrow 0$, which, in its turn, is consistent with the behaviour of the Ψ_u-function near the stationary state. All other variants of correlating (1) and (23) will not satisfy these properties of the Ψ_d- and Ψ_u -functions.

It is easy to show that correlations of the type (24) to (26) will also be valid if all the members of the Taylor series are accounted for in expansion (19). For instance, they are upheld for three members of the Taylor series (in this case $\Psi_{st} = \beta_1 + \beta_2 p_{st} + \beta_2 p_{st}^2$).

Two consequences follow from the above and both pertain to formula (26). Firstly it follows that:

$$\Psi_d\, p = \Psi_{st}\, p_{st} = z \tag{27}$$

where z, within some boundaries is a universal constant. Second, taking the derivative with respect to time of (26), we have

$$\frac{d\Psi_d}{dt} = -\frac{z}{p^2}\frac{dp}{dt}$$

and allowing that when the system approaches the stationary state the probability density is not diminished $dp/dt \geqslant 0$ we arrive at the evolution criterion:

$$d\Psi_d/dt \leq 0 \quad , \tag{28}$$

which has good physical sense and can be easily checked experimentally.

Constitutive Processes

Let us consider now the problem of the kinetics of an open system approaching the stationary state. At first let us confine ourselves to the case when equation (20) holds, i. e. the system is near the stationary state. Solving this equation we obtain:

$$p = p_{st}\,[1 - B\exp(-\alpha_1 t)] \;, \tag{29}$$

where B is a positive constant.
Substituting (29) into (17) we have:

$$\Psi = \Psi_{st} + \beta_1 \frac{B\exp(-\alpha_1 t)}{[1 - B\exp(-\alpha_1 t)]} \;. \tag{30}$$

As the system is near the stationary state, t ≫ 0, and $B\exp(-\alpha_1 t) \ll 1$ therefore equation (30) can be written in the form:

$$\Psi = \Psi_{st}\,[1 + A\exp(-\alpha_1 t)] \;, \tag{31}$$

where $A = \beta_1 B - \Psi_{st}$. This equation is in a good agreement with the relationship obtained from thermodynamic considerations of linea, irreversible processes from other considerations (see de Groot, Masur, 1969).

In the course of development and growth in organisms the living system approaches the final steady state, and, it is possible to say, that at sufficiently late stages of growth in the linear region, then equation (31) will be valid. It can be shown that changes in the respiratory intensity during growth of many animals are described by an equation of the type:

$$\dot{q}_{O_2} = \dot{q}_0\,[1 + A\exp(-\alpha_1 t)] \tag{32}$$

which is identical to equation (31). Therefore, Brody (1945) derived the following empirical equations which describe changes in basal metabolism during the growth of mammals viz.

for horses $\dot{q}_{O_2} = 22[1 + 7\,exp(-0.22t)]$
for cows $\dot{q}_{O_2} = 16[1 + 23\,exp(-0.096t)]$
for sheep $\dot{q}_{O_2} = 27[1 + 17\,exp(-0.22t)]$.

These equations are clearly based upon (32).

Equation (20) is inapplicable to nonlinear systems and expansion (19) must be used in its complete form. This sets great limitations to the derivation of kinetic properties, the equations for which describe a change in the dissipation function when the open system approaches the steady, or equilibrium state. However, confining oneself to the second and third terms in (19), such equations can be obtained in a rather simple form (Zotin, Zotina, 1977). Thus taking two terms of the Taylor series, we obtain equation (22) which can be written as:

$$dp/dt = ap^2 + bp + c \tag{33}$$

where $a = \alpha_2$, $b = -(\alpha_1 + 2\alpha_2 p_{st})$; $c = \alpha_1 p_{st} + \alpha_2 p_{st}^2$. This is the Riccati equation (see Kamke, 1951). Solving it we obtain

$$p = p_{st} + \frac{\alpha_1 exp(-\alpha_1 t)}{\alpha_2[c + exp(-\alpha_1 t)]} \tag{34}$$

Substituting (34) into (24) and (25) we obtain

$$\Psi_d = \frac{\alpha_2 c p_{st} \Psi_{st} + \alpha_2 p_{st} \Psi_{st}\, exp(-\alpha_1 t)}{\alpha_2 c p_{st} + (\alpha_1 + \alpha_2 p_{st})\, exp(-\alpha_1 t)}$$

$$\Psi_u = \frac{\beta_1}{1 + c\,exp(\alpha_1 t)}, \qquad \Psi = \Psi_d + \Psi_u . \tag{35}$$

As follows from these equations with $t \to \infty$, then $\Psi_d \to \Psi_{st}$ and $\Psi_u \to 0$ which is in conformity with the properties of these functions.

Now we shall make an attempt to use the relationships obtained to describe the kinetics of constitutive processes during development and growth in animals. For this purpose the signs of the constants α_1 and α_2 must be found. In equation (19) $f'(p_{st}) < 0$, since the change in the rate of probability density decreases as the system transfers to the stationary point. It follows that $\alpha_1 > 0$, since $\alpha_1 = -f'(p_{st})$ and similarly that $f''(p_{st}) < 0$ and $\alpha_2 < 0$, since $\alpha_2 = -f''(p_{st})/2!$. It follows that functions

Ψ_d and Ψ_u determined by equations (35) will continuously decrease with time. Taking all the constants except Ψ_{st} in (35) equal to 1 then,

$\Psi_d = \Psi_{st}(1 + e^{-t})$ and $\Psi_u = 1/(1 + e^{t})$,

i. e. both functions decrease with t increase. The Ψ_u -function decreases faster than Ψ_d , and intersection of these functions occurs at some point. These properties of the Ψ_u - and Ψ_d -functions are shown schematically in Fig. 62. Both functions decrease with time; before the intersection point the Ψ_u -function is located higher than the Ψ_d -function and after it lower. Correlations of the curves in Fig. 62 with real curves for the change of Ψ_d- and Ψ_u-functions during oogenesis (Fig. 59), development (Fig. 60) and growth (Figs. 61, 71 and 72) of animals shows that the kinetics of the constitutive approach to the final steady state in these periods of ontogenesis in animals can be described by equations (35). During oogenesis when the living system is far from the final steady state the Ψ_u -function is higher than the Ψ_d -function and they both decrease during oogenesis. During embryonic development and growth of the organisms the Ψ_u -function is lower than the Ψ_d -function, $\Psi_u \to 0$, $\Psi_d \to \Psi_{st}$.
The intersection point lies at the end of oogenesis or at the earliest stage of embryonic development.

Thus proceeding from a statistical interpretation of dissipative functions and assumptions of the form (17) and (18), it proves possible to obtain kinetic equations for the dissipation function, external, and bound dissipation functions and to give a quantitative description of the constitutive processes proceeding during development and growth in animals.

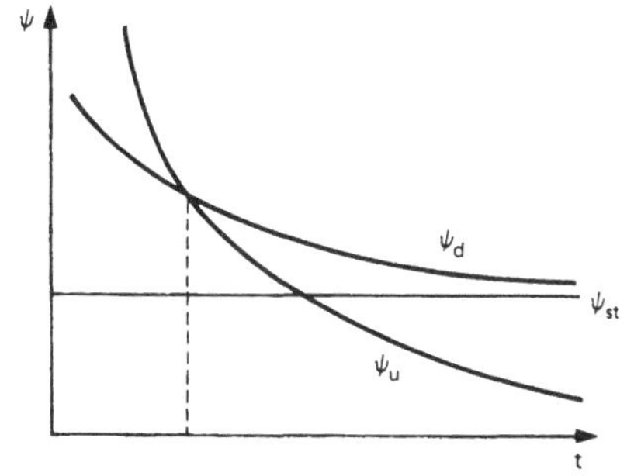

Fig. 62 Schematic picture of the change of Ψ_u - and Ψ_d -functions during development and growth of animals.

Stochastic Description of Growth and Aging in Animals

V. I. Timonin, R. S. Zotina

In earlier works (Zotina, Zotin, 1967, 1976; Zotin, 1972) aging of living systems was analyzed in terms of the thermodynamics of irreversible processes. To characterize these systems, functions were introduced, that depend on the rate of entropy production (dissipative functions); since functioning of any system involves a great number of irreversible physico-chemical processes all of which are accompanied by the production of dissipation heat. Consequently the dissipative functions serve as specific indicators of the intensity of processes proceeding within the system. In addition, according to Prigogine's theorem, the dissipative function of the system in a linear region continuously decreases as the system approaches the stationary, or equilibrium, state. Thus, the pattern of changes in the dissipative function gives a criterion for judging the evolution of the system.

Zotin (1974) proposed to subdivide the dissipative function into the sum of two dissipative functions:

$$\Psi = \Psi_u + \Psi_d \tag{1}$$

where Ψ_u is the bound dissipative function, Ψ_d the external dissipative function; he also suggested a total criterion for the evolution of thermodynamic systems

$$d\Psi_d / dt \leq 0 \tag{2}$$

which - with $\Psi_u = 0$ - is transformed into Prigogine's evolution criterion:

$$d\Psi / dt \leq 0 . \tag{3}$$

Thermodynamics and Kinetics of Biological Processes

Proceeding from these criteria it is clear that by observing the system's behaviour over time certain conclusions can be drawn as to the level of aging of the system; understanding aging as the system's approach to the stationary state.

Let us introduce a generalized measure of the aging of the system during time t in the form:

$$G(t) = \int_0^t \Psi \, dt . \tag{4}$$

Then, any parameter specifying the aging of the system can be presented as a function of the generalized measure of the aging of the system.

$$z_i = z_i \left[G(t) \right] . \tag{5}$$

Differentiating this parameter with respect to t we obtain:

$$\frac{dz_i}{dt} = \frac{dz_i}{dG(t)} \cdot \Psi \tag{6}$$

and, allowing the existence of z_i^{-1} we have:

$$\frac{dz_i}{dt} = \varphi_i(z_i) \cdot \Psi(t) \tag{7}$$

Since the dissipative function is presented as a sum according to (1):

then $$\frac{dz_i}{dt} = \varphi_1(z_i) \Psi_u(t) + \varphi_2(z_i) \Psi_d(t) . \tag{8}$$

This expression can be written in a more general form as:

$$\frac{dz_i}{dt} = \mathcal{F}_u \left[z_i, \Psi_u(t) \right] + \mathcal{F}_d \left[z_i, \Psi_d(t) \right] . \tag{9}$$

Thus, we obtain the result that at an arbitrary moment of time the change rate in the parameters determining the internal state of the system is composed of two parts. The first component depends on $\Psi_u(t)$ which tends to zero when the system tends to the local stationary state. Consequently, the first component describes the rate of system drift to a local stationary state. The second component depends on the function $\Psi_d(t)$. Hence, according to the evolution criterion (2) the second component specifies "pure" aging of the system. The first component in equation (9) must become zero, providing that $\Psi_u(t) = 0$. Assuming that when growth is terminated the living system is in the local stationary state, then equation (9) gives the most general form of the law governing a change of the aging parameter in living and in artificial systems.

It is usually believed that development, growth and aging are strictly determined processes disturbed only by random variations in the environment. However, according to Astaurov (1974) this process is also internally, a stochastic process, i. e. it is random in its essence. Therefore a change in time of the arbitrary parameters, $z_i(t)$, can be looked upon as a stochastic process.

Since a change in time of the parameters $z_i(t)$ takes place in a system far from equilibrium (all the living systems are far from equilibrium), the stochastic process, $z_i(t)$, can be regarded as a Markov process (this is one of the properties of highly dissipative systems - they rapidly forget the past). Assuming also that the local behaviour of the stochastic process, $z_i(t)$, can be described by Ito's stochastic differential (Venitzel, 1975) then

$$dz_i(t) = a(z_i, t)dt + b(z_i, t)d\eta(t), \qquad (10)$$

where $a(z_i, t)$ is the coefficient of the drift of the stochastic process $z_i(t)$, $B(z_i, t)$ the diffusion coefficient of the stochastic process $z_i(t)$, $n(t)$ the process of Brownian movement. Comparing (10) and (9) we see that $a(z_i, t)$ can be substituted for dz_i/dt and then we obtain

$$a(z_i, t) = \mathcal{F}_u[z_i, \Psi_u(t)] + \mathcal{F}_d[z_i, \Psi_d(t)]. \qquad (11)$$

The diffusion coefficient is regarded as being independent of z_i. It is known (Venitzel, 1975) that a Markov process whose local behaviour is described by Ito's stochastic differential is a diffusion process, and its transition probability density function satisfies the Fokker-Planck-Kolmogorov equation:

$$\frac{\partial f}{\partial t} = -\frac{\partial}{\partial z_i}[a(z_i, t)\cdot f] + \frac{1}{2}\frac{\partial^2}{\partial z^2}[b_i^2(z_i, t)\cdot f] \qquad (12)$$

It follows that knowing the functions $\mathcal{F}_u[z_i, \Psi_u(t)]$, $\mathcal{F}_d[z_i, \Psi_d(t)]$, $\Psi_u(t)$ and $\Psi_d(t)$ by solving equation (12) and integrating the solution over a tolerance region of values, z_i, it is possible to determine the cumulative distribution function of the process.

The theory presented can be used to describe the processes of organism development. As an example we can analyse the weight change during growth and aging in animals.

Case 1 Let $\varphi_z(z_i) = \alpha$, $\Psi_u = 0$, $\Psi_d = a e^{-\alpha t} + c$,

then

$$a(z, t) = \beta e^{-\alpha t} + \gamma \tag{13}$$

where $\beta = a\alpha$, $\gamma = c\alpha$, and B(z, t) are equal to the constant b. Then equation (12) will have the form:

$$\frac{\partial f}{\partial t} + (\beta e^{-\alpha t} + \gamma)\frac{\partial f}{\partial z} + \frac{1}{2} b \frac{\partial^2 f}{\partial z^2} = 0. \tag{14}$$

Initial condition: $f(z_0, t, z)|_{t=0} = \delta(z - z_0)$

Boundary conditions: $f(z_0, t, z)|_{z=\pm\infty} = 0$

To solve this equation let us introduce substitution of variables in the following manner:

$$\begin{aligned} t' &= bt \\ z' &= z - \int_0^t (\beta e^{-\alpha t} + \gamma)\, dt = z + \frac{\beta}{\alpha} e^{-\alpha t} - \frac{\beta}{\alpha} + \gamma t. \end{aligned} \tag{15}$$

In terms of the new variables equation (14) will have the form:

$$\frac{\partial f}{\partial t'} + \frac{1}{2}\frac{\partial^2 f}{\partial z'^2} = 0. \tag{16}$$

Solution of this equation yields the function

$$f'(t', z_0, z) = \frac{1}{\sqrt{2\pi t'}} \exp\left\{\frac{(z' - z_0')^2}{2t'}\right\}. \tag{17}$$

Hence, we have shown that the stochastic process of change of organism weight is a Gaussian stochastic process. Substituting t' and z' into (17) we obtain an expression for the solution in terms of the initial variables t and z i. e. the probability density of the particular process:

$$f(t, z_0, z) = \frac{\exp\{[z - (z_0 - \beta/\alpha\, e^{-\alpha t} + \beta/\alpha + \gamma t)]^2 / 2bt\}}{\sqrt{2\pi b t}} \tag{18}$$

A mathematical expectation of this process takes the form:

$$M(z) = z_0 - \frac{\beta}{\alpha} e^{-\alpha t} + \frac{\beta}{\alpha} + \gamma t, \tag{19}$$

while the variance: D(z) = bt, is in direct proportion to time.

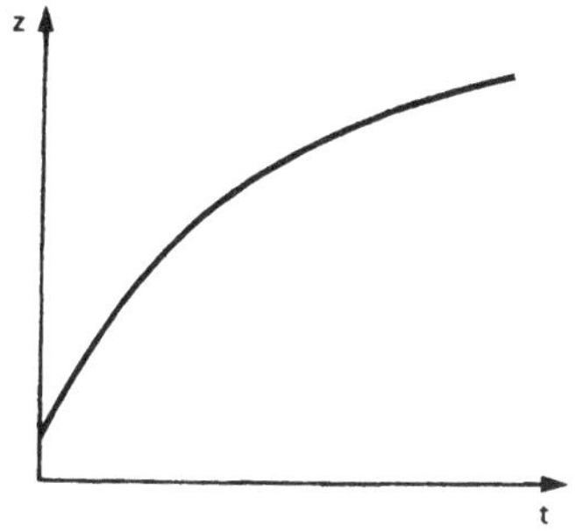

Fig. 63 - Change in body weight of animals during growth - case 1 (see text).

The mathematical expectation gives the average change in animal weight. In this example the weight rises constantly; the weight curve has neither a maximum, nor a point of inflection (Fig. 63). The growth of bird embryos (Terskova, 1976), fish (Hofman-Kadoshnikov, 1976), and of some insects (Walter, Lamprecht, 1978) follow this type of curve.

<u>Case 2</u> Let $\Psi_u = a_1 e^{-\alpha_1 t}$, $\Psi_d = a e^{-\alpha t} + c$, $\varphi_1(z) = z$

$\varphi_2(z) = d = \text{const}.$

Then $a(z,t) = \beta e^{-\alpha t} + \gamma + z a_1 e^{-\alpha_1 t}$, $\beta = ad$, $\gamma = cd$ (20)

b(z, t) assumed again equal to the constant b.

Then equation (12) will have the form:

$$\frac{\partial f}{\partial z} + \frac{\partial}{\partial z}\left\{\left[\beta e^{-\alpha t} + \gamma + z a_1 e^{-\alpha_1 t}\right] f\right\} - \frac{1}{2} b \frac{\partial^2 f}{\partial z^2} = 0. \quad (21)$$

Initial condition: $f(z_0, t, z)\big|_{t=0} = \delta(z - z_0)$

boundary conditions: $f(z_0, t, z)\big|_{z=\pm\infty} = 0$

To solve equation (21) let us introduce the following substitution of variables:

$$t' = \int_0^t b \exp\{-2\int_0^{t_2} a_1 e^{-\alpha_1 t_1} dt_1\} dt_2 = b\int_0^t \exp\{\frac{2a_1}{\alpha_1}(e^{-\alpha_1 t}-1)\} dt$$

$$z' = z\exp\{-\int_0^t a_1 e^{-\alpha_1 t_1} dt_1\} - \int_0^t (\beta e^{-\alpha t_2}+\gamma)\exp\{\int_0^{t_2} a_1 e^{-\alpha_1 t_1} dt_1\} dt_2 = \qquad (22)$$

$$= z\exp\{\frac{a_1}{\alpha_1}(e^{-\alpha_1 t}-1)\} - \int_0^t (\beta e^{-\alpha t}+\gamma)\exp\{\frac{a}{\alpha}(e^{-\alpha t}-1)\} dt.$$

In terms of the new variables the equation has the form:

$$\frac{\partial f}{\partial t} + \frac{1}{2}\frac{\partial^2 f}{\partial z^2} = 0.$$

This equation is exactly the same as (16), and its solution has the form (17). Hence, in this case also the process of weight change will be a Gaussian process.

Substituting t' and z' values into solution (17) for case 2 we obtain the probability density function for weight change during growth and aging in animals:

$$f(t, z_0, z) =$$

$$= \frac{\exp \dfrac{z - \exp\{-\frac{a_1}{\alpha_1}(e^{-\alpha_1 t}-1)[z_0 + \int_0^t (\beta e^{-\alpha t}+\gamma)\exp\{\frac{a_1}{\alpha_1}(e^{-\alpha_1 t}-1)\} dt]\}^2}{b_0 \int^t \exp\{\frac{2a}{\alpha}(e^{-\alpha t}-1)\} dt}}{\sqrt{2\pi b_0 \int^t e^{2a/\alpha(e^{-\alpha t}-1)} dt}}. \qquad (23)$$

The mathematical description of the weight change has the form:

$$M(z) = \exp\{-\frac{a_1}{\alpha_1}(e^{-\alpha_1 t}-1) z_0 + \int_0^t (\beta e^{-\alpha t}+\gamma)\exp\{\frac{a_1}{\alpha_1}(e^{-\alpha_1 t}-1)\} dt\} \qquad (24)$$

Variance:

$$D(z) = b\int_0^t \exp\{\frac{2a_1}{\alpha_1}(e^{-\alpha_1 t}-1)\} dt \qquad (25)$$

In expression (24) at small t, the main role is played by the first component

$$\exp\{-\frac{a_1}{\alpha_1}(e^{-\alpha_1 t}-1)\}$$

which is a Gompertz function. This function is widely applied in the theory of growth. It is used to describe tumour growth (Laird, 1964; Dronova et al., 1966; Erokhin, 1975; Emanuel, 1977), embryonic growth (Laird, 1966 a; Tyler et al., 1967), and the growth of animals and men (Laird et al., 1965; Laird, 1966 b, 1967; Walter, Lamprecht, 1978). The graph of this function at t = 0 passes through z_0, has one point of inflection and one maximum. Following the maximum the function decreases (Fig. 64).

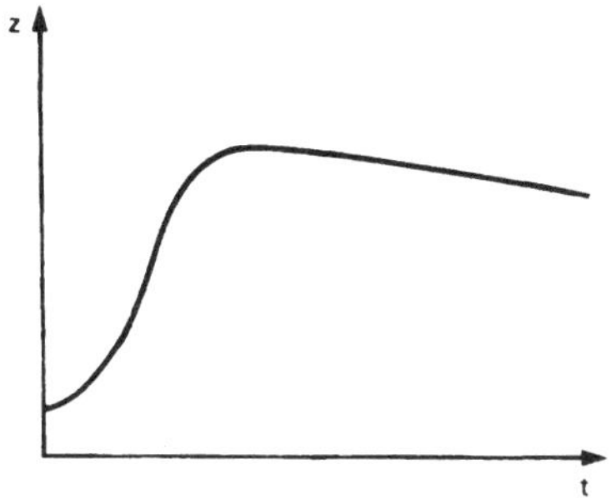

Fig. 64 Change in the body weight of animals during life - case 2 (see text).

Using asymptotes of the function reflecting the mathematical description of growth, i. e. the most probable growth curve, at $t \rightarrow \infty$ we obtain

$$M(z(t)) = ct + \mathfrak{D}$$
$$\mathfrak{D}(z(t)) = c_1 t + \mathfrak{D}_1 \qquad (26)$$

At large t the growth of animals decreases in an almost linear fashion.

Thus, through knowledge of the external and bound dissipation functions for a particular animal, the function describing the change of the animal's weight throughout its life can be found. Knowing that this weight decrease must not exceed 30 % the maximal life time of this animal can be calculated (Zotin, Zotina et al., 1978). The concepts presented here for the quantitative description of the changes observed during growth and aging in animals and men may find applications in other fields.

Thermodynamic Estimation of Development and Growth of Microorganisms

V. V. Koryagin, S. A. Konovalov

Determination of kinetic and thermodynamic characteristics over the course of development and growth in microorganisms is of primary importance, both for creating an experimental basis for the thermodynamics of biological processes, and for practical applications associated with microbiological cultures (particularly in industry and medicine).

The following methods have been employed in the study of microorganism. Heat production (direct calorimetry) and respiration (indirect calorimetry) were continously measured directly in a fermenter over the course of fermentation (Koryagin et al., 1974, 1976, 1977).

According to Zotin (1972, 1974) the dissipative function of living systems (Ψ), if the contribution from glycolysis is neglected, is equal to the respiration intensity:

$$\Psi \simeq \dot{q}_{O_2},$$

where $\dot{q}_{O_2}$ is the specific respiration rate.

The external dissipation function (Ψ_d) is proportional to the heat production of the system:

$$\Psi_d = \dot{q}$$

where $\dot{q}$ is the specific rate of heat production.

The bound dissipation function (Ψ_u) is proportional to the difference:

$$\Psi_u = \Psi - \Psi_d = \dot{q}_{O_2} - \dot{q}.$$

To increase the information content of thermodynamic data we suggested that, together with indices expressed as W/g determination of specific rates of heat production, respiration, and of the Ψ_u -function not only per weight unit, but also per volume unit, i. e. in W/l would be useful. The values to be estimated have the form: $\dot{q}^{\vee}$, $\dot{q}^{\vee}_{O_2}$ and $\Psi^{\vee}_u$.

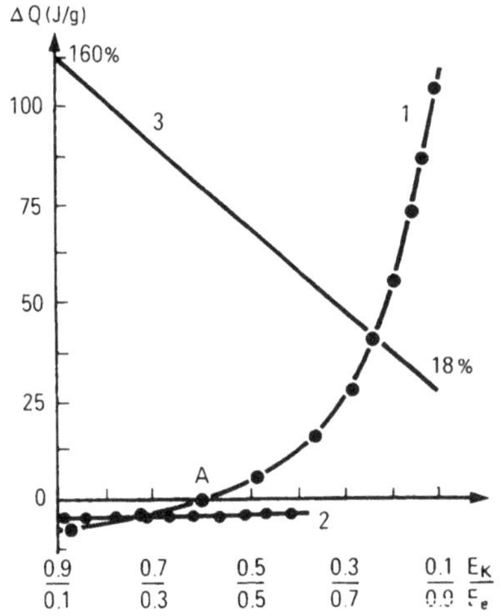

Fig. 65 - Difference between direct and indirect calorimetry determined from the data by Gololobov and Golubev (1970) (1) and Guenther (1965) (2) depending on the ratio between constructive (E_K) and energetic (E_e) metabolism coefficients and on the value of biomass yield (according to Gololobov and Golubev, 1970).
Point A is the intersection point of the curve (1) and the graph coinciding with abscissa (according to Minkevich and Eroshin, 1973, 1977).

The first stage was to use from the literature data for the amount of oxygen consumed and the expected level of heat production to calculate Q_{O_2} values (energy related to oxygen uptake adopting a thermal coefficient equal to 14.2 kJ/gO_2), Q_{sp} values (specific heat liberation in J/g of biomass), and their differences, ΔQ, which depend on differing physiological states. These observations are shown in Fig. 65. It is seen that with increase in biomass production the graph reflects differences between the data determined by direct and indirect calorimetry

$$\Delta Q = Q_{O_2} - Q_{sp} \ .$$

Curve 1 of Fig. 65 according to Gololobov and Golubev's data is of a non-linear character and even enters the negative region. The graph (curve 2), calculated after Guenther (1965), is practically parallel to the abscissa and is located in the negative region, while a graph plotted on the basis of Minkevich and Eroshin's (1977) data coincides with the abscissa. Point A is the intersection point calculated according to the principles of Gololobov and Golubev, and Minkevich and Eroshin. Point A in the graph 1 (Gololobov, Golubev, 1970) reflects a special case when $\Delta Q = 0$.

Thus a significant descrepancy has been revealed in the data of these authors on the respiration-heat production relationship in different metabolic states in microorganisms. It follows from the literature (see Zotin, Konoplev, Grudnitzky, 1978; Zotin, 1979) that the discrepancy between data from direct and indirect calorimetry was always observed when the size of organisms was less than 50 g.

The second state of the investigation involved kinetic and thermodynamic estimations of specific features of metabolism of microorganisms that belong to different taxonomic groups during the course of their development and growth. Fig. 66 shows kinetic and thermodynamic parameters for continuous cultivation of the yeast *Candida quilliermondii*. It is seen that the rhythmic character of population development and growth found in the yeast *Candida* has three activity extrema (as compared with the normal sigmoid growth curve) and is easily distinguished from thermogenesis, and has influenced the pattern of the thermodynamic parameters. The specific rate of heat production $\dot{q}$ (W/l) and specific rate of respiration $\dot{q}_{O_2}$ (W/l) have an almost linear, irreversible relationship with thermogenesis, $\dot{q}^{\vee}$, or, in other words, $\dot{q}$ (W/l) and the specific respiration rate per volume unit $\dot{q}^{\vee}_{O_2}$ (W/l) respectively. However, if the maximal value of the Ψ_u-function in the lag-phase agrees with data of Schaarschmidt and Brettel (1978) then the Ψ_u-function has a maximum in the exponential phase of growth, or more precisely, in the anabolic sub-phase of the second extremum in thermogenesis.

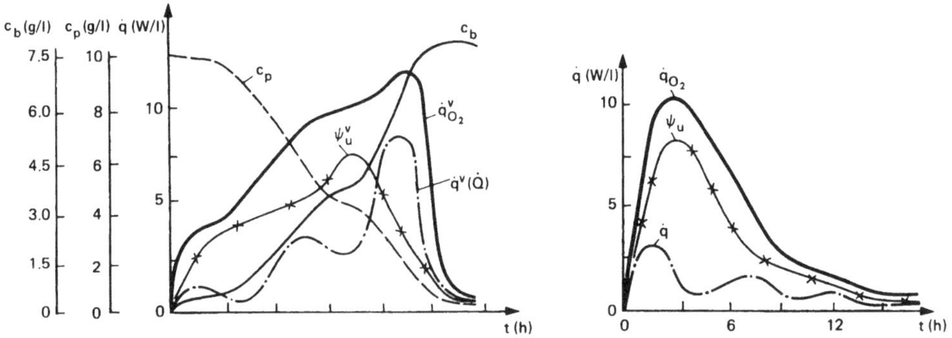

Fig. 66 Change of kinetic and thermodynamic parameters in the course of cultivation of the yeast *Candida quilliermondii*.
Notations: c_p, c_b parafin and biomass concentrations, respectively;
$\dot q_{O_2}, \dot q = \Psi_d, \Psi_u$ specific intensity of respiration, heat production and bound dissipation function;
$\dot q^V_{O_2}, \dot q^V = \Psi^V_d, \Psi^V_u$ the same per culture volume unit.

The physical meaning of the bound dissipation function, according to Zotin (1975), although not absolutely clear, lies in correlation with the rates of chemical reactions proceeding in an open system.

Certain specific feature of Ψ_u- and Ψ^V_u-functions are of interest. The Ψ_u-function has a maximum at the end of the anabolic sub-phase of the first extremum of thermogenetic rhythm (3 hours) and the Ψ^V_u-function at the end of the anabolic sub-phase of the second rhythmical extremum (10 hours). At theses times there is no uptake of substrate and no increase of biomass. More precisely, these times are the critical points between the absence, and renewal, of substrate uptake and the increase of biomass.

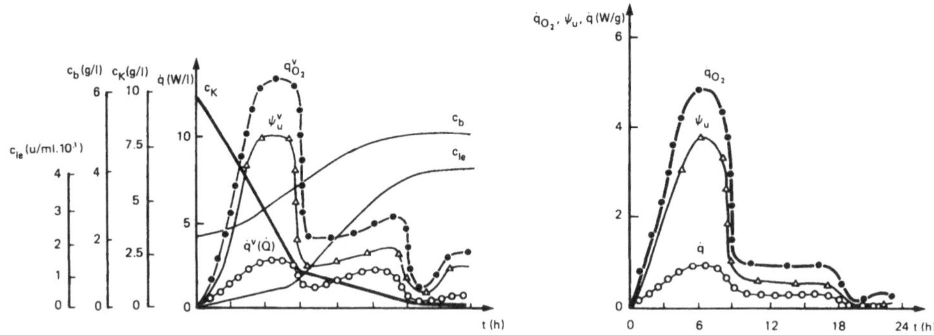

Fig. 67 Change in kinetic and thermodynamic parameters during cultivation of actinomycets. Notations: c_{le} concentration of lysing enzymes;
c_k, c_b concentration of starch and biomass, respectively; other same as in Fig. 66.

The changes in the kinetic and thermodynamic parameters in the course of continuous cultivation of *Actinomyces sp. II* are given in Fig. 67. This shows that this culture has two rhythm extrema in thermogenesis. The first maximum is concerned with the intensive uptake of the basic component of the substrate (starch), and this is characterized by a high rhythmic level in the catabolic sub-phase. The second rhythm maximum is concerned with the uptake of the second substrate component (addition of yeasts), for the utilization of which the culture appears to increase the production of lysing enzymes which are liberated into the environment to destroy the yeast cell walls. The thermodynamic parameters found by us are lower for the second maximum, and this seems to characterize the intensive production of secondary metabolites (complex of lysing enzymes). Thus the metabolism of a culture-producer of lysing enzymes can easily be described through kinetic and thermodynamic parameters.

Finally, it follows from Fig. 67 that as for yeasts (Fig. 66) a maximum in the $\Psi_{\check{u}}$ -function is observed at the end of a short, in this case anabolic, rhythmical sub-phase after the eighth hour, while the maximum in the thermodynamic parameters which characterize development and growth of the culture fall into the first rhythmical extreme, the second extreme, in terms of thermodynamics, is more weakly conditioned. The pattern of thermodynamic parameters in yeasts (Fig. 66) is different.

Changes of kinetic and thermodynamic parameters over the course of continuous cultivation during maturation (greater utilization of the substrate) of *Candida quilliermondii* is shown in Fig. 68. These data show only two rhythmical extrema, i. e. the second and third ones (compare with Fig. 66), because the first extreme is absent since there is no lag-phase in the given cultivation regime.

The first extremum (Fig. 68), as compared to the second one, is characterized by a lower uptake of oxygen, substrate, a lower level of heat production (thermogenesis), and by the presence of a significant amount of phosphorus. Evidently, the first extremum corresponds to one metabolic pathway (oxidative phosphorylation), and the second one to another metabolic pathway (free oxidation). Therefore the yeast culture apparently uses two differing metabolic pathways.

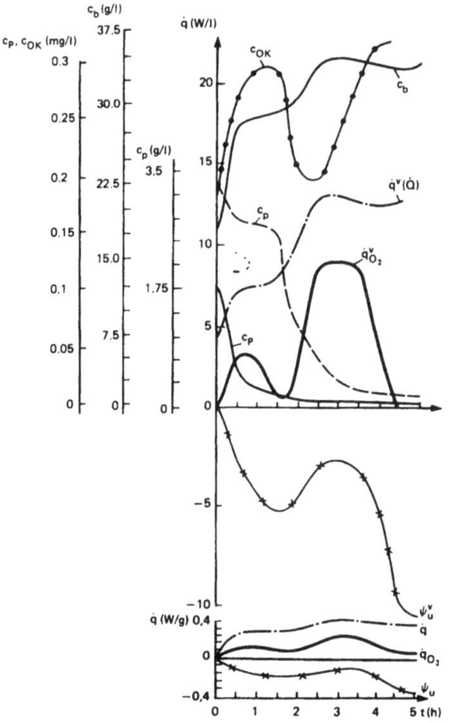

Fig. 68 - Change in kinetic and thermodynamic parameters during cultivation at dosing regime of the yeast *Candida quilliermondii*.
Notations: c_{p}, c_{oa}, c_{b}, c_{pa} concentrations of phosphorus, organic acids, biomass and paraffin respectively; other same as in Fig. 66.

The thermodynamic pattern of this process differs from the pattern presented in Fig. 66. In this case, although the Ψ_u - and Ψ_u^v -functions show differing rhythmical behaviour, they are located in the negative region. The negative value of Ψ_u - and Ψ_u^v -functions can be explained by the fact that differences between specific rates of respiration and heat production rather than between metabolism and heat production were determined.

It is seen from Fig. 68 that by accounting for contribution of other metabolic pathways (see the graph of organic acid concentration c_{oa}) the value of Ψ_u - and Ψ_u^v-function obtained are closer to the abscissa. The low values for the bound dissipation functions are also explained by a decrease in the energetic level of the yeast cell protoplasm or in the general case by a decrease in respiratory efficiency of the aging culture during maturation.

The Ψ_u -function (W/g) describes the rate of entropy production per unit mass, and consequently its numerator describes bound dissipation, and the denominator increase in biomass, i. e. biomass synthesis. Therefore this function is convenient for characterizing the maximum in developmental phases. The $\check{\Psi}_u$ -function (W/l) is concerned with the intensity of energy dissipation. However this energy is not distributed to the total biomass obtained, but suggests the absolute value of the dissipation level. It falls, as is seen, upon the inflection point in the growth curve. Therefore this function is convenient for characterizing the maxima in growth phases.

Thus, in all the figures presented the maxima of the developmental phases described by Ψ_u -functions are always in advance of the maxima in the growth phases characterised by $\check{\Psi}_u$ -function.

Transition Process in Ontogenesis as a Characteristics of Changes in the Stability of Animals to External Influences

V. V. Lapkin

A transition process in open systems usually means the establishment of a stationary state when the system deviates from this working regime. The stationary state for an equilibrium system is understood as that energy state in which the entropy production, or in other words, the dissipation function Ψ , takes the least positive value under stable environmental conditions (Prigogine's theorem). The condition for a steady nonequilibrium system is met by the minimum principle of one component of the dissipation function (the dissipation function Ψ consists of two components - the external dissipation function Ψ_d and the bound dissipation function Ψ_u), namely, the minimum in the external dissipation function Ψ_d corresponding to dissipation heat leaving the system (Zotin, 1974).

The dissipation heat corresponding to another component of the dissipation function, the bound dissipation function Ψ_u is left and realized within the system. When the open system approximates to the equilibrium state, this tends to zero. The principles of the minimum in the dissipation function and in the external dissipation function serve as thermodynamic criteria for the approximation of an open nonequilibrium system to the final stationary state, in which these functions take the smallest possible value.

The dissipation function Ψ for living systems is reflected in the specific rate of metabolism conditioned by the intensity of respiration, and glycolysis, the external dissipation function Ψ_d by the metabolic heat, leaving the system (heat production). The bound dissipation function Ψ_u is determined as the difference between the specific value of metabolism (indirect calorimetry) and heat production (direct calorimetry). The level of organism

Thermodynamics and Kinetics of Biological Processes

metabolism at rest (unexited state) or of basal metabolism defines the constitutive stationary state. Deviations from this state caused by internal and external factors are called inducible deviations. Deviations caused by oogenesis, regeneration, malignant growth, as well as, apparently, wound healing which rank among constitutive deviations (Zotin, 1972, 1974) are exceptions.

An analysis of experimental data on direct and indirect calorimetry carried out on different classes of animals (microorganisms, insects, fish, amphibians, mammals) has shown that ontogenesis is most correctly described by thermodynamic concepts which sub-divide the dissipation function into two components. At any rate, such an approach may explain by a decrease in the contribution of the bound dissipation function to Ψ -function some of the discrepancies between data from direct and indirect calorimetry obtained during early stages of ontogenesis and during growth and aging. Thereby possible applications of the Prigogine-Wiame theory to biological systems are extended and complemented. According to this theory the rate of entropy production (dissipation function) should spontaneously decrease with age (Zotin, 1974; Zotin, Zotina, 1978). When the system tends to the final stationary state the behaviour of the dissipation function and of its components indicates that the degree of nonequilibrium state in the living system does not remain constant, it decreases continuously. Along with studies of metabolic levels during the course of ontogenesis the problem of the organism's stability toward external influences is important in the thermodynamic characterisation of biosystems (Rubin, 1976). As Nikolaev (1976) notes, a decrease in the value of the entropy increment in open systems does not mean an increase in their stability toward external disturbances.

Regarding living organisms as totally nonequilibrium systems (Bauer, 1935; Nikolaev, 1976), which can exist only within a certain framework of nonequilibrium conditions we hope to find a correlation between the degree of the nonequilibrium biosystem and its stability to external effects which depends on age changes in the organism.

After Ashby (1956, 1960) and Shulman (1972) stability was taken as the field of states providing the integrity of the system, beyond the limits of which, it is destroyed. With this approach to system stability the problem is, practically, reduced to studies of age changes and tolerance limits (the environmental conditions providing for long survival) or of resistivity limits (the environmental conditions providing for a limited life time) of the organisms. In terms of thermodynamics the problem is to examine the age changes of maximal possible inducible deviations of the system from the constitutive stationary state under disturbing effects. Using Ashby's concepts, deviations of the system from some set of state, knowingly inherent to any age category of the organism should be discussed.

We have investigated age peculiarities in responses of fishes to such universal external disturbing factors as temperature and electric fields. We have also collected literature data relating to this problem for other animals.

The bream *Abramis brama L.* from the Rybinsk reservoir was taken as the object of investigation. In the Rybinsk latitude this species shows, in its life cycle, two stable stationary states (Lapkin et al., 1977; Poddubny et al., 1978), which are expressed by the fact that the fish can operate within two temperature ranges, 21 to 29 °C in summer, and 1 to 6 °C in winter. In adults, as compared to young fish, some trends toward a lower summer temperature is observed. Although there are not enough experimental data at present to give a quantitative estimate of this phenomenon, the decrease in the summer level of optimal temperature with age seems quite real. The fact is that basal metabolism in the region of temperature tolerance of the species monotonously increases with temperature rise (Winberg, 1956 a), while its value drops with the age of animals, fishes included (Zotin, 1972, 1974). Therefore a decrease in optimal temperature in a gradient environment with age is considered as a control mechanism of the optimal level of metabolism when poikilothermic (cold-blooded) organisms interact with the environment.

In this connection all the investigations of age dynamics of fish stability were carried out in summer on individuals acclimatised to 20 °C for 10 to 20 days, i. e. on those brought to the near-stationary summer state (corresponding to the tolerance zone for any age categories of the bream). Upper lethal temperature values at which the individual under study died whilst experiencing a constant rate of water heating equal to 1 to 2 °C per day were studied, i. e. particular temperature conditions of the environment were found which caused the death of individuals over a limited time (not more than 24 hours). Experiments were conducted on groups of 3 to 10 individuals (depending on their size) in tanks with high water volume (350 l) equipped with a system of positive aeration. This allowed the oxygen concentration in the water with equal temperature parameters to be regarded as equal (solubility of oxygen in water decreases linearly with temperature rise within the range 20 to 40 °C) for different age groups. Experiments with electric current were conducted on single individuals at a constant temperature of 20 °C in a 30 litre tank of aerated water. An electric field with uniform tension and a frequency of 50 Hz was generated. The individuals were placed along the lines of force of the field, and the minimal specific power inducing electronarcosis (complete immobilization of the individuals) was found (Lapkin, Zaytsev, 1976). The experimental results (Table 7) were evaluated by conventional methods (Kassandrova, Lebedev, 1970). The age of individuals was estimated from the sclerite structure of the scales (Chugunova, 1959; Pravdin, 1966).

If the lethal temperatures were the external factors under which the death of the organism occurred, then the electronarcosis data reflected death indirectly. The death of individuals under narcosis was only recorded in some experiments and, as a rule, it occurred 10 to 20 min after the application of the electric field. Consequently, over the course of the temperature as well as the current experiments individuals died in the same resistance zone, i. e. under experimental conditions providing for the limited life time of the animals. These values of the external stimulus directly or indirectly leading to the destruction of the system in the resistance zone were considered as the limiting possibilities for the system to deviate from the constitutive stationary state, or, as inducible transition processes quantitatively expressed in the stimulus values (°C and $\mu W/cm^3$).

Table 7

Size, age and values of lethal temperatures and electronarcosis reactions of the bream

Age	Number of individuals	Linear and weight composition		Lethal temperature	Electronarcosis
		Body length	Raw weight		
years		cm	g*	°C	μW/cm³
0.25	20	2.5 ± 0.1	0.25 ± 0.02	37.0 ± 0.3	260 ± 5
1	20	8.1 ± 0.7	10 ± 2	36.0 ± 0.4	40 ± 5
2	20	11.4 ± 0.6	25 ± 4	35.4 ± 0.7	27 ± 5
3	20	13.4 ± 0.4	52 ± 5	34.2 ± 0.8	18 ± 1
4	20	15.8 ± 0.7	81 ± 9	32.2 ± 0.5	14 ± 2
5	20	20.0 ± 0.9	162 ± 23	32.1 ± 0.6	9 ± 1
6	20	23.0 ± 0.4	245 ± 16	32.0 ± 0.6	9 ± 0.7
7	20	25.3 ± 0.8	313 ± 27	31.9 ± 0.7	8 ± 0.5
8	16	28.6 ± 0.9	485 ± 40	31.5 ± 0.6	6 ± 0.7
9	14	31.4 ± 0.5	630 ± 32	31.5 ± 0.5	5 ± 0.6
10	10	33.1 ± 0.6	720 ± 31	31.0 ± 0.5	5 ± 0.7
11	10	35.4 ± 0.8	855 ± 54	31.1 ± 0.6	5 ± 0.5
12	3	37.0 ± 2.4	910 ± 233	31.1 ± 1.4	4.7 ± 0.7
13	2	37.5 ± 6.2	1030 ± 502	30.7 ± 6.2	

* The specific weight of individuals is comparable with that of water, (Nikolsky, 1974) and the weight parameters reflect their body volume (cm³). The quoted confidence limits for Δl, ΔQ, $\Delta\Theta$ and ΔE are for $p = 0.95$

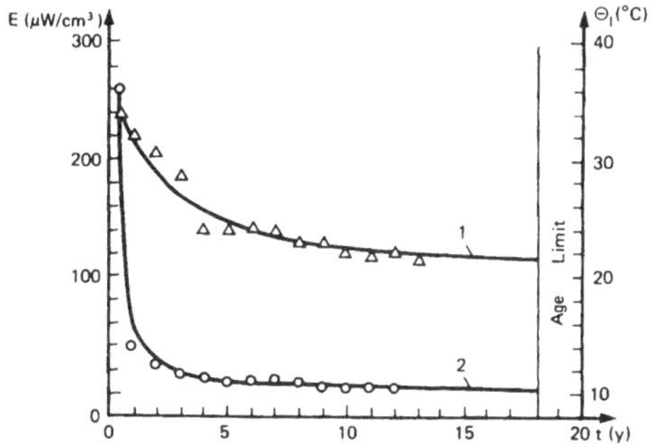

Fig. 69 Age dynamics of bream response to temperature and after electric current application: 1 lethal temperature Θ_l, 2 electronarcosis E.

The analysis of the experimental data (Fig. 69) has shown that, notwithstanding the stimulus, the age changes for maximum possible deviations in the bream resulting in its death, are of similar pattern. They are decreasing valued functions with time, since in young fish the lethal temperature is 37 °C and for electronarcosis 260 $\mu W/cm^3$ is required. In adults these parameters are stable at a fixed level and when extrapolated to the maximum age of Rybinsk reservoir bream, which is appraised as 18 to 20 years (Ostroumov, 1955; Butorin, 1978), they are 30.7 °C and 2.5 $\mu W/cm^3$, respectively. Such a range of extreme deviation within the system indicates that the tolerance zone of the organism becomes narrower with age, i. e. the number of states and standards of the system providing for its long term existence decreases (Fig. 70). With age, as follows from thermodynamic concepts, the rate of entropy production also decreases (Ψ - function) and the degree of nonequilibrium which derives from the sub-division of the species is determined by the bound dissipation function Ψ_u.

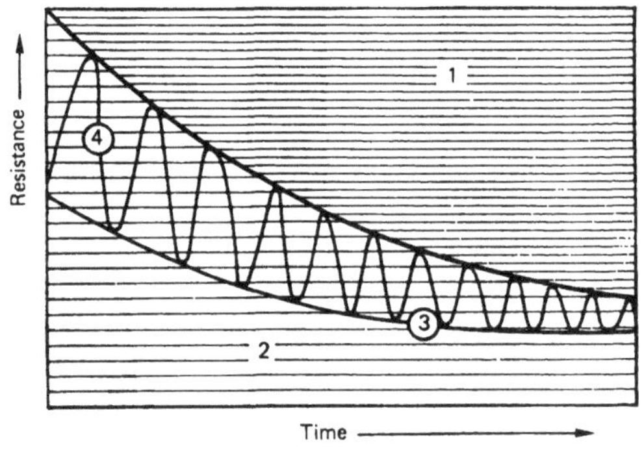

Fig. 70 Schematic picture of the change of the "field" of the biosystem stability depending on the organism age:
1 resistance zone providing the stability of the system during limited time; 2 tolerance zone or multitude of states providing the stability of the system during prolonged time; 3 constitutive approach of the system to the stationary state (or age changes of Ψ -function); 4 inducible deviations.

It follows that an increasingly nonequilibrium state or higher value of the Ψ_u-function conforms according to Ashby (1956) with higher stability and greater numbers of the system states.

The bound dissipation function has been determined for animals of different organizational levels: in microorganisms (Schaarschmidt et al., 1975; Schaarschmidt, Brettel, 1978), insects (Fourche, Lemaitre, 1974; Loehr et al., 1978), fishes (Grudnitzky, 1972 a, b, 1975) and higher vertebrates (Zotin, 1975, 1979). It has been found that the values of Ψ_u-function, as well as of Ψ- and Ψ_d-functions decrease. As an example (Fig. 71), let us take the data on the changes of these values in the cockroach *Blattella germanica L.* (Loehr et al., 1978). In addition a good correlation between Ψ_u-function and growth rate has been demonstrated for yeast cultures (Schaarschmidt, Brettel, 1978). It is noted that change in the bound dissipation function produces a better agreement with the number of formed cells than with a change in their mass. The physical meaning of the bound dissipation function is unclear at present. Zotin (1975) believes, however, that the heat determined by it may be utilized in the system in such ways as to increase local temperature, to accelerate chemical reactions, formation of dissipative structure, fixation of thermal radiation by macromolecules and intracellular structures. On account of the above it may be assumed that Ψ_u-function is responsible not only for growth and stability, but for the life span of the biological system as well. In fact, the value of this function decreases with age, and postulating that nonequilibrium and the existence of dissipative (Glansdorff, Prigogine, 1971) or dynamic (Nikolaev, 1976) structures are the necessary elements of living matter, it may be assumed that the biological death of an organism occurs when the Ψ_u-function equals zero, or, it acquires such a small value that further functioning of the organism becomes impossible. Evidently, this life time criterion ought to be valid for any open, nonequilibrium system.

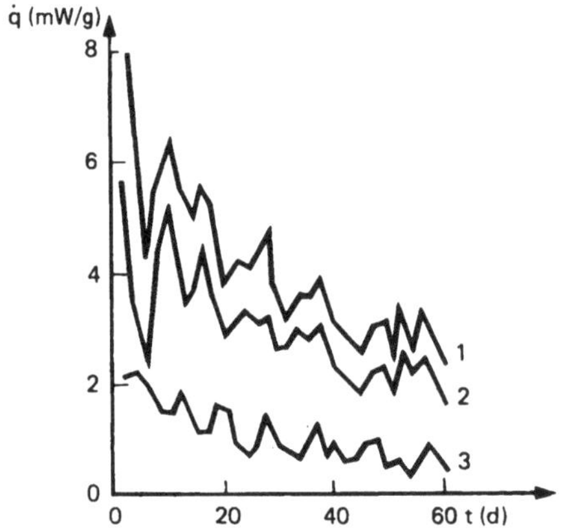

Fig. 71 Change of specific heat production q̇ during the growth of *Blattella germanica* males:
1 - dissipation function, 2 - bound dissipation function, 3 - external dissipation function (Loehr et al., 1978).

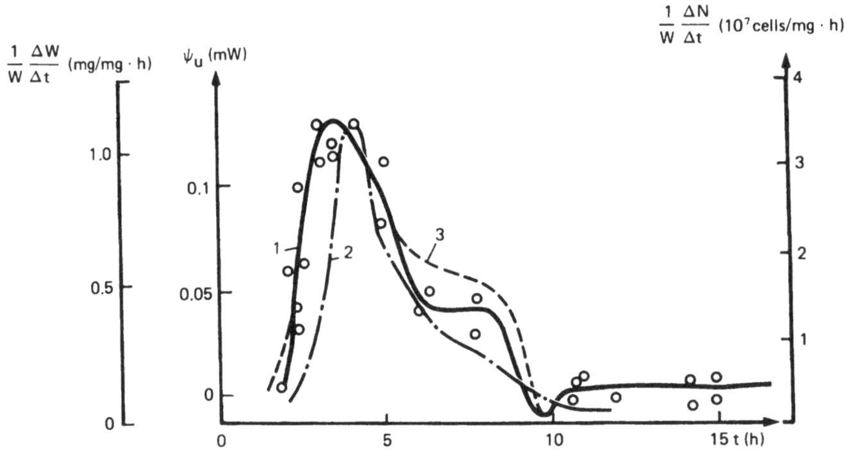

Fig. 72 Growth of a yeast culture under anaerobic conditions:
1 - change of Ψ_u-function, 2 - specific rate of the change in dry weight $1/W \cdot \Delta W/\Delta t$, 3 - specific rate of the change in cell number $\frac{1}{W}\frac{\Delta N}{\Delta t}$ (Schaarschmidt, Brettel, 1978).

On Specific Peculiarities of Transition Processes in Ontogenesis of Mammals

I. A. Arshavsky

In terms of our theory of individual development, which in contrast to the energetic rule of the surface, was named the energetic rule of skeletal muscles, specific features of transition processes in a living developing system, as compared to the non-living system (mechanical, electrical and others), have been described. The transition process in the latter is a transition from one established state (equilibrium or nonequilibrium one) to another state. Charging of condensers from a current source may be presented as an example of such a transition. Due to inertia, determined by the nature of substance and the structure pattern of the corresponding system, the transition process is realized here along some exponential curve, asymptotically approximating its limiting value. The system acquires the parameters of nonequilibrium state, the level of which is determined by the curve of the acquired energetic potential.

The change under study, i. e. the change in the value of the acquired potential (ε) when the system transfers from state 1 to state 2 can be written as a difference between the initial and final values of the corresponding function of the state (Fig. 73 a). The change of the same function (ε) under the reverse transition of the system from state 2 to state 1 will take place when the condenser is discharged. The reversed transition is also realized along some exponential curve asymptotically approximating the system to its initial state. The sum of both differences corresponds to the change of ε for a complete cycle; after its completion the sum appears to have a zero value. It follows that in a non-living system, after the completion of the whole cycle, this function (as apparently, is any other) is not changed in the cyclic process. As far as the system itself is concerned (its structural characteristics, rather than its state functions), it is evident

Thermodynamics and Kinetics of Biological Processes

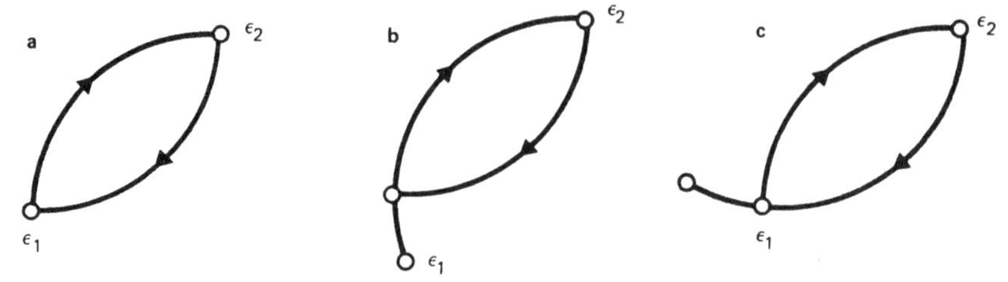

a: $\Delta\epsilon_{1\to2} = \epsilon_2 - \epsilon_1$; $\Delta\epsilon_{2\to1} = \epsilon_1 - \epsilon_2$

b: $\Delta\epsilon_{1\to2} = \epsilon_2 - \epsilon_1$; $\Delta\epsilon_{2\to1} = \epsilon_1 - \epsilon_2$; $\Delta\epsilon_{1\to2} + \Delta\epsilon_{2\to1} < 0$

c: $\Delta\epsilon_{1\to2} = \epsilon_2 - \epsilon_1$; $\Delta\epsilon_{2\to1} = \epsilon_1 - \epsilon_2$; $\Delta\epsilon_{1\to2} + \Delta\epsilon_{2\to1} > 0$

Fig. 73 Explained in the text.

that during each successive performance of the cyclic processes its dissipation, i. e. disturbance of some properties, will increase more and more. In the condenser this will be a slow decrease in its capacity, which becomes increasingly fast as the cycle is repeated. With stable quantitative parameters of the stimuli the initial, and final values of the functions of the system's states will become smaller and smaller. As the dissipation of the system proceeds the transition from state 2 to state 1 will be more and more incomplete with each successive cycle (Fig. 73 b) and the system gradually loses the ability to return to state 1.

A heat engine is another example. Like the living organism it exchanges substance and energy with the environment. The potential energy of the substances (fuel) supplied to the engine results in mechanical work. Beginning with the first movement of piston, i. e. from the moment of initiation of the first closed cycle, the heat engine is forced to a gradual, but more and more progressive dissipation-structural disorganization. The substances entering the heat engine, unlike those in living systems, are not used for the restoration of the disorganized structures forming its components. The work performed by the engine increases the entropy not only externally (liberating heat into the environment) but also internally disorganizing it as a system.

It is usually accepted that living systems behave in a similar fashion to non-living systems. The entire organism (and its elements, such as organs and cells) like the heat engine, works in a cyclic regime. The autooscillative strictly periodical work of the heat is usually presented as a classical example of a sub-system of the whole organism. It is believed that at the end of each working cycle of the heart the internal energy remains unchanged. This is how classical physiology describes each successive excitation or metabolic cycle, as a closed loop, a scheme for which is presented in Fig. 73 a.

If metabolic cycles in living systems (which in the heart consist of two phases - the catablic and anabolic phases) had the pattern of a completed loop, repeating periodically, there would be no increase in internal energy during the course of ontogenesis. This means that there would be no growth and development. Such circular metabolic cycles would be futile for the developing organism. What events are then responsible for the growth and development of living systems and what is the role of transition processes in these systems?

According to common concepts, growth is the result of the assimilation of nutrients derived from the environment. But why during growth does assimilation prevail over dissimilation, what mechanisms are involved in this process, and, finally, why growth is stopped at some definite instant? There are no answers to these questions as yet. To attempt explanations some authors compare these processes with those proceeding in non-living systems. More often changes of the organisms in the course of development and growth are associated with the exhaustion of some energetic sources; interpreting these sources in some general form. These are the so-called "theories of exhaustion" some of which are quite old, for example those due to Rubner (1908), Bauer (1935), Mainot (1914) and newer ones, such as those by Prigogine and Wiame (1946), Zotin (1972, 1974) and others. These theories consider each successive transition process, no matter whether it proceeds in macro time intervals as aperiodically realized complex behavioural reaction, or in micro time intervals as cyclic excitation processes, as a factor leading to a gradual exhaustion of the energetic fund in the course of individual development. Thereby, successive transition processes conduct the living cycle of individual development to its fatal completion in a pattern similar to that observed in non-living systems (see the scheme in Fig. 73 a, b). The data from long-term, comparative ontogenetic studies carried out in our laboratory on some species of mammals has made it possible to find some regularities which dispute this concept of ontogenesis (Arshavsky, 1967, 1971, 1972, 1974, 1975, 1976 a, b, 1977 a, b, c). According to the energetic rule for skeletal muscles substantiated in these studies, peculiarities in the energetics at various age periods, both in the entire organism, in the tissues, as well as in specific functioning of various vegetative systems of organs, depends directly on specific functioning peculiarities of skeletal muscle in each period, rather than on the surface rule, i. e. on the relationship between the mass and the surface of the body. The essence and physiological meaning of this rule lies in the fact that each successive motor activity, i. e. each successive excitation cycle, beginning with the zygote (cytoplasmic motor activity) up to establishing of the skeletal muscle is the factor which induces successive restoration (anabolism). As was shown by special analysis, restoration returns the living system not to its initial state but it proceeds through excess, bringing it in a spiral to a new level, at which its internal energy increases (Fig. 73 b). In the nerve cells this is expressed in trace hyper-polarization and trace hyper-relaxation. As found, the excess of anabolism is realized only in relation to activity. An experimental blocking of the latter, in spite of continuing food uptake brings about either a delay in growth and development or even its complete termination.

In mitotic cells, beginning with the zygote, as soon as the daughter cell is formed after division it exhibits cytoplasmic motor activity in an attempt because of endogenous deficiency to obtain growth substances and energy from the environment. This activity is produced by actin and myosin in the cytoplasm. They are already present in the zygote (Poglazov, 1966). Excess anabolism induced by each successive cytoplasmic motor activity leads to a gradual increase in the cytoplasmic mass. Due to the excess anabolism occurring during each periodically realized transition process the cytoplasm mass increases according to a third power function, the area of the diffusion surface of the cellular membrane by a second power function. The area of the membrane surface, A, for the cell is close to that of a sphere and equals, $A = 4\pi R^2$. The volume where metabolism is to be performed is equal to $V = \frac{4}{3}\pi R^3$. The volume-surface area relationship equals $V/A = R/3$. As a result substance and energy exchange between the cell and environment becomes less and less effective. A situation occurs when the increased mass cannot obtain sufficient of the growth substance and energy required for effective metabolism. And, what is more essential, the possibility of liberating metabolic products from the cell becomes more and more limited. Thus this contradiction induces aging and death of the cell. Nature has found a solution for this contradiction by division of the cell into two cells. Thus, even at the beginning of life discrete behaviour appears.

In a multicellular organism an increase in the mass (growth) of different systems of organs consisting of mitotic cells, and, thereby an increase of internal energy, is reached at the expense of an increase in the number of cell elements. It was found in investigations on the liver carried out in our laboratory that an important index of the level of structural energy, such as membrane potential, which determines the value of the free energy in the cell, is not changed in the course of ontogenesis (from birth to the adult state) and equals 40 mV. It is obvious then, that the working parameters of the organ as a whole are determined, not by intensive characteristics, but by extensive characteristics. It is clear therefore that notwithstanding the stability of the membrane potential, the working parameters of the entire organ (liver in this case) are also increased with the increase of its biomass (in an absolute expression).

In discussing systems consisting of postmitotic cells we shall limit ourselves mainly to the skeletal muscle system. As mentioned above, two forms of excess anabolism in motor activity should be distinguished. The first one is realized in the excess accumulation of mass, which conditions the growth process itself and thereby an increase of internal energy in the living system. The growth of skeletal muscles is carried out, not at the expense of hyperplasia of the cellular elements, but by their hypertrophy. The same is true for the cellular elements of the nervous system. The second form is expressed in the excess accumulation of structural-energetic potentials (or which is the same, the structural energy according to Bauer, 1935), increasing not the mass but its free energy in the course of ontogenesis.

One of the important unsolved problems in interpreting the basic regularities of ontogenesis, and the thermodynamic regularities specific to living systems in particular, is the problem concerning the factors and mechanisms responsible not only for the growth itself, but for its termination as well. Another unsolved problem is that concerning the mechanisms of aging, i. e. the mechanisms responsible for disorganization of living systems and their gradual transition to the equilibrium state.

Physiological data has shown that, while the body mass and its internal energy increase according to a third power law, the area of diffusion surface of transcapillary exchange through which the organism cellular elements exchange substance and energy with the internal medium, increases only by a second power law. This was first reported by Milman (1940) in connection with morphological studies.

A gradual decrease in the specific area of the diffusion surface brings about a decrease in the supply of tissues with the necessary amount of nutrients and oxygen. This factor seems to be the basic, limiting factor responsible for the arrest of further growth of skeletal muscle and of the mass of other organ systems. It is fixed by natural selection at the value which meets the requirements of valuable physiological functions during the birth period.

Thus in almost all the species of mammals studied the same value of the biomass (relatively expressed) is established in the adult state: in males 42 to 44 %, in females 37 to 38 %. At birth this value is about 20 %.

But how can we interpret the fact that in spite of a progressing decrease in the rate of blood supply to cellular elements whose mass increases by a third power law, the ability to carry out work becomes greater and greater? To answer this question we consider specific developmental features of skeletal muscles in ontogenesis.

In the pre-natal period the membrane potential of skeletal muscles (measurements were conducted on gastrocnemius) is 16 to 18 mV, the protein content is 6 to 8 %. In the first days after birth these values are 25 to 30 mV, and 13 to 15 %, respectively. In adults they are 80 to 85 mV, and 22 to 23 %, respectively. In stenobiontic mammals (rats, rabbits) the membrane potential is 2 to 4 mV and the protein content 2 to 3 %. This is less than in eurybiontic animals (squirrels, hares, dogs). In extremes expression of these differences is naturally well pronounced. In animals developing under hypodynamic conditions these values are essentially lower.

The membrane, or resting potential, is determined by the number of charges formed on the cellular membrane during its repolarization with some excess of the initial state (hyperpolarization), in case of the depolarized cell, return from state 2 to state 1 (Fig. 73 b). The number of charges producing some potential differences on the cellular membrane is determined by its capacity. Special measurements of cellular membrane capacity in postnatal ontogenesis showed a gradual increase with age (Fig. 74). Reduction of resistance with age results from a gradual increase in muscle fibre diameter.

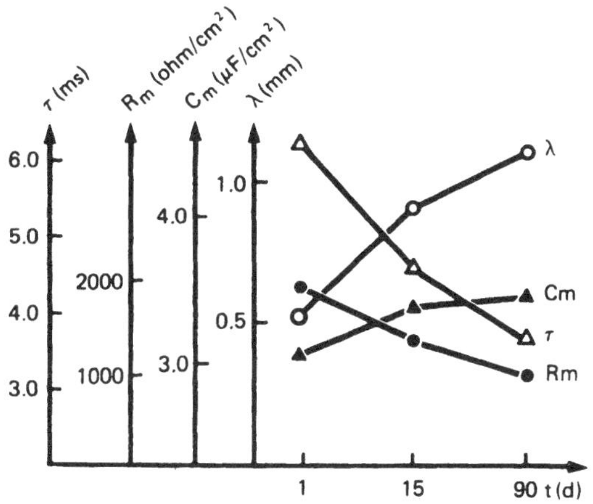

Fig. 74 Change of passive electrical characteristics of muscle membrane in rats at different age periods: τ time constant, R_m membrane resistance, C_m membrane capacity, λ space constant.

As mentioned above, a standard condenser gradually decreases its capacity due to periodically repeated charging and discharging of the condenser. This is conditioned by a gradual dissipation resulting from breakdown in structure, bringing it closer and closer to the equilibrium state. The muscle cells membrane, as a condenser with repeated transition process due to periodically repeated depolarization (discharging) and repolarization (charging), gradually increases its capacity. This results from gradual increase in complexity in the cell membrane structure and the level of its order during the course of ontogenesis. This increases the nonequilibrium nature of the developing system. The heat engine operating as an open system becomes, due to periodical production of work, increasingly disorganised and, eventually, approaches equilibrium. The living system can grow and develop because in the repeated transition processes it carried out work due to which its nonequilibrium nature increases more and more. In non-living, physical systems the transition process can be well described by differential equations. The transition process in living systems which are nonlinear cannot be expressed by differential equations. Nonequilibrium in the living system is maintained, not because it is fed by negative entropy. Nutrients contain only plastic materials (proteins and lipids) and potential energy is contained in hydrocarbons and lipids. During oxidation of the latter the liberated energy is transferred to ATP molecules for storage until required. We cannot give a detailed account of this problem but we would only like to note that the energy liberated in ATP hydrolysis is utilized not only for the performance of contractile activity but, what is more essential, for reconstruction (with some excess) of the initial structure. Corresponding values of the membrane potential and corresponding relaxation level of the actomyosin complex (sarcomere) are meant. Membrane relaxation potentials determine the value of the structural energy which is a specific expression of free energy in living systems, and at the expense of which work is performed. Living developing systems unlike non-living systems escape the death penalty of the Second Law of thermodynamics not only because they are open systems exchanging substance and energy with the environment. Firstly they create their own negentropy (structural energy). Due to the fact that their development is concerned with the environment by increasing their own negentropy they, respectively, increase entropy in the environment. In this sense they are also subject to the Second Law of thermodynamics.

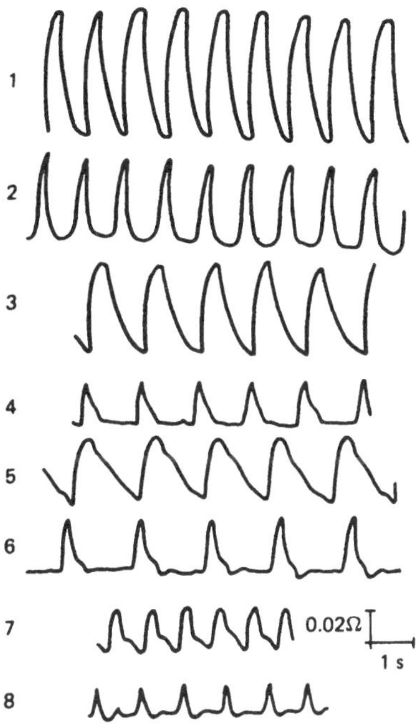

Fig. 75 Rheograms of blood current in dog hind leg at different age periods: 1-5 days, 3-12 days, 5-18 days, 7-1.5 months (2, 4, 6, 8, - differential constituents). Calibration: vertical line - 0.01 Ohm, horizontal ones - 1 s.

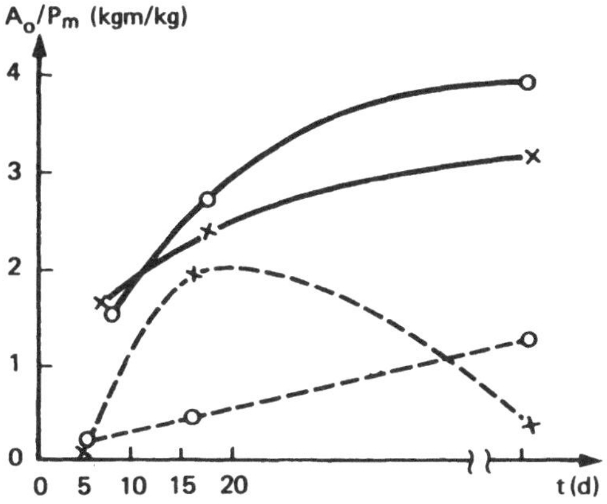

Fig. 76 Characteristics of optimal work in kgm per muscle mass unit (kg) as function of age: full lines - dogs, dotted - rabbits, crosses - gastrocnemius, circles - anterior tibial muscle.

During the course of post-natal ontogenesis the specific area of the diffusion surface for transcapillar exchange gets smaller at the instant when antigravitation reactions occur (Fig. 75). After these in association with the acquired ability to move in the gravitational forcefield, the initiated phase-tetanic activity of skeletal muscles begins to induce a larger contribution from the second form of anabolism. This induction is expressed differently in steno- and eurybiontic mammals. A unit of muscle mass in eurybiontic mammals can perform greater work (kg/m) than can stenobiontic animals.

Fig. 76 presents a similar instance for post-natally developing rabbits and dogs. Rabbits after achieving the standing posture under normal conditions comparatively soon adapt the sitting posture. The specific value for work activities, particularly for gastrocnemius, necessarily decreases.

Our data has established the higher energetic cost necessary in providing the excess of the first form of anabolism and lower one for the second form. Chemical work in restoring structural energy in living systems after the performance of antigravitation reactions becomes more efficient and economical. It is then quite understandable why, in spite of a pronounced increase of specific area of diffusion surface at the moment of performance of antigravitation reactions, that the work potential of skeletal muscles and of the entire organism does not decrease, but even increases significantly.

The specific, characteristic decrease in the diffusion surface area of transcapillary exchange in post-natal ontogenesis acquires some stable and morphologically constant value in the so-called stationary (adult) period. It seems that this value should remain constant in the course of ontogenesis and due to the performance of successive transition processes the living system itself should have been a long-term value. However, the specific diffusion surface area begins to decrease even in the so-called "stationary state" in the course of childbirth, and whilst as a result of alterations in the walls of the diffusion surface for transcapillary exchange, i. e. arteriosclerotic and aterosclerotic changes. As a result, not only oxygen (and nutrients) supply becomes limiting, which leads to the development of tissue hypoxia, but, what is more important, liberation of metabolic products, acid products in particular, also become limited. Connective-tissue formations of the vessel system walls, i. e. collagen and elastic fibres, are altered. As they are products formed by living cells connective-tissue elements are non-living matter, therefore they can only gradually dissipate, and cannot show negentropic tendencies. It has been stated that in special experiments the artery walls when influenced by deforming forces (systolic ejection) are incapable of excess formation of additional elastic forces. It is this phenomenon that differentiates them from living cellular formations. Together with sclerotic changes in large arterial vessels resulting in their greater rigidity in old age, thickening and atrophy of the capillary walls (pericapillary sclerosis) takes place.

The following conclusion can be drawn. Aging is the result of the fact that from the defined stage of the so-called stationary period the specific area of the diffusion surface for transcapillary exchange begins to decrease due to sclerotic changes, particularly in the walls of brain vessels. As a result of acidotic alteration of the nerve cells (apparently, not only of the cytoplasm, but of their nuclear apparatus as well) the nerve-trophic regulation of metabolism is disturbed first of all in the skeletal muscles and then in other organs. Due to these causes the ability to induce excess of the second form of anabolism in motor activity declines more and more. The transition process acquires the pattern presented in Fig. 73 b, and therefore protein content in the skeletal muscles decreases (up to 18 %), the membrane potential in skeletal muscle cells is reduced by up to 60 to 50 mV, and in liver cells by up to 30 mV. The amount of total muscle mass decreases (in absolute and relative terms). As a result the living system approaches closer and closer to the equilibrium state.

It should be noted in conclusion that in the course of ontogenesis other forms of transition processes are also observed which are performed not only in micro-, but in macrointervals. They are represented at critical stages of individual development; in view of this ontogenesis is divided into several periods differing only in their morphophysiological content.

The paper summarizes the data obtained by V. D. Rozanova, T. A. Adzhimolaeva, S. I. Enikeeva, S. S. Solomatin, G. G. Savkin, L. A. Li, N. V. Darinsky, B. A. Elantsev, G. B. Levitsky, L. A. Siryk, V. A. Elin, M. S. Musin, D. U. Ermatova, M. I. Trigulov, E. Z. Rabinovich, P. V. Glazirina and by the author of this contribution.

Stable Orbitals and Transition Processes in the Models of Biological Systems

V. A. Vasiliev, Yu. M. Romanovsky

When studying biological objects, as a rule, one has to analyse open systems in a state far from thermodynamic equilibrium. At the same time each system can be in a state of dynamic equilibrium with the environment. In this sense stationary states of such systems are discussed. A transition process in an open system means the establishment of the stationary state following a deviation; the simulated and established stationary states may disagree. Before giving a detailed account of transition processes, the term stationary state should be specified.

The simplest and most thoroughly investigated, both theoretically and experimentally, example of a stationary state is the established regime in the continuous cultivation of microorganisms - a chemostat culture (Jerusalimsky, 1963; Romanovsky et al., 1975). There is a single specific point, a steady mode corresponding to the stationary state in the simplest dynamic model of the chemostat. When the flow rates of the limiting substrate are above the critical values, the biomass concentration in the stationary state is zero, when these rates are below the critical values, the biomass is greater than zero. In simulation of the stationary state transition processes proceed which bring the system back to the initial state. The transition processes described in the simplest model are of an aperiodic character and differ little from relaxation processes in thermodynamic equilibrium systems. However, more complex types of transition processes are observed in these experiments.

Complex models take into account such variables as concentrations of life products; the number of ribosomes in the cells: interactions of several microorganisms species and the age structure of populations. Firstly, this

Thermodynamics and Kinetics of Biological Processes

increases the dimensions of the phase area in the various models, and secondly it changes, qualitatively, the character of the stationary states and their number. Transitions processes in such models are not necessarily of an exponential type; they may be oscillatory; almost instantaneous transitions from the vicinity of a stationary state into another stationary state are also possible.

Such models also describe the autooscillation regimes of the chemostat (in the phase space of steady limiting cycles). The limiting cycles should be considered as a stationary state, or more exactly, as a dynamic, stationary state. Transitions processes are defined as the establishment of the stationary variation amplitudes of the variables, their mean values (in the variation period), and mean values of the yield of the corresponding products. It is noteworthy that if the area of oscillation amplitudes is considered as a phase area an analogy is established between the transition processes leading to dynamic stationary states and to standard stationary states.

When describing the stationary state of the dynamic model it is imperative to find out whether it is stable or not, i. e. to define whether it is observed experimentally in a real system. Usually investigations of stationary states use the stability definitions of Lyapunov (Glansdorff, Prigogine, 1971; Romanovsky et al., 1975). Such an interpretation of stability underlies the methods of thermodynamics of irreversible processes developed by Prigogine's group (Glansdorff, Prigogine, 1971). However, when modelling many biological processes, according to Lyapunov, the requirement of stability is overrated and it is expedient to confine ourselves to the requirement for the orbital stability of the process. For instance, the dynamic stationary state, the relaxation limiting cycle conforming with an experimentally observed event, can only be orbitally steady, but is unsteady according to Lyapunov. Another example of this process specific to biology is the exponential growth of variables within the transition process. Evidently, such motion is unsteady according to Lyapunov and may only be orbitally steady. Thus the concept "orbital stability" permits the extension of the stationary state to limit cycles, and the introduction of an important characteristics of the transition process trajectory.

Biological systems function in changing external conditions: the environment is changed, the organisms grow, develop, age and so on. In many cases the characteristic times for such changes are many-fold larger than those of the processes under study, information being available only as to the direction of such changes. In a similar situation it is not advisable to construct mathematical models consisting of equations with variable coefficients, it is sufficient to confine oneself to the analysis of simpler models with fixed coefficients. These models should be analyzed with sufficient exactness to show how the state of the studied system changes with slow change in external conditions. Having performed such an analysis and knowing details of the evolution of the external conditions a trajectory close to the true motion of the studied biological system can be plotted. An important state in the solution of this problem is to find the system of orbitally stable trajectories involving both stationary states and transition processes. Such trajectories will be designated as stable orbitals.

Modelling of living objects would be pointless and hopeless if they did not possess stable orbitals from the moment of their origination and on through growth and aging. The transition process is initiated by the actions of external or casual internal factors. After the transition process the system can return to the stationary state without undergoing qualitative changes. Transitions processes can bring the system to a new stationary state or even to a new steady orbital. It is because of such transition processes that new qualities appear in the system. It is worth noting that there is a similarity between such transition processes in the systems under consideration (thermodynamically open) and phase transition in equilibrium systems (Nitzan et al., 1974). As in the case of phase transitions, transition processes can be divided into those of types I and II. In a transition process of type I the extensive parameters of the system (reagent concentration, entropy production etc.) are changed abruptly, in that of the type II only symmetrical properties of the system are changed abruptly. Near the critical points the behaviour of nonequilibrium systems is formally described in a similar fashion to that for systems at equilibrium (Nitzan et al., 1974).

Consider the model of the chemostat; this describes a set of steady orbitals (Romanovsky et al., 1975). It is evident that in many tasks the living cell may be considered as a chemostat. The most important technological problem is the creation of control programs for chemostat production. The desired effect is maximal product yield. An orbital model has therefore to be found, where some function is maximized. Having found such an orbital it is easy to construct the optimal program for chemostat regulation.

The immune system in an organism has many orbitals. Transition processes between different orbitals proceed under the influence of antigens. Consider a simple variant of the punctate dynamic model proposed by Stepanova for describing the struggle of an organism immune system against cancer cells (Stepanova, Romanovsky, 1975). The number of malignant cells, X, and the number of lymphocytes, Y, serve as variables. It is known that in case of transformation there appears at first a significant number of lymphocytes, but then, with tumour growth, the activity of the immune system decreases. Accounting for the lymphocytes and cancer cells a model can be described as follows:

$$\frac{d}{dt}X = \alpha X - \beta XY \qquad (1)$$
$$\frac{d}{dt}Y = aXY - bX^2Y - cY + \varkappa - \beta XY .$$

The model assumes:

1. tumour cells are reproduced exponentially; 2. interaction between lymphocytes and cancer cells results in cancer cell death; 3. the specific rate of reproduction, Y, is $\mu = aX - bX^2$ (i. e. at low X the lymphocyte reproduction is timulated by the tumour, at high X it is inhibited), 4. there exists a smal stable inflow of lymphocytes with the rate cY.

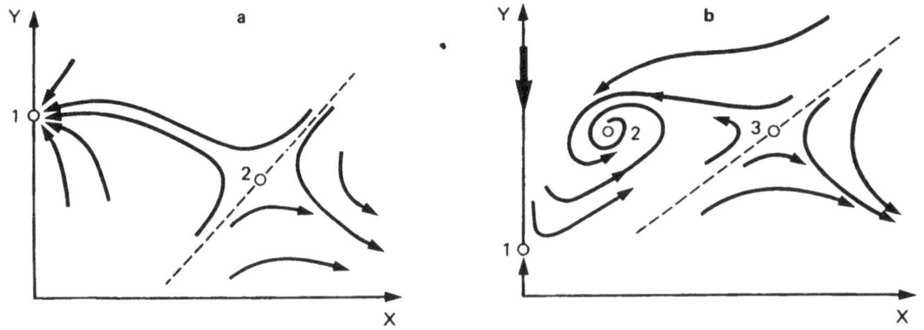

Fig. 77 Phase pattern of the system (1) for aperiodic transition processes (a) and for oscillative transition processes (b).

If X > some constant the equation system (1) determines two final stationary states (its phase pattern is presented in Fig. 77 a). The integral curves either approach a fixed point "1", or tend to infinity ($X \rightarrow \infty$, $Y \rightarrow \infty$). In the first case all the integral curves above the separatrix are stable orbitals resulting in recovery ($X \rightarrow 0$). All the trajectories below the separation are stable orbitals of "death". Such interlocation of the orbitals explains why, occasionally, stimulation of the immune apparatus with some stimulator inhibits tumour development. However, it can help only in the early stages when the X value is small. Such a cure provides a transition process bringing to the orbital which faster leads to the recovery or to the transition process through the separatrix which qualitatively changes the situation.

With some combination of the parameters another variant of the model phase pattern is possible (1) (Fig. 77 b). There are three special points: point "1" on the ordinate is unstable, the integral curves are screwed at the stable point "2". There is no complete recovery in this case - at the end of the fluctuating transition process an equilibrium between the immune forces and tumour is established. Although the phase pattern differs from the previous situation, the system also has two qualitatively different orbitals, which lead to death or partial recovery. Thus even this simple model enables us to show the possibility of initiating transition processes resulting in different outcomes in the struggle against tumours.

To illustrate the problem of stable orbitals in growing objects let us analyse the properties of stationary dissipative structures (Turing, 1952; Romanovsky et al., 1975; Vasiliev, 1978). This example is also interesting because a distributed, i. e. infinite-dimensional, system is modelled. Turing (1952) was the first to pay attention to dissipative structures in relation to the theory of morphogenesis. Now dissipative structures are experimental observables in both chemical and biological objects. For example, dissipative structures are initiated in a cultivator without complete internal mixing (Zarubin et al., 1976; Levandowsky et al., 1976).

Let us consider the critical points and transition processes in a distributed system modelled by Turing's equation system (Romanovsky et al., 1975; Vasiliev, 1978).

$$\frac{\partial}{\partial t} x = A + x^2 Y - (B+1)x + D_x \frac{\partial^2}{\partial r^2} x$$
$$\frac{\partial}{\partial t} Y = Bx - x^2 Y + D_y \frac{\partial^2}{\partial r^2} Y, \tag{2}$$

where t is the time, r the space coordinate, D_x and D_y diffusion coefficients. Let us put the second boundary value for this system adding boundary conditions of reactor impermeability at its end plane

$$\frac{\partial}{\partial r} x(t,0) = \frac{\partial}{\partial r} x(t,L) = \frac{\partial}{\partial r} Y(t,0) = \frac{\partial}{\partial r} Y(t,L) = 0 \tag{3}$$

where L is the system length.

For boundary conditions of (3) the solutions of (2) can be represented as the following

$$x(t,r) = x_0 + \sum_{k=0} P_k(t) \cos(\pi k r/L);$$
$$Y(t,r) = Y_0 + \sum_{k=0} P_k(t) \varkappa_k \cos(\pi k r/L), \tag{4}$$

where X = A and Y_0 = B/A are coordinates of the single finite point of the local model corresponding to system (2): $\varkappa_k$ are constants depending on the system parameters (A, B, D_x/D_y, L and k). The trivial solution of the boundary problem of (2) and (3) is the uniform stationary state $\{X(t,r) = X_0;\ Y(t,r) = Y_0\}$. The problem also has spatially non-uniform stationary solutions corresponding to dissipative structures. The existence conditions of dissipative structures in the distributed system are mostly similar to those of trigger concentrated systems. For a'broad class of systems sufficient conditions for the existence of dissipative structures can be obtained from an analysis of the stability of the stationary uniform solution. To analyze the stability toward small disturbances the following procedure is adopted: 1. in equation (4) $P_k(t)$ is presented as $P_k(t) = P_k \exp\{\lambda_k t\}$; 2. (4) is substituted into equations (2) and linearized relative to P_k; 3. a characteristic equation determining the connection between the spatial period 2L/k and λ_k is obtained from the existence conditons of nontrivial solutions of the derived algebraic equation system; 4. if the characteristic equation for some k has even one root of $\widetilde{\lambda}_k$ with a positive actual part, the studied uniform solution is unstable (Romanovsky et al., 1975; Vasiliev, 1978). If the free member of the characteristic equation is negative (i. e. there is an odd member of λ_k with positive

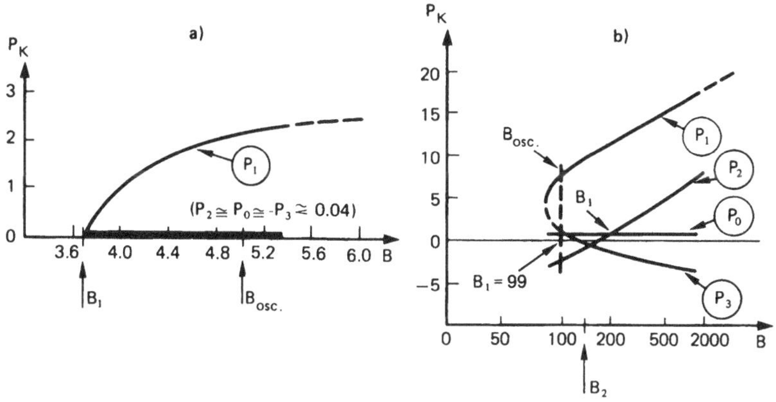

Fig. 78 Diagram of states of the equation system (2): a) A = 1.7; D_x = 0.08; D_y = 0.4; L = 1; b) A = 10; D_x = 0.75; D_y = 1; L = 1.

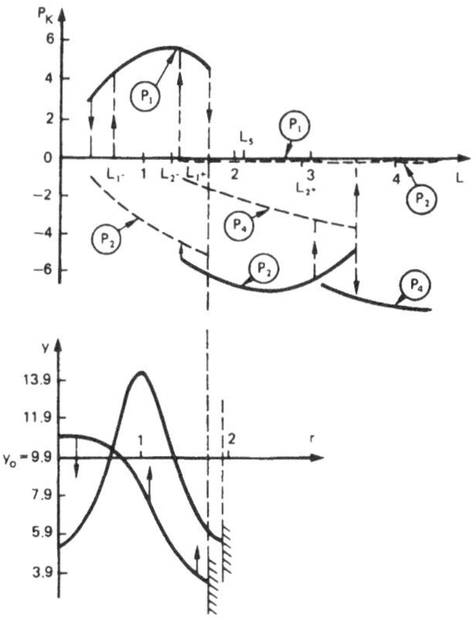

Fig. 79 Diagram of states of the equation system (2): A = 10; B = 99; D_x = 0.75; D_y = 1.

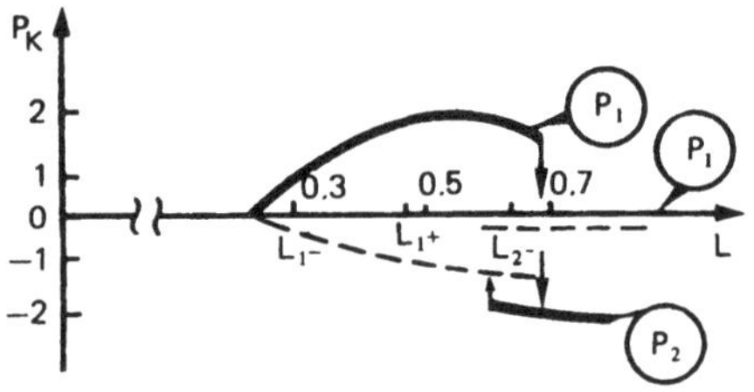

Fig. 80 Diagram of states of the equation system (2): A = 5; B = 18; D_x = 0.04; D_y = 1.

actual parts), the distributed system (2) has stationary spacial periodic solutions of the period 2L/k. The system (2) can have several such solutions differing in periods within space; autooscillation solutions are also possible.

Figs. 77 to 80 show the diagrams for stationary states of the system (2) and (3), for the parameters B and L. Fig. 79 also gives the stationary distributions of the variable Y for k = 1 and k = 2. With $B > B_k$ or ($L > L_{k^-}$ and $L < L_{k^+}$) the free member of the characteristic equation for the given k value is negative. If $B > B_{osc} = 1 + A^2$, then the corresponding local model has a limit cycle, i. e. the conditions of autoexcitation of symphase fluctuations (k = 0) of the system (2) are met. P_k, the amplitude of harmonic components of the stationary solution (4) is plotted against the ordinate axes.

There are two types of orbitals derived for changes in parameter B, that with hysteresis loop (Fig. 78 a) and without hysteresis (Fig. 78 b). With $B < B_1$ (Fig. 78 a) any deviation from the uniform stationary state is extinguished exponentially. If $B_1 < B < B_{osc}$ any excitation of the uniform stationary state initiates dissipative structures, the transition process being monotonous. If $B > B_{osc}$ another stable orbital of synphase autooscillations exists. These orbitals intersect at B ≅ 5.25, and at B > 5.25 only synphase oscillations are stable. Fig. 78 b presents a case when a dissipative structure within period 2 also exists with $B < B_1$. The transition process from the uniform state to the dissipative structure results from great disturbances and is monotonous. If at the same values of D_x, D_y, and A, L is increased n times, the dissipation structure will have n half-periods, (in Fig. 78 b, B_1 should be substituted for B_n, P_1 for P_n, P_2 for P_{2n} and so on - the graph remains unchanged); this property is called translation invariance of dissipative structures (Romanovsky et al., 1975). With over-threshold localized disturbances of the uniform stationary state the transition process proceeds by additional self-construction in the dissipative structure: in the spatial distribution of the components there first appears the first peak, then the second one etc., up to n/2. Note also that as B increases (Fig. 78 b) the system behaves in a similar manner to the previous case.

Figs. 79 and 80 show the relationship between the amplitudes of the dissipation structures with periods 2L and L (k = 1 and k = 2) and the system length. They are of special interest in view of the application of dissipation structure theory to the problems of morphogenesis, cell divisions etc. Hysteresis loops are seen in these state diagrams. The biological meaning of hysteresis consists in the fact that transition from a dissipation structure (stationary state) with k = 1 to that with k = 2 proceeds to the end; in other words, if, with small increase in L the system passed to the state with k = 2, it cannot return to the previous state (k = 1). With a further increase in L the system passes to the state with k = 4, k = 8 etc. Thus a stable orbital k = 1 → k = 2 → k = 4 → k = 8 ..... is distinguished. The transition process on this orbital is monotonous; an example of initial and final states of a transition is presented in Fig. 79. In addition to the stable orbital indicated there are others in the system, such as for example k = 3 → k = 6 → k = 12 ..... . Transition from one orbital to another results only from sufficiently large specific stimulation of the system.

The orbitals in Figs. 78 a to 80 are plotted for cases involving slow changes in the parameters B and L. With changes in these parameters the movement of the system deviates from the orbital. Slow changes are those for which the time of the transition process returning the system to the orbital is much less than the times characterizing the velocities of the changes in parameters values.

If B and L values increase not slowly all the points of transition on the orbital graphs are shifted to the left. Thus if B is rapidly increased from 3.7 to 4.6 the system will pass to the regime of synphase autooscillations, although with slow increase in B this would take place at B = 5.25. Thus the critical points of the orbitals depend upon the velocity of the parameter changes, but the existence of the points themselves continues. This means that if only a qualitative analysis of the orbital of a developing object is required, one can confine oneself to the investigation of simple partial models.

The equation system (2) describes a model of an active medium. However, there arises real noise in the object. Therefore the experience of stochastic modelling of dissipative structure is interesting. Without dwelling upon the details, we should like to point out that even with relatively high levels of fluctuations the shape of the orbitals is not qualitatively changed, but even for small fluctuations there is a decrease in the time of transition from one state into another. This indicates that the transition process in the system is unsteady, according to Lyapunov, although the orbital itself is stable.

Using the graphs presented it is possible to calculate the relationship P(S)/L (the rate of entropy production, according to Prigogine, related to the length unit of the system) from the system length. Such a relationship will increase monotonously with leaps at the transition points (Figs. 79 and 80); during transition from the state k = 1, into the state k = 2. Now let us assume that formation of dissipative structures in a cell or in the whole organism is possible. Then during growth with the attainment of $L = L_1$ the dissipative structure (k = 1) is generated; with further increase in L P(S)/L increases; it then rises abruptly after the transition into the state with k = 2, and it is useful to the organism to divide into two organisms. The scheme presented is indirectly supported by the data on the changes in heat production rate during the development of hen eggs (Zotin, Zotina, 1967; Zotin, 1972). Note also that the existence of orbitals of the type analyzed (Figs. 79 and 80) could explain such phenomena as competence, initiation and induction. However, it should be noted that not only chemical dissipative structures give such orbitals. Only some further accumulation of data and the analysis of the finer properties of the model can elucidate the physical basis of these phenomena.

Note in conclusion that the investigation of the orbitals described with those models includes the description of transition processes. These may be of importance in the identification of the models. A complete description of transition processes can be rarely obtained, but often detailed analysis is not required as, for practical purposes (e. g. for identification), it is sufficient only to find the types of transition processes and the critical values of the parameters, at which transitions of the first or second order take place. These points are the most indicative parameters of the system, as are the transition points of liquid-vapour, liquid-crystal, Curie-points; they are of primary importance in the study of equilibrium systems. We believe that special attention should be paid to such phenomena when analyzing experimental data.

Concluding Remarks

As declared in our earlier book (Lamprecht, Zotin, 1978), the thermodynamics of biological processes ought to describe all fundamental biological phenomena from a single standpoint. Only then can it form a reliable basis for theoretical biology. The present monograph gives an account (naturally an incidental and incomplete one) of ecological problems, reactions of living systems to external effects, of adaptation of organisms and of the kinetics of ontogenesis. Together with the problems touched upon in the previous book (development, growth and aging of organism, dissipative structures and orderliness of the living system, some problems of evolution) they outline the scope of the problems which need to be solved by thermodynamics for biological processes. Such specific and important phenomena, as control processes, have not been considered as yet. We hope that these phenomena will be dealt with, in detail, in a subsequent book in the Series "Thermodynamics of Biological Processes", which is to be published in the near future.

As in the previous book, it should be stated that at present the main difficulties in the formulation of a system of thermodynamics for biological processes are concerned with the difficulties of thermodynamics itself. Therefore in this book the main attention has been paid to purely thermodynamic problems. The first chapters showed how many unsolved problems are to be found in the thermodynamics of nonlinear, irreversible processes, and how unreliable, therefore are the foundations of the thermodynamics of biological processes. At the same time it is evident that the purely physical and mathematical problems which were dealt in the first chapters of the book, touch only upon limited aspects of biological phenomena. In view of this biologists may be disappointed and sceptical as to the possibilities inherent in modern thermodynamics and consequently doubt whether recent developments in such fields as the thermodynamics of living processes (which is not new for biology) are well-timed.

Indeed, all the powerful mathematical apparatus, difficult to understand, which is used in modern thermodynamic studies is aimed at solving the problems which are, in practice, not raised in biology (except in ecological phenomena), or are presented in some extreme case when the living system confronts a disaster. Only in these cases do situations arise in which the apparatus of nonlinear theory is required. Usually the organisms function normally, and their functioning is described with linear thermodynamic relationships. This has been substantiated in work by Lurie and Wagensberg (1979) and also follows from the material presented in the biological chapters of this book. For example, the kinetics of transition processes during the course of intensive muscle work in man, even under extreme conditions is an oscillatory regime initiated which is specific to nonlinear irreversible processes. This is concerned with the fact that processes in the living system are regulated by many control systems preventing the organism from entering nonlinear regions. Evidently, in the course of biological evolution the most reliable organizational forms for living processes were selected which allowed the organisms to remain within the realm of linear reactions and only in extraordinary cases move out into the nonlinear region. From the standpoint of the biologist modern thermodynamics with its more and more complex mathematical apparatus deals with the initial stages of the arising life when the first structures resembling living systems were initiated from a primary broth. Giant fluctuations occurred at those states, and nonlinear processes played a leading role, both in the initiation and functioning of protobionts. After the origination of true living systems evolution began, and natural selection eliminated everything unreliable and irregular. A perfect control system appeared and developed, such as basic organism reactions, and the processes involved returned to those of a linear type.

All the above, certainly, requires accurate experimental proof, but at the same time if it is valid, biologists can be saved from the difficulties arising in the thermodynamics of nonlinear, irreversible processes. This does exclude the possibility that many biological phenomena may be described in terms of the thermodynamics of linear irreversible processes, and some of the most developed parts of the nonlinear theory. Researchers in the field of thermodynamics have to solve many problems to explain the linear behaviour of such obviously nonlinear systems as living systems. They must rapidly go the way which biological evolution followed for many millions of years controlling strongly nonlinear systems and turning them into ordinary organisms.

References

Adjemian, L. C., Grinin, A. P., Kuni, F. M. 1975. Time asymptotics of kinetic nuclei of linear hydrodynamics. Theor. math. Phys., 24, 255-264 (Russian).

Adjemian, L. C., Kuni, F. M. 1975. Non-cumulative projecting and elimination of time derivatives from nonequilibrium distribution function. Theor. math. Phys. 24, 368-381 (Russian).

Adjemian, L. C., Kuni, F. M., Novozhilova, T. Yu. 1974 a. Mori method of projecting operators in nonlinear hydrodynamics. Theor. math. Phys., 18, 125-136 (Russian).

Adjemian, L. C., Kuni, F. M., Novozhilova, T. Yu. 1974 b. Bogolyubov method of kinetic equation in nonlinear statistical hydrodynamics. Theor. math. Phys., 19, 383-394 (Russian).

Adjemian, L. C., Kuni, F. M., Novozhilova, T. Yu. 1974 c. Absence of hydrodynamical-pole contributions to the nonlinear hydrodynamical functional. Lett. Nuovo Cimento, 10, 78-80

Afanasyeva-Ehrenfest, T. A. 1928. Irreversibility, unilaterality and Second Law of thermodynamics. J. Appl. Phys., 5, 1-30 (Russian).

Agarwal, G. S., Narducci, L. M., Gilmore, R., Feng, D. H. 1978. Optical bistability: a self-consistent analysis of fluctuations and the spectrum of scattered light. Phys. Rev., A 18, 620-634.

Akcasu, Z. A. 1977. Fluctuations of macrovariables in nonlinear systems: Langevin equation approach. J. stat. Phys., 16, 33-58.

Alekseev, V. V. 1976. Biogeocenoses - autogenerators and triggers. J. gen. Biol., 37, 738-743 (Russian).

Alekseev, V. V., Kryshev, I. I., Polyakova, M. S., Sazykina, T. G. 1978. Dynamics and statistical mechanics of biogeocenoses with the fixed mass of limiting biogenic element. In: The Man and the Biosphere. v. 2, 40-102, Moscow, Univ. Press (Russian).

Alekseeva, K. D. 1967. Influence of muscles work on fishes energetic metabolism. In: Metabolism and Biochemistry of Fishes. Moscow, Nauka, 193-198 (Russian).

Alimov, A. F. 1975. Intensity of metabolism in fresh-water Bivalvia. Ecology, 1, 10-20 (Russian).

Anderson, M. L., Boyd, R. K. 1971. Nonequilibrium thermodynamics in chemical kinetics. Canad. J. Chem., 49, 1001-1007.

Andronov, A. A., Leontovich, E. A., Gordon, I. I., Mayer, A. G. 1967. Bifurcation Theory of Dynamic Systems on the Plane. Moscow, Nauka (Russian).

Andronov, A. A., Vitt, A. A., Khaykin, S. E. 1959. Oscillation Theory. Moscow-Leningrad, Physmathgiz (Russian).

Arnold, L. 1973. Stochastische Differentialgleichungen. München, Oldenburg.

Arnold, L., Horsthemke, W., Lefever, R. 1978. White and coloured external noise and transition phenomena in nonlinear systems. Z. Physik, B 29, 367-373.

Arnold, V. I. 1972. Normal forms of functions near the confluent critical points, Weil's groups A_K, D_K, E_K and Lagrange peculiarities. Funct. Analysis and its Appl., 6, 3-25 (Russian).

Arnold, V. I. 1978. Additional Chapters of the Theory of Ordinary Differential Equations. Moscow, Nauka (Russian).

Arshavsky, I. A. 1967. Essays on Age Physiology. Moscow, Medizina (Russian).

Arshavsky, I. A. 1971. Physiological mechanisms of some basic regularities of ontogenesis. Adv. physiol. Sci., 2, 100-141 (Russian).

Arshavsky, I. A. 1972. Musculo-skeletal activity and rate of entropy in mammals. In: Advances in Psychobiology. New York, Wiley, 1-52.

Arshavsky, I. A. 1974. Focus for development: the energy rule of skeletal muscles. Editor. Developm. Psychobiol., 7, 291-295.

Arshavsky, I. A. 1975. Factors determining growth (physiological mechanisms of growth). In: Quantitative Aspects of Organism Growth. Moscow, Nauka, 147-161 (Russian).

Arshavsky, I. A. 1976 a. Biological and medical aspects of adaptation and stress in terms of the data of ontogenesis physiology. In: Urgent Problems of Modern Physiology. Moscow, Nauka, 144-192 (Russian).

Arshavsky, I. A. 1976 b. Ontogenesis and aging. In: Biological Possibilities of Life Prolongation. Kiev, Naukova Dumka, 29-39 (Russian).

Arshavsky, I. A. 1977 a. On paradoxes in organism individual development and mechanism of gerontogenesis. In: Physiological and Molecular Aspects of Ontogenesis. Kiev, Naukova Dumka, 186-200 (Russian).

Arshavsky, I. A. 1977 b. Physiological mechnisms of life span of mammals. Adv. modern Biol., 83, 287-304 (Russian).

Arshavsky, I. A. 1977 c. Changes of the vascular system with age and mechanisms of gerontogenesis. In: Functional Peculiarities of the Heart under Physical Loads in Age Aspect. 2, 118-141, Stavropol (Russian).

Aschoff, J., Pohl, H. 1970. Der Ruheumsatz von Vögeln als Funktion der Tageszeit und der Körpergröße. J. Ornithol., 111, 38-47.

Ashby, W. R. 1956. An Introduction to Cybernetics. London, Chapman & Hall.

Ashby, W. R. 1960. Design for a Brain. London, Chapman & Hall.

Asplund, K. K. 1970. Metabolic scope and body temperatures of whiptail lizards (Cnemidophorus). Herpetologica, 26, 403-411.

Astaurov, B. L. 1974. Heredity and Development. Moscow, Nauka (Russian).

Balescu, R. 1975. Equilibrium and Nonequilibrium Statistical Mechanics. New York, Wiley.

Barott, H. G. 1937. Effect of temperature, humidity, and other factors on hatch of hen's eggs and energy metabolism of chick embryos. US Dept. agr. tech. Bull, 553, 1-45.

Bartholomew, G. A., Tucker, V. A. 1964. Size, body temperature, thermal conductance, oxygen consumption, and heart rate in Australian varanid lizards. Physiol. Zool., 37, 341-354.

Bashamohideen, M., Parvatheswararao, V. 1976. Size-metabolism relation in animals. Zool. Anz., 196, 333-377.

Bashkirov, A. G., Zubarev, D. N. 1970. The generalized Kramers-Fokker-Planck equation. Physica 48, 137-145.

Bataille, J., Edelen, D. G. B., Kestin, J. 1978. Nonequilibrium thermodynamics of the nonlinear equations of chemical kinetics. J. Non-Equilib. Thermodyn., 3, 153-168.

Bataille, J., Kestin, J. 1976. General forms of the dissipation inequality. J. Non-Equilib. Thermodyn. 1, 25-31.

Bataille, J., Kestin, J. 1977. Irreversible thermodynamics of continua and internal variables. In: Proc. 2nd Intern. Symp. Continuum Models of Discrete Systems. Montreal.

Bataille, J., Kestin, J. 1979. Irreversible processes and physical interpretation of rational thermodynamics. J. Non-Equilib. Thermodyn., 4, 229-258.

Bauer, E. S. 1935. Theoretical Biology. Moscow-Leningrad, Inst. exp. Med. Press (Russian).

Bedeaux, D. 1977. Equivalence of the master equation and the Langevin equation. Phys. Lett., 62 A, 10-12.

Belokopytin, Yu. S. 1968. The level of basal metabolism in some marine fishes. Problems Ichthyol., 8, 382-385 (Russian).

Benedict, F. G. 1938. Vital Energetics. Washington, Carnegie Inst.

Bennett, A. F. 1972. The effect of activity on oxygen consumption, oxygen debt. and heart rate in the lizards *Varanus gouldii* and *Sauromalus hispidus*. J. comp. Physiol., 79, 259-280.

Bennett, A. F., Dawson, W. R. 1976. Metabolism. In: Biology of Reptilia, 5, 127-223, New York, Acad. Press.

Bennett, A. F., Dawson, W. R., Bartholomew, G. A. 1975. Effects of activity and temperature on aerobic and anaerobic metabolism in the Galapagos marine iguana. J. comp. Physiol., B 100, 317-329.

Bennett, A. F., Gleeson, T. T. 1976. Activity metabolism in the lizard *Sceloporus occidentalis*. Physiol. Zool., 49, 65-76.

Bennett, A. G., Licht, P. 1973. Relative contributions of anaerobic and aerobic energy production during activity in amphibia. J. comp. Physiol., 87, 351-360.

Berger, M., Hart, J. S., Roy, O. Z. 1970. Respiration, oxygen consumption, and heart rate in some birds during rest and flight. Z. vergl. Physiol., 66, 201-214.

Biot, M. A. 1970. Variational Principles in Heat Transfer. Oxford, Univ. Press.

Blazka, P. 1955. Temperaturadaptation des Gesamtmetabolismus bei der Weinbergschnecke *Helix pomatia L.* - Zool. Jahrb., Abt. I, Physiol., 65, 430-438.

Blum, E. H., Mehra, R. K. 1975. Chemical kinetics and catastrophe theory. Res. Report NASA, No. 41.

Blumenfeld, L. A. 1974. Problems of Biological Physics. Moscow, Nauka (Russian).

Bochkov, G. N., Kuzolev, O. E. 1977. On general theory of heat fluctuations in nonlinear systems. J. exp. theor. Phys., 1(72), 238-245 (Russian).

Bogolyubov, N. N. 1946. Problems of Dynamic Theory in Statistical Physics. Moscow, Gostekhizdat (Russian).

Bogolyubov, N. N. 1970. Selected Works., v. 2, Kiev, Naukova Dumka (Russian).

Bogolyubov, N. N. 1971. Selected Works., v. 3, Kiev, Naukova Dumka (Russian).

Bogolyubov, N. N. 1973. Quasimeans in the problems of statistical mechanics. In: Statistical Physics and Quantum Theory of Field. Moscow (Russian).

Bogolyubov, N. N. 1978. On stochastic processes in dynamic systems. Phys. Elementary Particles and Atomic Nucleus, 9, 501-579 (Russian).

Bogolyubov, N. N., Zubarev, D. N., Tserkovnikov, Yu. A. 1960. Asymptotically correct solution for slow Hamiltonian of the superconductivity theory. J. exp. theor. Phys., 39, 120-122 (Russian).

Bogolyubov, N. N. (Jr) 1974. Method for Studying Model Hamiltonians. Moscow, Nauka (Russian).

Boiteux, A., Goldbeter, A., Hess, B. 1975. Control of oscillating glycolysis of yeast by stochastic, periodic, and steady source of substrate: a model and experimental study. Proc. nat. Acad. Sci. USA, 72, 3829-3833.

Boltzmann, L. 1896. Vorlesungen über Gastheorie. Leipzig, Barth.

Bom, D. 1963. Quantum Theory. New York, Prentice-Hall.

Bonifacio, R., Gronchi, M., Lugiato, L. A. 1978. Photon statistics of a bistable absorber. Phys. Rev., 18 A, 2266-2279.

Boyling, J. B. 1972. An axiomatic approach to classical thermodynamics. Proc. Roy. Soc. London, Ser. A., 329, 35-70.

Brett, J. R. 1965. The relation of size to rate of oxygen consumption and sustained swimming speed of sockeye salmon *(Oncorhynchus nerka)*. J. Fish. Res. Bd. Canada, 22, 1491-1501.

Brett, J. R. 1973. Energy expenditure of sockeye salmon, *Oncorhynchus nerka*, during sustained performance. J. Fish. Res. Bd. Canada, 30, 1799-1809.

Brett, J. R., Sutherland, D. B. 1965. Respiratory metabolism of pumpkinseed *(Lepomis gibbosus)* in relation to swimming speed. J. Fish. Res. Bd. Canada, 22, 405-409.

Bröcker, Th., Lander, L. 1975. Differentiable Germs and Catastrophes. Cambridge, Univ. Press.

Brody, S. 1945. Bioenergetics and Growth. New York, Reinhold.

Bulsara, A. R., Schieve, W. C., Gracc, R. F. 1978. Phase transitions induced by white noise in bistable optical systems. Phys. Lett., 68 A, 294-296.

Bunimovich, L. A. 1970. About one transformation of circumference. Math. Notes, 8, 205-216 (Russian).

Burton, A. C. 1939. The properties of steady state compared to these of equilibrium as shown in characteristic biological behavior. J. cell. comp. Physiol., 14, 327-349.

Butler, P. J., West, N. H., Jones, D. R. 1977. Respiratory and cardio-vascular responses of the pigeon to sustained, level flight in a wind-tunnel. J. exp. Biol., 71, 7-26.

Butorin, N. V. (Ed.). 1978. Ivankovskoe Reservoir and its Life. Trans. Inst. Biol. Inner Waters, v. 34, Leningrad, Nauka (Russian).

Byzova, Yu. B. 1970. Oxygen uptake and temperature in some millipeds (Julidae, Diplopoda). Ann. Zool. (Preprints Symp. Paris), 34-40.

Byzova, Yu. B. 1972. Respiration of soil invertebrates. In: Ecology of Soil Invertebrates. Moscow, Nauka, 3-39 (Russian).

Byzova, Yu. B. 1977. Habitation temperature and metabolism intensity of soil invertebrates. In: Adaptation of Soil Animals to Environmental Condition. Moscow, Nauka, 3-44 (Russian).

Byzova, Yu. B., Zalesskaya, N. T. 1980. Effect of temperature on oxygen consumption in Monotarsobius curtipes C. Koch (Chilopoda, Lithobiidae) from different parts of the areal. J. gen. Biol. 41, 80-87 (Russian).

Callahan, J. 1974. Singularities and plane maps. Amer. math. Monthly 81, 211-240.

Callahan, J. 1977. Singularities and plane maps. II. Sketching catastrophes. Amer. math. Monthly, 84, 765-803.

Calvao, P. E., Tarasantchi, J., Guertzenstein, P. 1965. Heat production of tropical snakes in relation to body weight and body surface. Amer. J. Physiol., 209, 501-506.

Caratheodory, C. 1909. Untersuchungen über die Grundlagen der Thermo-dynamik. Math. Annalen 67, 355-386.

Carpenter, R. E. 1974. Flight metabolism of flying foxes. In: Symp. Swimm. Flying Nature, Calif., Pap. Abstr., p. 77.

Casey, T. M. 1976. Flight energetics in sphinx moths: heat production and heat loss in *Hyles lineata* during free flight. J. exp. Biol., 64, 545-560.

Chadwick, L. E. 1947. The respiratory quotient of Drosophila during flight. Biol. Bull. 93, 229-239.

Chadwick, L. E., Gilmour, D. 1940. Respiration during flight in *Drosophila repleta Wollaston*: the oxygen consumption considered in relation to the wing-rate. Physiol. Zool. 13, 398-410.

Chakrabarti, C. G. 1975. Theory of thermodynamic fluctuations. Indian J. Phys. 49, 691-695.

Chandrasekar, S. 1943. Stochastic problems in physics and astronomy. Rev. mod. Phys. 15, 1-89.

Chaturvedi, S., Gardiner, C. W., Matheson, I. S., Walls, D. F. 1977. Stochastic analysis of a chemical reaction with spatial and temporal structures. J. stat. Phys., 17, 469-489.

Chernavsky, D. S. 1973. On dynamic and statistical theories. Proc. 8th Winter School LIYuF, 255-278 (Russian).

Chernavsky, D. S., Puchov, N. M. 1971. Unstable processes in quantum mechanics. Theor. math. Phys., 7, 219-231 (Russian).

Chugunova, N. I. 1959. Handbook for Investigations of Age and Length of Fishes. Moscow, Acad. Sci. Publ. (Russian).

Cleland, W. W. 1963 a. The kinetics of enzyme-catalyzed reactions with two or more substrates or products. I. Nomenclature and rate equations. Biochem. biophys. Acta, 67, 104-137.

Cleland, W. W. 1963 b. The kinetics of enzyme-catalyzed reactions with two or more substrates or products. II. Inhibition: nomenclature and theory. Biochem. biophys. Acta, 67, 173-187.

Cleland, W. W. 1963 c. The kinetics of enzyme-catalyzed reactions with two or more substrates or products. III. Prediction of initial velocity and inhibition patterns by inspection. Biochem. biophys. Acta, 67, 188-196.

Cody, M. L. 1974. Optimization in ecology. Science, 183, 1156-1168.

Cohen, D. 1966. Optimizing reproduction in a randomly varying environment. J. theor. Biol., 12, 119-129.

Cohen, E. G., Berlin, T. M. 1960. Note on the derivation of the Boltzmann equation from the Liouville equation. Physica, 26, 717-729.

Cohn, A. 1955. Optimal systems: vascular system. Bull. math. Biophys., 16, 316-342.

Conrad, M. 1973. Thermodynamic extremal principles in evolution. Biophysik, 9, 191-196.

Conzalez-Gascon, F. 1977. On a new first integral of certain dynamical systems. Phys. Lett., A 61, 375-376.

Cooper, S. C., Ferguson, J. H. 1973. The effects of cold acclimation upon the oxygen consumption of two species free-living nematodes. J. Nematodology, 5, 241-245.

Crell, B., de Pali, T., Uhlmann, A. 1978. H-Theoreme für die Fokker-Planck-Gleichung. Wiss. Z. Karl-Marx-Univ. Leipzig, Math. Naturwiss., 27, 229-234.

Cunningham, D. J. C. 1974. Integrative aspects of the regulation of breathing: a personal view. In: Physiology, Ser. I. v. 2. Respiratory Physiology, Baltimore, Butterworths, 303-363.

Czajkowski, G., Ebeling, W. 1977. Nonequilibrium phase transitions in enzyme reaction systems. J. Non-equil. Thermodyn., 2, 1-16.

Day, W. A. 1972. The Thermodynamics of Simple Materials with Fading Memory. Berlin, Springer.

Davies, P. C. W. 1978. Thermodynamics of black holes. Rep. Prog. Phys., 41, 1313-1355.

Dawson, W. R., Hudson, J. W. 1970. Birds. In: Comparative Physiology of Thermoregulation., v. 1, 224-310, New York, Acad. Press.

Delgado Domingos, J. J., Nina, M. N. R., Whitelaw, J. H. (Eds.). 1974. Foundations of Continuum Thermodynamics. London, Macmillan.

de Groot, S. R., Mazur, P. 1969. Non-Equilibrium Thermodynamics. Amsterdam, North-Holland.

Dejours, P. 1967. Neurogenic factors in the control of ventilation during exercise. Circulat. Res., 20, Suppl. I, 146-153.

DelleDonne, M., Ortoleva, P. 1978. Generalized nonlinear master equations for local fluctuations: dimensionality expansious and augmental mean field theory. J. stat. Phys., 18, 319-331.

Dewel, G., Walgraef, D., Borckmans, P. 1977. Renormalization group approach to chemical instabilities. Z. Physik, B 28, 235-237.

Ditkin, V. A., Kuznetzov, P. I. 1951. Handbook for Operation Calculation. Moscow-Leningrad, Gos. Izd. Techn. Theor. Lit. (Russian).

Dmi'el, R., Rappoport, D. 1976. Effect of temperature on metabolism during running in the lizard *Uromastix aegyptius*. Physiol. Zool., 49, 77-84.

Dolnik, V. P. 1968. Energy metabolism and animal evolution. Adv. modern Biol., 66, 276-293.

Don Stevens, E., Josephson, R. K. 1977. Metabolic rate and body temperature in singing katydids. Physiol. Zool., 50, 31-42.

Dorfman, J., Cohen, E. 1973. Velocity-correlation functions in two and three dimensions. Phys. Rev., A 6, 776-790.

Dorfman, N. L., Ovsyannikov, L. L. 1978. Birth rate and weight of newborn mammals optimal from evolutionary viewpoint. In: Dynamics of Biological Systems, v. 2, 60-75, Gorky, Univ. Press (Russian).

Dronova, L. M., Belich, E. I., Erokhin, V. N., Emanuel, N. M. 1966. Kinetic regularities of erythromyelosis development in rats. Proc. Acad. Sci. USSR, Ser. biol., 5, 743-750 (Russian).

Dufty, J. 1971. Long-time behavior of space- and time-dependent correlation functions. Phys. Rev., A 5, 2247-2255.

Durnin, J. V. G. A., Passmore, R. 1967. Energy, Work and Leisure. London, Heinemann.

Ebeling, W. 1976. Strukturbildung bei Irreversiblen Prozessen. Eine Einführung in die Theorie Dissipativer Strukturen. Leipzig, Teubner.

Ebeling, W. 1978. Non-equilibrium transitions and stationary probability distributions of stochastic processes. Phys. Lett., 68 A, 430-432.

Ebeling, W., Engel-Herbert, H. 1977. Strukturelle Instabilitäten nichtlinearer irreversibler Prozesses. In: Rostocker Physikalische Manuskripte 2, 23-43.

Ebeling, W., Malchow, H. 1979. Bifurcations in a bistable reaction-diffusion system. Ann. Physik, 36, 121-134.

Edelen, D. G. B. 1973 a. On the existence of symmetry relations and dissipation potentials. Arch. Rat. Mech. Anal., 51, 218-227.

Edelen, D. G. B. 1973 b. Asymptotic stability, Onsager fluxes and reaction kinetics. Intern. J. Eng. Sci., 11, 819-832.

Edelen, D. G. B. 1974. Primitive thermodynamics: a new look at the Clausius-Duhem inequality. Intern. J. Eng. Sci., 12, 121-141.

Edelen, D. G. B. 1975. On the thermodynamics of evolving chemical systems and the approach to equilibrium. Adv. Chem. Phys., 33, 399-442.

Edelen, D. G. B. 1976. A thermodynamics with internal degree of freedom and nonconservative forces. Intern. J. Eng. Sci., 14, 1013-1032.

Edelen, D. G. B. 1977. Lagrangian mechanics of nonconservative nonholonomic systems. In: Noordhoff, Leyden, Sect. 9, 279-283.

Edney, E. B. 1964 a. Acclimation to temperature in terrestrial isopods. I. Lethal temperature. Physiol. Zool., 37, 364-377.

Edney, E. B. 1964 b. Acclimation to temperature in terrestrial isopods. II. Heart rate and standard metabolic rate. Physiol. Zool., 37, 378-393.

Edwards, G. A. 1953. Respiratory metabolism. In: Insect Physiology, New York, Wiley.

Efremov, G. F. 1968. Fluctuation-dissipative theorem for nonlinear media. J. exp. theor. Phys., 55, 2322-2333 (Russian).

Efremov, G. F. 1972. On the theory of heat fluctuation in nonequilibrium systems. Proc. Higher Educ. Establ., Radiophys., 15, 1207-1219 (Russian).

Eigen, M. 1971. Selforganization of Matter and the Evolution of Biological Macromolecules. Berlin, Springer.

Emanuel, N. M. 1977. Kinetics of Experimental Tumour Processes. Moscow, Nauka (Russian).

Emelyanov, V. I., Klimontovich, Yu. L. 1976. Quantum electronic, v. 3, 848.

Erneux, T., Herschkowitz-Kaufman, M. 1977. Rotating waves as asymptotic solutions of a model chemical reaction. J. chem. Phys., 66, 248-250.

Ernst, M. H., Hauge, E. H., Leeuwen, J. M. J. 1971. Asymptotic time behaviour of correlation functions. Phys. Rev., A 5, 2055-2065.

Erokhin, V. N. 1975. Kinetic regularities of development of experimental malformations (mathematical laws of tumour growth). In: Quantitative Aspects of Organism Growth, Moscow, Nauka, 89-111 (Russian).

Farmanfarmaian, A. 1966. The respiratory physiology of Echinoderms. In: Physiology of Echinodermata, New York, Wiley, 245-265.

Farmer, G. J., Beamish, F. W. H. 1969. Oxygen consumption of *Tilpadia milotica* in relation to swimming speed and salinity. J. Fish. Res. Bd. Canada, 26, 2807-2821.

Feinn, D., Ortoleva, P. 1977. Catastrophe and propagation in chemical reactions. J. chem. Phys., 67, 2119-2131.

Feistel, R., Ebeling, W. 1978. Deterministic and stochastic theory of sustained oscillations in autocatalytic reaction systems. Physica, 93 A, 114-137.

Felderhof, B. U. 1961. Eigenschappen van de master equation. Nederlands Tijdsch. Natuurkunde, 27, 232-250.

Feller, W. 1968. An Introduction to Probability Theory and its Application., v. 1, New York, Wiley.

Feller, W. 1971. An Introduction to Probability Theory and its Application., v. 2, New York, Wiley.

Fesenko, A. I., Glushko, A. A. 1976. Device for measuring the rate of temperature changes. Discover and Invention, 34, 94 (Russian).

Feynman, R. P. 1972. Statistical Mechanics. Mass., Benjamin.

Fisher, M. E. 1974. The renormalization group in the theory of critical behavior. Rev. mod. Phys., 46, 597-616.

Fisher, R. A. 1958. The Genetical Theory of Natural Selection. New York, Dover.

Fourche, J., Lemaitre, C. 1974. Le métabolisme énergetique au cours des métamorphoses de *Ceratitis capitata* - relations thermogénése - respirations. Bull. Biol. France et Belgique, 108, 97-102.

Fox, R. F. 1978. Fluctuation theories and Gaussian stochastic processes. J. math. Phys., 19, 127-130.

Fox, R. F., Kac, M. A. 1977. A remark on the theory of fluctuations far from thermodynamic equilibrium. Biosystems, 8, 187-191.

Frolov, V. P. 1978. Black holes: quantum processes, thermodynamics, astrophysics. In: Black Holes., Moscow, Mir, 5-30 (Russian).

Fry, F. E. G. 1957. The aquatic respiration of fish. In: The Physiology of Fishes, v. I, 1-63, New York, Acad. Press.

Furukawa, H. 1976. Equations of motion for fluctuation and variance in nonequilibrium state. Progr. theor. Phys., 56, 464-476.

Furukawa, H. 1977. Langevin type equations in nonequilibrium state. Progr. theor. Phys., 58, 1127-1141.

Galkovskaya, T. A., Sushchenya, L. M. 1978. Growth of Aquatic Animals under Varied Temperature., Minsk, Nauka i Tehnika (Russian).

Gardiner, C. W. 1976. A comment on chemical Langevin equations. J. stat. Phys., 15, 451-454.

Gardiner, C. W., McNeil, K. J., Walls, D. F., Matheson, J. S. 1976. Correlations in stochastic theories of chemical reactions. J. stat. Phys., 14, 307-331.

Gardiner, C. W., Walls, D. F. 1978. A generalised Landau theory for chemical instabilities. J. Phys., A 11, 161-171.

Gatten, R. E. 1974. Effects of temperature and activity on aerobic and anaerobic metabolism and heart rate in the turtles *Pseudemys scripta* and *Terrapene ornata*. Comp. biochem. Physiol. 48 A, 619-648.

Gershuni, G. Z., Zhukhovitsky, E. M. 1972. Convective stability of Incompressible Liquid., Moscow, Nauka (Russian).

Ghiretti, F. 1966. Respiration. In: Physiology of Mollusca, v. 2, 175-208, New York, Acad. Press.

Gierer, A., Meinhardt, H. 1974. Biological pattern formation involving lateral inhibition. In: Lectures on Mathematics in Life Science, 7, 163, Berlin, Springer.

Gikhman, I., Skorokhod, A. W. 1971. Stochastische Differentialgleichungen. Berlin, Akademie-Verlag.

Gillespie, D. T. 1976. A general method for numerically simulating the stochastic time evolution of coupled chemical reactions. J. comp. Physiol., 22, 403-434.

Ghilarov, M. S. 1957. Julidae and their role in soil formation. Soil Science, 6, 74-80 (Russian).

Glansdorff, P., Nicolis, G., Prigogine, I. 1974. The thermodynamic stability theory of non-equilibrium states. Proc. nat. Acad. Sci. USA, 71, 197-199.

Glansdorff, P., Prigogine, I. 1970. Non-equilibrium stability theory. Physica, 46, 344-366.

Glansdorff, P., Prigogine, I. 1971. Thermodynamic Theory of Structure, Stability and Fluctuations. New York, Wiley.

Glushko, A. A. 1976 a. Moisture-sensitive element. Discover and Invention, 33, 127 (Russian).

Glushko, A. A. 1976 b. Heat model of man. Discover and Invention, 36, 160 (Russian).

Glushko, A. A., Selivanov, V. V. 1974. Heat flowmeter of flowing media. Discover and Invention, 14, 131 (Russian).

Glushko, A. A., Gorodinsky, S. M., Smirnov, L. S. 1971 a. Generator of heat current for studying heat release of man. Discover and Invention, 11, 20 (Russian).

Glushko, A. A., Gorodinsky, S. M., Smirnov, L. S. 1971 b. Generator of temperature. Discover and Invention, 12, 7 (Russian).

Glushko, A. A., Smirnov, L. S., Arbitan, A. A. 1974. Thermometric element. Discover and Invention, 18, 87 (Russian).

Gnedenko, B. V. 1962. The Theory of Probability. New York, Chelsea.

Goel, N. S., Maitra, S. C., Montroll, E. W. 1971. On the Volterra and other nonlinear models of interacting populations. Rev. mod. Phys., 43, 231-276.

Goldbeter, A., Lefever, R. 1972. Dissipative structures for an allosteric model. Application to glycolytic oscillations. Biophys. J., 12, 1302-1315.

Gololobov, A. D., Golubev, I. N. 1970. Microkinetics of hydrocarbon type of nutrition of microorganisms. Microbiol. Industry, 3, 14-22 (Russian).

Golubitsky, M. 1978. An introduction to catastrophe theory and its applications. Siam Rev., 20, 352-387.

Golubitsky, M., Guillemin, V. 1973. Stable Mappings and their Singularities., Berlin, Springer.

Gonzalez-Gascon. 1977. On a new frist integral of certain dynamical systems. Phys. Lett., A 61, 375-376.

Gorodinsky, S. M., Glushko, A. A., Orekhov, B. V. 1976. Calorimetry in Isolating Prophylactics of Man. Moscow, Mashinostroenie (Russian).

Grad, H. 1949. On the kinetic theory of rarefied gases. Comm. Pure Appl. Math., 2, 331-407.

Graham, R. 1973. Statistical theory of instabilities in stationary nonequilibrium systems with applications to laser and nonlinear optics. In: Quantum Statistics in Optics and Solid State Physics. Ergebn. exakte Naturw., 66, 1-97, Berlin, Springer.

Graham, R. 1978. Path-integral methods in nonequilibrium thermodynamics and statistics. In: Stochastic Processes in Nonequilibrium systems. 83-138, Berlin Springer.

Graham, R., Haken, H. 1968 a. The quantum-fluctuations of the optical parametric oscillator. Z. Physik, 210, 276-302.

Graham, R., Haken, H. 1968 b. Quantum theory of light propagation in fluctuating lase-active medium. Z. Physik, 213, 420-450.

Grainger, J. N. R. 1956. Effects of changes of temperature on the respiration of certain Crustacea. Nature. London, 178, 930-450.

Green, M. S. 1952. Markoff random processes and the statistical mechanics of time-dependent phenomena. J. chem. Phys., 20, 1281-1295.

Greenwald, O. E. 1971. The effect of body temperature on oxygen consumption and heart rate in the Sonora gopher snake, *Pituophis catenifer affinis*. Copeia, 1, 98-106.

Grinin, A. P. 1975. Calculation of asymptotics of kinetic nuclei of linear hydrodynamics at high times in extended hydrodynamic description. Proc. Leningrad Univ., N22, 17-23 (Russian).

Grinin, A. P. 1976. Sufficient bilinear hydrodynamic description in the problem about time asymptotics of kinetic nuclei of linear hydrodynamics. Proc. Leningrad Univ., N4, 43-50 (Russian).

Grodins, F. S. 1963. Control Theory and Biological Systems. New York, Columbia Univ. Press.

Gromysz-Kalkowska, K. 1976. The influence of temperature and duration of acclimation on the metabolic rate in *Orthomorpha gracilis (C. L. Koch)* (Diplopoda). Folia Biol., 24, 55-64.

Grudnitzky, V. A.† 1972 a. Dependence of heat production on weight during the growth of *Xiphophorus helleri H.* C. R. Acad. Sci. USSR, 202, 1435-1437 (Russian).

Grudnitzky, V. A.† 1972 b. Energetic regularities during the growth of *Xiphophorus helleri H.* In: Energetic Aspects of the Growth and Metabolism in Aquatic Animals. Kiev, Naukov Dumka, 40-41 (Russian).

Grudnitzky, V. A.† 1975. Energetic regularities during growth of fishes. In: Quantitative Aspects of Organism Growth. Moscow, Nauka, 48-53.

Guenther, K. P. 1965. Aerobic hydrocarbon fermentation. Biotechn. Bioeng. 7, 445-446.

Gurov, K. P. 1978. Phenomenological Thermodynamics of Irreversible Processes. Moscow, Nauka (Russian).

Güttinger, W. 1974. Catastrophe Geometry in Physics and Biology. Lecture Notes in Biomathematics, 4, 1-30, Berlin, Springer.

Gyarmati, J. 1970. Nonequilibrium Thermodynamics. Field Theory and Variational Principles. Berlin, Springer.

Haase, R. 1963. Thermodynamik der Irreversiblen Prozesse. Darmstadt, Steinkopff.

Haken, H. 1970. Laserlicht - ein neues Beispiel für eine Phasenumwandlung? In: Festkörperprobleme, Bd. X, 351-365, Marburg, Pergamon Press.

Haken, H. (Ed.). 1973 a. Synergetics., Proc. Intern. Symp. on Synergetics. Elmau 1972. Stuttgart, Teubner.

Haken, H. 1973 b. Distribution function for classical and quantum systems far from thermal equilibrium. Z. Physik, 263, 267-282.

Haken, H. (Ed.). 1974. Cooperative Effects, Progress in Synergetics. Proc. Summerschool at Erice, Sicily 1974, Amsterdam, North-Holland.

Haken, H. 1975 a. Cooperative phenomena in systems far from thermal equilibrium and in nonphysical systems. Rev. mod. Phys., 47, 67-121.

Haken, H. 1975 b. Statistical physics of bifurcation, spatial structures, and fluctuations of chemical reactions. Z. Physik, B 20, 413-420.

Haken, H. (Ed.). 1977. Synergetics. Proc. Intern. Workshop on Synergetics at Schloss Elmau 1977, Berlin, Springer.

Haken, H. 1978. Synergetics. An Introduction to Nonequilibrium Phase Transitions and Self-Organisation in Physics, Chemistry and Biology. Berlin, Springer.

Haldane, J. B. S. 1956. The relation between density regulation and natural selection. Proc. roy. Soc., Ser. B, 145, 306-308.

Hanusse, P. 1977. Monte-Carlo simulation of finite fluctuations. J. chem. Phys., 67, 1282-1284.

Harris, S. 1967. Proof that successive derivatives of Boltzmann's H-function for a discrete velocity gas alternate in sign. J. math. Phys., 8, 2407-2410.

Hart, J. S., Berger, M. 1972. Energetics, water economy and temperature regulation during flight. In: Proc. 15th Intern. Ornithol. Congr., Hague 1970, Leiden, Brill, 189-199.

Hasegawa, H. 1976 a. On the construction of a time-reversed Markoff Process. Progr. theor. Phys., 55, 90-105.

Hasegawa, H. 1976 b. Variational principles for nonequilibrium states and the Onsager-Machlup formula. Progr. theor. Phys., 56, 44-60.

Hasegawa, H. 1977 a. Thermodynamic properties of non-equilibrium states subject to Fokker-Planck equations. Progr. theor. Phys., 57, 1523-1537.

Hasegawa, H. 1977 b. Variational approach in studies with Fokker-Planck equations. Progr. theor. Phys., 58, 128-146.

Hasegawa, H., Nakagomi, T. 1979 a. Optical instabilities from the entropy production point of view. In: Suppl. Progr. theor. Phys., No. 64, 321-338.

Hasegawa, H., Nakagomi, T. 1979 b. Semiclassical laser theory in the stochastic and thermodynamic frameworks. J. stat. Phys. 21, 191-214.

Hasegawa, H., Nakagomi, T. 1980. On the characterization of the stationary state of a class of dynamical semigroups. J. stat. Phys. 23, 639-652

Hasegawa, H., Nakagomi, T., Mabuchi, M., Kondo, K. 1980. Nonequilibrium thermodynamics of easing and bistable optical systems. J. stat. Phys. 23, 281-313

Hashitsume, H. 1952. A statistical theory of linear dissipative systems. Progr. theor. Phys. 8, 461-478.

Heath, J. E., Adams, P. A. 1967. Regulation of heat production by large moths. J. exp. Biol. 47, 21-31.

Heinrich, B. 1971. Temperature regulation of the spinx moth, *Manduca sexta*. I. Flight energetics and body temperature during free and tethered fligth. J. exp. Biol. 54, 141-152.

Heinrich, B. 1975. Thermoregulation in bumblebees. II. Energetics of warm-up and free flight. J. comp. Physiol. 96, 155-166.

Heinrich, B., Casey, T. M. 1973. Metabolic rate and endothermy in sphinx moths. J. comp. Physiol. 82, 195-206.

Hemmingsen, A. H. 1960. Energy metabolism as related to body size and respiratory surface, and its evolution. Rep. Steno memor. Hospital Nordisk Insulinlabor 9, 7-110.

Hepp, K., Lieb, E. H. 1973. On superradiance phase transition for molecules in a quantined radiation field: the Dicke model. Ann. Phys. 86, 360-

Hermann, R. 1973. The mathematics of thermodynamics. In: Geometry, Physics and Systems. New York, Dekker, 259-287.

Hess, W. R. 1914. Das Prinzip des Kleinsten Kraftverbrauches im Dienste Haemodynamischer Forschung. Arch. Anat. Physiol., 1/2, 1-62.

Hoare, M. R., Rahman, M. 1973. On the spectral theory of Rayleigh's piston. I. The discrete spectrum. J. Phys. (London). A Math. nucl. Gen. 6, 1461-1478.

Hochachka, P. W., Somero, G. N. 1973. Strategies of Biochemical Adaptation Philadelphia, Saunders.

Hocking, B. 1953. The intrinsic range and speed of flight of insects. Trans. roy. entom. Soc. London 104, 223-345.

Hofmann-Kadoshnikov, P. B. 1975. Molecular-genetical theory of programming growth and its role as amethod for studying of growth. In: Quantitative Aspects of Organism Growth., Moscow, Nauka, 189-202 (Russian).

Hohenberg, P. C., Halperin, B. I. 1977. Theory of dynamic critical phenomena. Rev. Modern Phys. 49, 435-479.

Horsfild, K. I., Comming, G. 1967. Angles of branching and diameters of branches in the human bronchial tree. Bull. math. Biol. 29, 245-259.

Horsthemke, W., Brenig, L. 1977. Non-linear Fokker-Planck equation and asymptotic representation of the master equation. Z. Physik B 27, 341-348.

Horsthemke, W., Lefever, R. 1977. Phase transition induced by external noise. Phys. Lett. 64 A, 19-21.

Horsthemke, W., Malek-Mansour, M. 1976. The influence of external noise on non-equilibrium phase transitions. Z. Physik B 24, 307-313.

Husimi, K. 1942. Theory of Probability and Statistics. Tokyo, Kawade Shobo (Japanese).

Ierusalimsky, N. D. 1963. Bases of Physiology of Microbes. Moscow, Acad. Sci. Press (Russian).

Insarov, G. E. 1975. Stepwise model of the growth and development of organisms. In: Quantitative Aspects of Organism Growth, Moscow, Nauka, 114-121 (Russian).

Insarov, G. E., Semenov, S. M. 1977. Modelling of growth and reproduction of organisms based on Haldane-Semevsky optimum principle. In: Theor. exp. Biophys., Kaliningrad 7, 117-124 (Russian).

Ivlev, V. S. 1954. Relationship between metabolism intensity in fishes and their body weight. Physiol. J. USSR 40, 717-721 (Russian).

Ivlev, V. S. 1959. From the experience of estimation of the evolutionary significance of level in the energy metabolism. J. gen. Biol. 20, 94-103 (Russian).

Ivlev, V. S. 1962. Methods for measuring of active metabolism. In: Guidebook for Investigation Methods of Fishes Physiology. Moscow, Acad. Press (Russian).

Ivlev, V. S. 1963. On parameters characterizing the level of energy metabolism in animals. In: Application of Mathematical Methods in Biology. 2, 146-151, Leningrad Univ. Press (Russian).

Ivleva, I. V. 1972. Effect of temperature on metabolism rate in poikilothermic animals. Adv. modern Biol. 73, 134-155 (Russian).

Ivleva, I. V. 1974. Metabolic rate in Crustacea living under low temperature conditions. In: Hydrobiology and Biogeography of Shelves in Cold and Moderate Waters of the World Ocean. Leningrad, Nauka, 132-133 (Russian).

Janes, E. T. 1963. Information theory and statistical mechanics. In: Statistical Physics., Brandeis Lectures, 3, 160-185.

Jansky, L. 1966. Body organ thermoregulation of the rat during exposure to cold and at maximal metabolic rate. Fed. Proc. 25, No 4, part I, 1297-1302.

Jauch, J. M. 1975. Analytical thermodynamics. I. Thermostatistics - general theory. Found. Phys. 5, 111-132.

Job, S. 1955. The oxygen consumption of *Salvelinus fontinalis*. Univ. Toronto Biol. Ser., 61, 1-39.

Kabashima, S. 1978. Oscillatory to non-oscillatory transition due to noise in a parametric oscillator. Ann. Israel Phys. Soc., Stat. Phys. 2, 710-713.

Kabashima, S., Kawakubo, T. 1979. Observation of a noise-induced phase transition in a parametric oscillator. Phys. Lett. 70 A, 375-376.

Kabashima, S., Kogure, S., Kawakubo, T., Okada, T. 1979. J. phys. Soc. Japan.

Kac, M. 1959. Probability and Related Topics in Physical Sciences. London, Interscience Publ.

Kamke, E. 1959. Gewöhnliche Differentialgleichungen. Leipzig.

Karlin, S., Taylor, H. M. 1975. A First Course in Stochastic Processes. New York, Acad. Press.

Kassiandrova, O. N., Lebedev, V. V. 1970. Processing of Observations Results. Moscow, Nauka (Russian).

Kawasaki, K. 1970. Kinetic equations and time-correlation functions of critical fluctuations. Ann. Phys. 61, 1-56.

Kawasaki, K. 1971. Dynamical theory of fluctuations near the critical points. In: Critical Phenomena. Proc. Intern. School Phys. "Enrico Fermi", New York, Acad. Press, 342-379.

Kawasaki, K., Gunton, J. D. 1972. A note on projection operators in non-equilibrium statistical mechanics. Phys. Lett. 40 A, 35-36.

Kawasaki, K., Gunton, J. D. 1973. Theory of non-linear transport processes. Phys. Rev., A 8, 2048-2064.

Kazakov, I. E. 1975. Statistical Theory of Control Systems in the State Space. Moscow, Nauka (Russian).

Keizer, J. 1975 a. A theory of spontaneous fluctuations in macroscopic systems. J. chem. Phys. 63, 398-403.

Keizer, J. 1975 b. Concentration fluctuations in chemical reactions. J. chem. Phys. 63, 5037-5043.

Keizer, J. 1977. Master equations, Langevin equations, and the effect of diffusion on concentration fluctuations. J. chem. Phys. 67, 1473-1476.

Keizer, J., Fox, R. F. 1974. Qualms regarding the range of validity of Glansdorff-Prigogine criterion for stability of non-equilibrium states. Proc. nat. Acad. Sci. USA 71, 192-196.

Khanin, M. A., Bukharov, I. B. 1970. On phenomenological theory of functional parameters of oxygen transport system. Physiol. J. USSR 56, 1801-1807 (Russian).

Khanin, M. A., Dorfman, N. L. 1973. Evolutional mathematical theory of growth. J. gen. Biol. 34, 294-304 (Russian).

Khanin, M. A., Dorfman, N. L., Bukharov, I. B., Levadny, V. G. 1978. Extremal Principles in Biology and Physiology, Moscow, Nauka (Russian).

Khinchin, A. Ya. 1943. Mathematical Foundations of Statistical Mechanics. Moscow, Gostekhizdat (Russian).

King, J. K., Farner, D. S. 1961. Energy metabolism, thermoregulation and body temperature. In: Biology and Comparative Physiology of Birds. v. 2, 215-288, New York, Acad. Press.

Kleiber, M. 1947. Body size and metabolic rate. Physiol. Rev., 27, 511-541.

Kleiber, M. 1961. The Fire of Life. An Introduction to Animal Energetics. New York, Wiley.

Klein, M. J. 1956. Entropy and the Ehrenfest urn-model. Physica 22, 569-575.

Klimontovich, Yu. L. 1980. Kinetic Theory of Electromagnetic Processes. Moscow, Nauka (Russian).

Kogan, A. M. 1965. Deduction of Grad type equations and study of their relaxation properties by entropy maximization methods. Appl. Math. Mech. 29, 122-133 (Russian).

Kogan, A. M., Rozonoer, L. I. 1964. On macroscopic description of kinetic processes. C. R. Acad. Sci. USSR 158, 566-569 (Russian).

Kolmogorov, A. 1931. Über die analytischen Methoden in der Wahrscheinlichkeitsrechnung. Math. Annalen 104, 415-458.

Kamiya, A. T., Togawa, T., Yamamoto, A. 1974. Theoretical relationship between the optimal model of the vascular tree. Bull. math. Biophys. 36, 311-323.

Konoplev, V. A., Sokolov, V. E., Zotin, A. I. 1978. Criterion of orderliness and some problems of taxonomy. In: Thermodynamics of Biological Processes. Berlin, de Gruyter, 349-359.

Koryagin, V. V., Vorotilo, S. P., Konovalov, S. A., Chirkov, I. M. 1974. Calorimetric studies of microorganisms-producers of enzymes destroying yeast cellular walls. Appl. biochem. Microbiol. 10, 621-623 (Russian).

Koryagin, V. V., Konovalov, S. A., Chirkov, I. M., Vorotilo, S. P., Zvereva, T. A. 1976. Thermogenesis of mesophile, thermotolerant and thermophile strains of microorganisms-producers of enzymes destroying cellular walls of the yeasts. Appl. biochem. Microbiol. 12, 505-507 (Russian).

Koryagin, V. V., Zvereva, G. A., Chirkov, I. M., Bravicheva, R. N. 1977. On the problem of thermogenesis of yeasts *Candida quilliermondii* strain H-542, grown on paraffin in flowing cultivation regime. Appl. biochem. Microbiol. 13, 33-35 (Russian).

Krasovsky, A. A. 1974. Phase Space and Statistical Theory of Dynamic Systems. Moscow, Nauka (Russian).

Krogh, A. 1916. The Respiratory Exchange of Animals and Man. London

Krogh, A., Weis-Fogh, T. 1951. Respiration exchange of the desert locust *(Schistocerca gregaria)* before, during and after flight. J. exp. Biol. 28, 344-357.

Krylov, N. S. 1950. Works for Substantiating Statistical Physics. Moscow-Leningrad, Acad. Sci. Press (Russian).

Kubo, R. 1957. Statistical-mechanical theory of irreversible processes. I. General theory and simple applications to magnetic and conduction problems. J. Phys. Soc. Japan 12, 570-586.

Kubo, R. 1966. The fluctuation-dissipation theorem. Rep. Progr. Phys 29, 255-284.

Kubo, R., Matsuo, K., Kitahara, K. 1973. Fluctuation and relaxation of macrovariables. J. stat. Phys. 9, 51-96.

Kulakov, Yu. I. 1969. Elements of the Theory of Physical Structures. Novosibirsk, Univ. Press (Russian).

Kuni, F. M. 1974. Statistical theory of viscous-elastic properties of liquid. Theor. math. Phys. 21, 233-246 (Russian).

Kuni, F. M., Adjemian, L. C., Novozhilova, T. Yu. 1973 a. Non-linear generalization of the method of projecting Mori operators. Leningrad Univ. Press (Russian).

Kuni, F. M., Adjemian, L. C., Novozhilova, T. Yu. 1973 b. Formalism of memory function for the systems in exterior fields. Leningrad Univ. Press (Russian).

Kuni, F. M., Adjemian, L. C., Novozhilova, T. Yu. 1975. Cumulent decompositions of kinetic coefficients of nonlinear hydrodynamics. Proc. Leningrad Univ. N4, 8-13.

Kuni, F. M., Novozhilova, T. Yu., Adjemian, L. C. 1973 a. Mori's memory function formalism in nonlinear statistical hydrodynamics. Ann. Inst. Henri Poincaré 19, 375-386.

Kuni, F. M., Novozhilova, T. Yu., Adjemian, L. C. 1973 b. Nonlinear nonlocal statistical hydrodynamics. Lett. Nuovo Cimento 8, 708-712.

Kuni, F. M., Novozhilova, T. Yu., Adjemian, L. C. 1974. Nonlinear and nonlocal in time and space statistical hydrodynamics. Proc. Leningrad Univ. N 10, 14-19 (Russian).

Kuni, F. M., Novozhilova, T. Yu., Adjemian, L. C. 1977. Nonlinear hydrodynamic equations in exterior fields and generalized Einstein ratios for kinetic coefficients. Proc. Leningrad Univ. N 10, 141-142 (Russian).

Kuni, F. M., Storonkin, B. A. 1971. Nonequilibrium statistical means in nonlinear by thermodynamic parameters gradients approximation. Theor. math. Phys. 9, 124-136 (Russian).

Kuni, F. M. Storonkin, B. A. 1972 a. Statistical thermodynamics of strongly nonequilibrium systems. I. Statistical and dynamic correlations. Proc. Leningrad Univ. N 4, 13-23 (Russian).

Kuni, F. M., Storonkin, B. A. 1972 b. Statistical thermodynamics of strongly nonlinear systems. II. Motion equations. Proc. Leningrad Univ. N 10, 7-14 (Russian).

Kuni, F. M., Storonkin, B. A. 1973. Nonlinear nonequilibrium statistical thermodynamics. Theor. math. Phys. 14, 220-234 (Russian).

Kuramoto, Y. 1975. Hydrodynamic fluctuation around nonequilibrium steady state. Progr. theor. Phys. 53, 1302-1319.

Kurtz, T. G. 1976. Limit theorems and diffusion approximation for density dependent Markov chains. Math. Programming Study 5, 67-78.

Kurtz, T. G. 1978. Strong approximation theorems for density dependent Markov chains. Stoch. Proc. Appl. 6, 223-240.

Labas, Yu. A., Khlebovich, V. V. 1976. "Phenotypical window" of the genome and progressive evolution. In: Saline Adaptations of Aquatic Organisms. Study of Sea Fauna 17, 4-25, Leningrad, Nauka (Russian).

Lack, D. 1954. The Natural Regulation of Animal Numbers. Oxford.

Laird, A. K. 1964. Dynamics of tumor growth. Brit. J. Cancer 18, 490-502.

Laird, A. K. 1966 a. Dynamics of embryonic growth. Growth 30, 263-275.

Laird, A. K. 1966 b. Postnatal growth of birds and mammals. Growth 30, 349-363.

Laird, A. K. 1967. Evolution of the human growth curve. Growth 31, 345-355.

Laird, A. K., Tyler, S. A., Barton, A. D. 1965. Dynamics of normal growth. Growth 29, 233-248.

Lamprecht, I., Zotin, A. I. (Eds.). 1978. Thermodynamics of Biological Processes. Berlin, de Gruyter.

Landau, L. D., Lifshitz, E. M. 1952. Course of Theoretical Physics. V. Statistical Physics. London, Pergamon Press.

Landau, L. D., Lifshitz, E. M. 1957. Electrodynamics of Continua. Moscow, Gostekhizdat (Russian).

Landau, L., Lifshitz, E. 1967. Physique Statistique. Moscow, Mir.

Landau, L. D., Lifshitz, E. M. 1974. Quantum Mechanics. Moscow, Fizmatgiz (Russian).

Landau, L. D., Lifshitz, E. M. 1976. Statistical Physics. Moscow, Nauka (Russian).

Landauer, R. 1973. Entropy changes for steady-state fluctuations. J. stat. Phys. 9, 351-371.

Landauer, R. 1975. Inadequacy of entropy and entropy derivatives in characterizing the steady state. Phys. Rev., Gen. Phys. 12, 636-638.

Landauer, R. 1979. The role of fluctuations in multistable systems and in the transition to multistability. Ann. N. Y. Acad. Sci. 316, 433-452.

Langer, J. S. 1975. New computational method in the theory of spinodal decompostion. Phys. Rev. A 11, 1417-1429.

Langevin, P. 1908. Sur la théorie du mouvement brownien. C. R. Acad. Sci. Paris 146, 530-533.

Lapkin, V. V., Poddubny, A. G., Golovanov, V. K. 1977. Seasonal variability of behaviour of eurytherm fishes under thermogradient conditions of the medium. In: Regulation of Animal Behaviour. Moscow, Nauka, 177-178 (Russian).

Lapkin, V. V., Zaytsev, K. N. 1976. Reaction of fishes to electric field of different frequency. Information Bull. Inst. Biol. Inner Waters 30, 14-17 (Russian).

Larkin, A. I., Khmelnitsky, D. E. 1969. Phase transition in uniaxial segnetoelectrics. J. exp. theor. Phys. 56, 2087-2098 (Russian).

Lasiewski, R. C. 1963. Oxygen consumption of torpid, resting, active and flying hummingbirds. Physiol. Zool. 36, 122-140.

Lasiewski, R. C., Dawson, W. R. 1967. A re-examination of the relation between standard metabolic rate and body weight in birds. Condor 69, 13-23.

Lavrentev, A. V., Hreshchik, I. G. 1978. Generalised minimum production entropy, Virodov principle and deduction of transfer equation from variational principle. VINITI, Depon. No. 2128 (Russian).

Lebowitz, J. L., Bergmann, P. G. 1957. Irreversible Gibbsian ensembles. Ann. Phys., 1, 1-23.

Leech, J. W. 1958. Classical Mechanics. London, Methuen.

Lefever, R., Horsthemke, W. 1979. Multiple transitions induced by light intensity fluctuations in illuminated chemical systems. Proc. nat. Acad. Sci. USA 76, 2490-2494.

Lemarchand, H., Nicolis, G. 1976. Long range correlations and the onset of chemical instabilities. Physica 82 A, 521-542.

Leon, J. A. 1976. Life histories as adaptive strategy. J. theor. Biol. 60, 301-336.

Levandowsky, M., Childress, W. S., Spiegel, E. A., Hutner, S. H. 1975. A mathematical model of pattern formation by swimming microorganisms. J. Protozool. 22, 296-306.

Levontin, R. C., Cohen, D. 1969. On population growth in a randomly varying environment. Proc. nat. Acad. Sci 62, 1056-1060.

Li, J. C. N. 1962. Caratheodory's principle and the thermokinetic potential in irreversible thermodynamics. J. phys. Chem. 66, 1414-1420.

Lifshitz, E. M., Pitaevsky, L. P. 1978. Statistical Physics, v. 2, Moscow, Nauka (Russian).

Lindblad, G. 1974 a. Expectations and entropy inequalities for finite quantum systems. Comm. math. Phys. 39, 111-119.

Lindblad, G. 1974 b. Completely positive maps and entropy inequalities. Comm. math. Phys. 40, 147-151.

Lindblad, G. 1976. On the generators of quantum dynamical semigroups. Comm. math. Phys. 48, 119-130.

Loehr, K. D., Sayyadi, P., Lamprecht, I. 1976. Thermodynamic aspects of development for *Tenebrio molitor L.* Experientia 32, 1002-1003.

Lomov, I. A. 1970 a. Rubner law, contradictions and hypotheses. VINITI, Depon. No. 1492 (Russian).

Lomov, I. A. 1970 b. Relative power and size of animal. VINITI, Depon. No. 1489 (Russian).

Lomov, I. A. 1975. Relative power and basal metabolism. In: Quantitative Aspects of Organism Growth. Moscow, Nauka, 244-248 (Russian).

Lotka, A. J. 1925. Elements of Physical Biology. Baltimore, Williams & Wilkins.

Ludwig, H. W. 1963. O_2-Konsum und RQ fliegender Insekten. Zool. Anz., Suppl. 26, 355-359.

Lurie, D. 1978. Stochastic and thermodynamic Liapunov function. Preprint Departamento de Fisica Teoreica Univ. de Barcelona.

Lurie, D., Wagensberg, J. 1979 a. Entropy balance in biological development and heat dissipation in embryogenesis. J. Non-Equilib. Thermodyn. 4, 127-130.

Lurie, D., Wagensberg, J. 1979 b. Non-equilibrium thermodynamics and biological growth and development. J. theor. Biol. 78, 241-250.

Ma, S. K. 1976. Modern Theory of Critical Phenomena. Mass., Wiley.

Machlup, S., Onsager, L. 1953. Fluctuations and irreversible processes. II. Systems with kinetic energy. Phys. Rev. 91, 1512-1515.

Malek-Mansour, M., Houard, J. 1979. A new approximation scheme for the study of fluctuations in nonuniform nonequilibrium systems. Phys. Lett. 70 A, 366-368.

Malek-Mansour, M., Nicolis, G. 1975. A master equation description of local fluctuations. J. stat. Phys 13, 197-217.

Margalef, R. 1968. Perspectives in ecological theory. Illinois, Univ. Chicago Press.

Marshak, M. E. 1961. Regulation of Respiration in Man. Moscow, Medezina (Russian).

Marshall, S. M. 1973. Respiration and feeding on copepods. In: Adv. Marine Biol. 11, 57-120, London.

Matsuno, K. 1978. Evolution of dissipative system: a theoretical basis of Margalef's principle on ecosystem. J. theor. Biol. 70, 21-23.

Matsuo, K. 1977. Relaxation mode analysis of nonlinear birth and death processes. J. stat. Phys. 16, 169-195.

Matykhin, V. A. 1973. Bioenergetics and Physiology of Fishes Swimming. Novosibirsk, Nauka (Russian).

Matyukhin, V. A., Stolbov, A. Ya. 1970. Oxygen consumption by *Coregonus autumnalis migratorius* and *Thymallus arcticus P.* in relation to swimming velocity. In: Adaptation of Aquatic Animals. Novosibirsk, Nauka, 147-149 (Russian).

May, R. M. 1973. Stability and Complexity in Model Ecosystems. Princeton, Univ. Press.

May, R. M. 1974. Biological populations with non-overlapping generations: stable points, stable cycles and chaos. Science 186, 645-647.

McCann, F. V., Boettiger, E. G. 1952. Relation between oxygen consumption and wing load in flies. Fed. Proc. 11, 102.

McKean, H. P. (Jr.). 1966. Speed of approach to equilibrium for Kac's caricature of a Maxwellian gas. Arch. rat. mech. Analysis 21, 343-367.

McNeil, D. R. 1972. On the simple stochastic epidemic. Biometrika 59, 494-497.

McQuarrie, D. 1967. Stochastic approach to chemical kinetics. Suppl. Rev. Ser. Appl. Prob., London, Methuen.

Meixner, J. 1969. Processes in simple thermodynamic materials. Arch. rat. mech. Analysis 33, 33-53.

Meixner, J. 1973. The entropy problem in thermodynamics of processes. Rheol. Acta, Germany, 12, 465-467.

Meixner, J., Reik, H. G. 1959. Thermodynamik der Irreversiblen Prozesse. In: Handbuch der Physik III/2, Berlin, Springer.

Merkin, D. R. 1975. On the problem of the structure of forces. Appl. math. Mech. 39, 929-932 (Russian).

Merkin, D. R. 1976. Introduction into the Theory of Movement Stability. Moscow, Nauka (Russian).

Mesarovic, M. D. (Ed.). 1968. Systems theory and biology. Proc. III Systems Symp. Case Inst. Technology, Berlin, Springer.

Milman, M. S. 1940. Genesis of old age. In: Old Age. Kiev, Acad. Sci. Press, 41-52 (Russian).

Minkevich, I. G., Eroshin, V. K. 1977. On inverse ratio of heat liberation and oxygen consumption by microorganism cultures. In: Theses 8th All-Union Conf. on Calorimetry, Chernogolovky, p. 429 (Russian).

Minot, C. S. 1914. On immortality and on the development of death. In: New Ideas in Biology, v. 3, 67-104 (Russian).

Mishnev, O. G. 1977. Formulation of the dynamic theory of fluctuation by the method of nonequilibrium statistical operator. Proc. Higher Educ. Inst., Ser. Phys. 5, 145-147.

Misra, B., Prigogine, I., Courbage, M. 1979. From deterministic dynamics to probabilistic descriptions. Proc. nat. Acad. Sci. 76, 3607-3611.

Miyashita, K. 1969. Effects of locomotory activity, temperature and hunger on the respiratory rate of *Lycosa T-insignita Boes. et Str.* (Araneae: Lycosidae). Appl. Entomol. Zool. 4, 105-113.

Moberly, W. R. 1968. The metabolic responses of the common iguana, *Iguana iguana*, to walking and diving. Comp. biochem. Physiol. 27, 21-32.

Montroll, E. W., Shuler, K. E. 1958. The application of the theory of stochastic processes to chemical kinetics. Adv. chem. Phys. 1, 361-399.

Moran, P. A. P. 1961. Entropy, Markov processes and Boltzmann's H-theorem. Proc. Camb. Phil. Soc. 57, 833-842.

Mori, H. 1965. Transport, collective motion and Brownian motion. Progr. theor. Phys. 33, 423-455.

Mori, H. 1975. Stochastic processes of macroscopic variables. Progr. theor. Phys. 53, 1617-1640.

Morimoto, T. 1963. Markov processes and the H-theorem. J. phys. Soc. Japan 18, 328-331.

Mortensen, R. E. 1969. Mathematical problems of modeling stochastic non-linear dynamic systems. J. stat. Phys. 1, 271-296.

Muravskaya, Z. A., Belokopytin, Yu. S. 1975. Influence of active movement on metabolism of nitrogen and respiration of fish. Biology of Sea, N5, 39-44 (Russian).

Murray, B. 1926. The physiological principle of minimum work. Proc. nat. Acad. Sci. 12, 207-214.

Nakano, H. 1973. Notes on variation principles in the quantum and classical theories of transport processes. Prog. theor. Phys. 49, 1503-1515.

Nakoryakova, S. B. 1964. Study of the stability of mechanical equilibrium of unevenly heated binary system. Proc. Perm. Polytechn. Inst. 15, 113-120 (Russian).

Nicolis, G. 1972. Fluctuations around non-equilibrium states in open nonlinear systems. J. stat. Phys. 6, 195-222.

Nicolis, G., Allen, P., van Nypelseer, A. 1974. Some remarks on the theory of fluctuations around nonequilibrium states. Progr. theor. Phys. 52, 1481-1497.

Nicolis, G., Babloyantz, A. 1969. Fluctuation in open systems. J. chem. Phys. 51, 2632-2637.

Nicolis, G., Malek-Mansour, M. 1978. Nonequilibrium phase transitions and chemical reactions. Suppl. Progr. theor. Phys. No 64, 249-266.

Nicolis, G., Malek-Mansour, M. 1980. (to appear).

Nicolis, G., Prigogine, I. 1971. Fluctuations in nonequilibrium systems. Proc. nat. Acad. Sci. USA 68, 2102-2107.

Nicolis, G., Prigogine, I. 1977. Self-Organization in Nonequilibrium Systems. From Dissipative Structures to Order through Fluctuations. New York, Wiley.

Nicolis, G., Turner, J. W. 1977. Stochastic analysis of a nonequilibrium phase transition: some exact results. Physica 89 A, 326-338.

Nicolis, G., Turner, J. W. 1979. Effect of fluctuation on bifurcation phenomena. In: Proc. Confer. on Bifurcation Theory and Application in Scientific Disciplines. Ann. N. Y. Acad. Sci 316, 251-262.

Nikolaev, L. A. 1976. Bases of Physical Chemsitry of Biological Processes. Moscow, Visshaya Shkola (Russian).

Nikolsky, G. V. 1974. Ecology of Fishes. Moscow, Visshaya Shkola (Russian).

Nitzan, A., Ortoleva, P., Deutch, J., Ross, T. 1974. Fluctuations and transitions at chemical instabilities; the analog to phase transition. J. chem. Phys. 61, 1056-1073.

Nitzan, A., Ross, J. 1974. A comment on fluctuations around nonequilibrium steady states. J. stat. Phys. 10, 379-390.

Obukhov, A. M. On symmetrized nonlinear systems. C. R. Acad. Sci. USSR 233, 35-38 (Russian).

Onsager, L. 1931 a. Reciprocal relations in irreversible processes. I. Phys. Rev. 37, 405-426.

Onsager, L. 1931 b. Reciprocal relations in irreversible processes. II. Phys. Rev. 38, 2265-2279.

Onsager, L., Machlup, S. 1953. Fluctuations and irreversible processes. Phys. Rev. 91, 1505-1512.

Onuki, A. 1978. Langevin equation with multipoissonian noise. J. stat. Phys. 19, 325-332.

Oppenheim, I., Shuler, K. E., Weiss, G. H. 1977. Stochastic theory of non-linear rate processes with multiple stationary states. Physica 88 A, 191-214.

Oster, G. F., Perelson, A. S. 1974. Chemical reaction dynamics. I. Geometrical structure. Arch. rat. Mech. Analysis 55, 230-274.

Oster, G. F., Perelson, A. S., Katchalsky, A. 1973. Network thermodynamics: dynamic modelling of biophysical systems. Quart. Rev. Biophys. 6, 1-134.

Ostroumov, A. A. 1955. On age composition of shoal and the growth of bream of Rybinsk reservoire. In: Proc. Biol. Sta. "Borok" 2, 166-183 (Russian).

Pasquis, P., Lacaisse, A., Dejours, P. 1970. Maximal oxygen uptake in four species of small mammals. Respir. Physiol. 9, 298-309.

Pasynsky, A. G. 1957. Theory of open systems and its significance for biochemistry. Adv. modern Biol. 43, 263-279 (Russian).

Pasynsky, A. G. 1968. Biophysical Chemistry. Moscow, Visshaya Shkola (Russian).

Patashinsky, A. Z., Pokrovsky, V. L. 1975. Fluctuation Theory of Phase Transitions. Moscow, Nauka (Russian).

Patashinsky, A. Z., Pokrovsky, V. L. 1977. Renorm-group method in the theory of phase transitions. Adv. phys. Sci. 121, 55-96 (Russian).

Pauli, W. 1928. Über das H-Theorem vom Anwachsen der Entropie vom Standpunkt der neuen Quantenmechanik. In: Problem der Modernen Physik. Leipzig, Hirzel, 39-45.

Pawula, R. F. 1967 a. Approximation of the linear Boltzmann equation by the Fokker-Planck equation. Phys. Rev. 162, 186-188.

Pawula, R. F. 1967 b. Generalizations and extensions of the Fokker-Planck-Kolmogorow equations. IEEE Trans. Inform. Theory 13, 33-41.

Pearson, O. P. 1950. The metabolism of hummingbirds. Condor 52, 145-152.

Penrose, O. 1970. Foundations of Statistical Mechanics. Oxford. Pergamon Press.

Pfaffelhuber, E. 1977. Information-theoretic stability and evolution criteria in irreversible thermodynamics. J. stat. Phys. 16, 69-90.

Pianigiani, G. 1976. Existence of invariant measures for piecewise continuous transformations. Preprint Inst. Phys. Sci. Technol., Univ. Maryland.

Pianigiani, G. 1977. Absolutely continuous invariant measures for the process $x_{n+1}=Ax_n(1-x_n)$. Preprint Inst. Phys. Sci. Technol., Univ. Maryland.

Pimm, S. L. 1976. Existence metabolism. Condor 78, 121-124.

Poddubny, A. G., Golovanov, V. K., Lapkin, V. V. 1978. Seasonal dynamics of selected temperature of fishes. In: Theoretical Aspects of Fish Hatchery Studies of Reservoires. Leningrad, Nauka, 151-167 (Russian).

Poglazov, B. V. 1966. The Structure and Functions of Contractile Proteins. New York, Acad. Press.

Pomeau, Y. 1971. Long-time behavior of transport coefficients in fluids. Phys. Rev. A 5, 2569-2587.

Pontrjagin, L. S. 1934. Über Autoschwingungssysteme, die den Hamiltonschen nahe liegen. Phys. Z. Sowjetunion 6, 25-28.

Prange, H. D. 1976. Energetics of swimming of a sea turtle. J. exp. Biol. 64, 1-12.

Prange, H. D., Schmidt-Nielsen, K. 1970. The metabolic cost of swimming in ducks. J. exp. Biol. 53, 763-777.

Presnov, E. V. 1973. Potential character of the evolution criterium of thermodynamics of irreversible processes. J. phys. Chem. 47, 2902-2904 (Russian).

Presnov, E. V. 1978. Formalism of non-equilibrium phenomenological thermodynamics. In: Thermodynamics of Biological Processes. Berlin, de Gruyter, 31-59.

Presnov, E. V., Kofman, G. B., Khlebopros, R. G. 1977. Change of entropy in the open system. In: Theor. exp. Biophys., Kaliningrad, 7, 111-116 (Russian).

Presnov, E. V., Zotina, R. S. 1978. Stochastic consideration of constitutive processes and of the evolution criterion. In: Thermodynamics of Biological Processes. Berlin, de Gruyter, 99-104.

Prigogine, I. 1963. Non-Equilibrium Statistical Mechanics. New York, Wiley.

Prigogine, I. 1967. Introduction to Thermodynamics of Irreversible Processes. New York, Wiley.

Prigogine, I. 1979. Proc. XVII Solvay Confer. on Physics, New York, Wiley.

Prigogine, I., Defay, R. 1951. Thermodynamique Chimique. Liege, Desoer.

Prigogine, I., Wiame, J. M. 1946. Biologie et thermodynamique des phénomènes irréversibles. Experientia 2, 451-453.

Pritchard, A. W., Florey, E., Martin, A. W. 1958. Relationship between metabolic rate and body size in an elasmobranch *(Squalus suckleyi)* and in a teleost *(Ophiodon elongatus)*. J. mar. Res. 17, 403-411.

Prits, A. K. 1974. Principle of Stationary States of Open Systems and Population Dynamics. Kaliningrad, Univ. Press (Russian).

Prokhorov, Yu. V., Rozanov, Yu. A. 1973. Theory of Probabilities. Moscow, Nauka (Russian).

Prosser, C. L. (Ed.). 1973. Comparative Animal Physiology. v. 1, 349-429, Philadelphia, Saunders.

Pugachev, V. S., Khazanov, I. E., Evlanov, L. G. 1974. Foundations of Statistical Theory of Control Systems. Moscow, Mashinostroenie (Russian).

Raab, J. L., Eng, P., Waschler, R. A. 1976. Metabolic cost of grade running in dogs. J. appl. Physiol. 41, 532-535.

Rao, G. M. M. 1968. Oxygen consumption of rainbow trout *(Salmo gairdneri)* in relation to activity and salinity. Canad. J. Zool. 46, 781-786.

Rashevsky, N. 1973. The principle of adequate design. In: Foundations of Mathematical Biology., v. 3, 143-175, New York, Acad. Press.

Rayleigh Lord. 1891. Dynamical problems in illustration of the theory of gases. gases. Phil. Mag. 32, 538-445

Reichle, D. E. 1968. Relation of body size to food intake, oxygen consumption, and trace element metabolism in forest floor arthropods. Ecology 49, 538-542.

Rockefeller, R. T. 1970. Convex Analysis. New Jersey, Princeton Univ. Press.

Romanoff, A. L. 1967. Biochemistry of the Avian Embryo. New York, Wiley.

Romanovsky, Yu. M., Stepanova, N. V., Chernavsky, D. S. 1975. Mathematical Modelling in Biophysics. Moscow, Nauka (Russian).

Rosen, R. 1967. Optimality Principles in Biology. London, Butterworths.

Rosenmann, M., Morrison, P. R. 1974. Maximum oxygen consumption and heat loss facilitation in small homeotherms by He-O_2. Amer. J. Physiol. 226, 490-495.

Rosenstein, R., McCarthy, L. E., Borisch, H. L. 1974. Influence of hypoxia on tidal volume response to CO_2 in decerebrate cats. Respir. Physiol. 20, 239-250.

Ruben, J. A. 1976. Aerobic and anaerobic metabolism during activity in snakes. J. comp. Physiol. B 109, 147-157.

Rubin, A. B. 1976. Thermodynamics of Biological Processes. Moscow, Univ. Press (Russian).

Rubner, M. 1908. Das Problem der Lebensdauer und seine Beziehungen zu Wachstum and Ernährung. München und Berlin.

Ruelle, D. 1973. Dissipative systems and differential analysis. In: Mathematical Methods in Theoretical Physics. Colorado Ass. Univ. Press, 461-482.

Ruelle, D. 1977. Applications conservant une mesure absolument continue par rapport a dx gun (0,1). Comm. math. Phys. 55, 47-51.

Schaarschmidt, B., Brettel, R. 1978. The change of the Ψ_u-function during the growth of microbial cultures. In: Thermodynamics of Biological Processes. Berlin, de Gruyter, 181-189.

Schaarschmidt, B., Zotin, A. I., Brettel, R., Lamprecht, I. 1975. Experimental investigations of the bound dissipation function. Change of the Ψ_u -function during the growth of yeast. Arch. Microbiol. 105, 13-16.

Schiller, N. N. 1902. Basic laws of thermodynamics. J. russian Phys. Chem. Soc. 34, 377-426 (Russian).

Schlögl, F. 1971 a. Stability of steady state. Z. Phys. 243, 303-309.

Schlögl, F. 1971 b. On thermodynamics near a steady state. Z. Phys. 248, 446-458.

Schlögl, F. 1972. Chemical reaction models for non-equilibrium phase transitions. Z. Phys. 253, 147-161.

Schlögl, F. 1973. Stability criteria in non-equilibrium thermodynamics. In: Synergetics. Stuttgart, Teubner, 21-27.

Schlögl, F. 1974. Reversibility invariants and non-linear thermodynamics. In:Cooperative Effects. Amsterdam, North-Holland, 149-169.

Schmidt-Nielsen, K. 1970. Energy metabolism, body size and problems of scaling. Fed. Proc. 29, 1524-1532.

Schmidt-Nielsen, K. 1975. Animal Physiology. Adaptation and Environment. Cambridge Univ. Press.

Schmidt-Nielsen, K., Kanwisher, J., Lasiewski, R. C., Cohn, J. E., Bretz, W. L. 1969. Temperature regulation and respiration in the ostrich. Condor 71, 341-352.

Schmutzer, E. 1973. Grundprinzipien der Klassischen Mechanik und der Klassischen Feldtheorie. Berlin, Deutscher Verl. Wiss.

Schnakenberg, J. 1977. Thermodynamic Network Analysis of Biological Systems. Berlin, Springer.

Segrem, N. P., Hart, J. S. 1967. Oxygen supply and performance in Peromyscus: metabolic and circulatory responses to exercise. Canad. J. Physiol. Pharm. 45, 531-542.

Sel'kov, E. E. 1968. Self-oscillations in glycolysis. I. A simple kinetic model. Europ. J. Biochem. 4, 79-86.

Sel'kov, E. E. 1972. Nonlinearity of multienzyme systems. In: Analysis and Simulation of Biochemical Systems. Amsterdam, North-Holland, 145-161.

Semevskaya, V. A., Semevsky, F. N. 1977. Problem on stability of ecological systems of the type "parasite-host", "predator-prey" in terms of optimality principle. Zool. J. 56, 5-9 (Russian).

Sergeev, M. V. 1974. Generalized equation of transfer in the theory of irreversible processes. Theor. math. Phys. 21, 402-414 (Russian).

Sewel, G. L. 1965. Quantum-statistical theory of irreversible processes: dynamics of gross variables. Physica 31, 1520-1536.

Shristian, J. J. 1963. Endocrine adaptive mechanisms and the physiological regulation of population growth. In: Physiological Mammalogy, v. 1, 189-353, New York, Acad. Press.

Shulman, G. E. 1972. Physiological-Biochemical Peculiarities of Annual Cycles of Fishes. Moscow, Pishchevaya Promyshlennost (Russian).

Shumakov, V. I., Novoseltsev, V. N., Sakharov, M. P., Stengold, E. Sh. 1971. Modelling of Physiological Systems of Organisms. Moscow, Medizina (Russian).

Siegel, A. 1960. Differential operator approximations to the linear Boltzmann equation. J. math. Phys. 1, 378-390.

Simons, S. 1969. On the time derivatives of Boltzmann's H-function. J. Phys. A 2, 12-13.

Sinay, Ya. G. 1963. On properties of the spectra of ergodic dynamical systems. C. R. Acad. Sci. USSR 153, 1261-1263 (Russian).

Sinay, Ya. G. 1970. Several accurate results on correlation decrease. Addition to G. M. Zaslavsky's book 1970 (Russian).

Slebodziński, W. 1970. Exterior Forms and their Applications. Warsawa, Polish Sci. Publ.

Slonim. A. D. 1952. Animal Heat and its Regulation in the Organism of Animals. Moscow-Leningrad, Acad. Sci. Press (Russian).

Slymour, W. G. 1973. Physiological correlates of forced activity and burrowing in the spadefoot toad, *Scaphiopus hammondii*. Copeia No. 1, 103-115.

Smit, H., Amelink-Koutstaal, J. M., Vijverberg, J., Vaupel-Klein, J. C. 1971. Oxygen consumption and efficiency of swimming goldfish. Comp. biochem. Physiol. 39, 1-28.

Smith, J. M. 1974. Models in Ecology. Cambridge Univ. Press.

Sorokin, V. S. 1953. Variation method in convection theory. Appl. math. Mech. 17, 39-48 (Russian).

Sorokin, V. S., Sushkin, I. V. 1960. Equilibrium stability of conducting liquid heated from the bottom in a magnetic field. J. exp. theor. Phys. 39, 612-620 (Russian).

Spohn, H. 1978. Entropy production for quantum dynamical semigroups. J. math. Phys. 19, 1227-1230.

Spohn, H., Lebowitz, J. L. 1978. Irreversible thermodynamics for quantum systems weakly coupled to thermal reservoirs. Adv. phys. Chem. 38, 109-142

Stepanova, N. V., Romanovsky, Yu. M. 1975. Classification of mathematical models in microbiology and methods of their investigation. In: Application of Mathematical Methods in Microbiology. Pushchino, Nauka, 3-26 (Russian).

Sternberg, S. 1964. Lectures on Differential Geometry. New York, Prentice Hall.

Stolbov, A. Ya. 1972. Effect of dosed loads on the intensity of gas exchange in some species of Baykal fishes. In: Energetic Aspects of Growth and Metabolism of Aquatic Animals. Kiev, Naukova Dumka, 209-211 (Russian).

Storonkin, B. A., Kuni, F. M. 1973. Statistical thermodynamics of strongly nonequilibrium systems. III; Curie relation. Proc. Leningrad Univ., No. 22, 18-24 (Russian).

Stratonovich, R. L. 1961. Selected Problems of Fluctuation Theory in Radio Industry. Moscow, Sovetkoye radio (Russian).

Stratonovich, R. L. 1963. Topics in the Theory of Random Noise. v. I, New York, Gordon & Breach.

Stratonovich, R. L. 1966. A new representation for stochastic integrals and equations. SIAM J. Control 4, 362-371.

Stratonovich, R. L. 1967. About major correlations of nonlinear thermodynamics of irreversible processes. Proc. Moscow Univ. Ser. Physics and Astronomy, No. 4, 84-89 (Russian).

Stratonovich, R. L. 1969. On nonlinear thermodynamics of irreversible processes. Proc. Moscow Univ. Ser. Physics and Astronomy, No. 1, 40-46 (Russian).

Stratonovich, R. L. 1970 a. Fluctuation-dissipative thermodynamics with time-even and time-odd variables. Proc. Moscow Univ. Ser. Physics and Astronomy, No. 5, 479-486 (Russian).

Stratonovich, R. L. 1970 b. On nonlinear quantum theory of heat fluctuations. J. exp. theor. Phys. 58, 1612-1622 (Russian).

Stratonovich, R. L. 1970 c. Heat noises of nonlinear resistances. Proc. Higher Educ. Inst., Radiophys. 13, 1512-1522 (Russian).

Stratonovich, R. L. 1971. Application of linear and nonlinear fluctuation-dissipative thermodynamics to mechanics. Proc. Moscow Univ., Ser. Physics and Astronomy, No. 1, 3-10 (Russian).

Stratonovich, R. L. 1978. On deriving temporal irreversibility of thermodynamic processes from microscopic reversibility. Theor. math. Phys. 36, 74-85 (Russian).

Stratonovich, R. L., Platonov, A. A. 1972. Homogenous isotropic Brownian movement in linear-cubical approximation. Proc. Moscow Univ. Ser. Physics and Astronomy, No. 6, 671-678 (Russian).

Stratonovich, R. L., Platonov, A. A. 1973. Heat fluctuations in nonlinear macroscopic thermodynamics. Radiotechn. and Electr. 18, 326-334 (Russian).

Stueckelberg, E. G. G. 1952. Théorème H et unitarité de S. Helv. Phys. Acta 25, 577-580.

Sushchenya, L. M. 1972. Respiration Intensity of Crustaces. Kiev, Naukova Dumka (Russian).

Suzuki, M. 1977. Anomalous fluctuation and relaxation in unstable systems. J. stat. Phys. 16, 477-504.

Suzuki, M. 1978. Scaling theory of transient nonlinear fluctuation and formation of macroscopic order. In: Suppl. Progr. theort. Phys. No. 64, 402-422.

Suzuki, M. 1979. Theory of instability, non-linear Brownian motion and formation of macroscopic order. In: Proc. XVII Solvay Confer. Phys., New York, Wiley.

Taylor, C. R., Dmi'el, R., Fedak, M., Schmidt-Nielsen, K. 1971. Energetic cost of running and heat balance in a large bird, the rhea. Amer. J. Physiol. 221, 597-601.

Taylor, C. R., Schmidt-Nielsen, K., Raab, J. L. 1970. Scaling of energetic cost of running to body size in mammals. Amer. J. Physiol. 219, 1104-1107.

Teal, J. M. 1969. Direct measurement of CO_2 production during flight in small birds. Zoologica 54, 17-23.

Temkin, I. B. 1974. Physical Exercise and Cardio-Vascular System. Moscow, Visshaya shkola (Russian).

Terskova, M. I. 1975. On application of parabolic law to describing parameters characterizing growth of bird embryos. In: Quantitative Aspects of Organism Growth. Moscow, Nauka, 70-80 (Russian).

Thom, R. 1969. Topological models in biology. Topology 8, 313-336.

Thom, R. 1972. Statbilité Structurelle et Morphogénèse. New York, Benjamin.

Thom, R. 1975. Structural Stability and Morphogenesis. An Outline of a General Theory of Models. New York, Wiley.

Thomas, H. 1973. Dynamics of current instabilities. In: Synergetics. Stuttgart, Teubner.

Thomas, S. P. 1975. Metabolism during flight in two species of bats, *Phyllostomus hastatus* and *Pteropus gouldii*. J. exp. Biol. 63, 273-293.

Thomas, S. P., Suthers, R. A. 1972. The physiology and energetics of bat flight. J. exp. Biol. 57, 317-335.

Tikhonov, V. M., Mironov, M. A. 1977. Markov Processes. Moscow, Sovetskoye Radio (Russian).

Tishchenko, S. V. 1976. Construction of generalized hydrodynamics by the method of nonequilibrium statistical operator. Theor. math. Phys. 26, 96-108 (Russian).

Truesdell, C. 1972. A First Course in Rational Continuum Mechanics. Baltimore, John Hopkins Univ.

Tucker, V. A. 1965. Oxygen consumption, thermal conductance, and torpor in the California pocket mouse, *Perognathus californicus*. J. cell. comp. Physiol. 65, 393-404.

Tucker, V. A. 1968. Respiratory exchange and evaporative water loss in the flying budgerigar. J. exp. Biol. 48, 67-87.

Tucker, V. A. 1969. The energetics of bird flight. Sci. Amer. 220, 70-78.

Tucker, V. A. 1970. Energetic cost of locomotion in animals. Comp. biochem. Physiol. 34, 841-846.

Tucker, V. A. 1972. Metabolism during flight in the laughing gull, *Larus atricilla*. Amer. J. Physiol. 222, 237-245.

Tucker, V. A. 1973. Bird metabolism during flight: evaluation of a theory. J. exp. Biol. 58, 689-709.

Turing, A. M. 1952. Chemical basis of morphogenesis. Phil. roy. Soc. London, Ser. B, 237, 37-72.

Turner, J. S. 1977. Discrete simulation methods for chemical kinetics. J. phys. Chem. 81, 2379-2408.

Turpaeva, E. P., Soldatova, I. N., Simkina, R. G. 1970. Metabolism intensity in Black Sea bivalvia. Zool. J. 49, 1571-1572 (Russian).

Tyler, S. A., Stearner, S. Ph. Muller, H. D. 1967. Radiation effect on growth of the chicken X-irradiated during embryonic development. Growth 31, 105-117.

Uhlmann, A. 1977. Markov master equation and the behaviour of some entropy-like quantities. Rostocker Physik. Manuskripte 2, 45-54.

Uylings, H. B. M. 1977. Optimization of diameters and bifurcation angles in lung and vascular tree structure. Bull. math. Biophys 39, 245-257.

van den Broek, C., Houard, J., Malek-Mansour, M. 1980. Physica (submitted).

van Kampen, N. G. 1954. Quantum statistics of irreversible processes. Physica 20, 603-622.

van Kampen, N. G. 1957 a. Derivation of the phenomenological equations from the master equation. I. Even variables only. Physica 23, 707-719.

van Kampen, N. G. 1957 b. Derivation of the phenomenological equations from the master equation. II. Even and odd variables. Physica 23, 816-824. 816-824.

van Kampen, N. G. 1960. Non-linear thermal fluctuations in a diode. Physica 26, 585-604.

van Kampen, N. G. 1961 a. A power series expansion of the master equation. Canad. J. Phys. 39, 551-567.

van Kampen, N. G. 1961 b. Exact calculation of the fluctuation spectrum for a nonlinear model system. J. math. Phys. 2, 592-601.

van Kampen, N. G. 1965. Fluctuations in non-linear systems. In: Fluctuations Phenomena in Solids. New York, Acad. Press, 139-177.

van Kampen, N. G. 1970. Stochastic Processes in Physics. Lecture Notes, Utrecht.

van Kampen, N. G. 1976. The expansion of the master equation. Adv. chem. Phys. 34, 245-309.

van Kampen, N. G. 1977 a. The two expansion of the master equation. Phys. Lett. 62 A, 383-385.

van Kampen, N. G. 1977 b. A soluble model for diffusion in a bistable potential. J. stat. Phys. 17, 71-88.

van Kampen, N. G. 1978. Stochastic phenomena. In: Fundamental Problems in Statistical Mechanics. Warsawa, Polish Acad. Sci., 223-263.

Vasiliev, V. A. 1978. Stationary dissipative structures. In: Thermodynamics of Biological Processes. Berlin, de Gruyter, 277-292.

Ventzel, A. D. 1975. Lecture of the Theory of Casual Processes. Moscow, Nauka (Russian).

Vernberg, F. J. 1959. Studies on the physiological variation between tropical and temperature zone fiddler crabs of the genus Uca. III. The influence of temperature acclimation on oxygen consumption of whole organisms. Biol. Bull. 117, 582-593.

Virodov, I. P. 1977. Basic variational principles of irreversible thermodynamics and construction of their formalism. VINITI, Depon. No. 3982 (Russian).

Virodov, I. P. 1978. Axiomatics of phenomenological irreversible thermodynamics and the evolution criterium of thermodynamical systems. VINITI, Depon. No. 2127 (Russian).

Virodov, I. P. On evolution criterium of thermodynamical systems and relativity covariant aspects of irreversible thermodynamics. In: Questions of Physics of Pattern Formation and Phase Changes. Kalinin, 3-15 (Russian).

Vojta, G. 1967. Variationsprinzipien und Schwankungstheorie für Nichtgleichgewichtssysteme im Funktionalformalismus. Abhandl. DAW zu Berlin (M. P.) 1, 183-195.

Volkenstein, M. V., Chernavsky, D. S. 1978. Information and biology. J. social. Biol. Struct., 95-108.

Volterra, V. 1931. Leçons sur la Theorie Mathematique de la Lutte pour la Vie. Paris.

von Neumann, J. V. 1932. Mathematische Grundlagen der Quantenmechanik. Berlin, Springer.

Wagner, H. 1966. Long-wavelength exitations and the Goldstone theorem in many-particle systems with "broken symmetries". Z. Physik 195, 273-299.

Walgraef, D., Dewel, G., Borckmans, P. 1980. Fluctuations near nonequilibrium phase transitions to new uniform states. Phys. Rev.

Walter, Ch. 1973. Stability properties and periodic behaviour of controlled biochemical systems. In: Nonlinear Problems in the Physical Sciences and Biology. Berlin, Springer, 343-357.

Walter, R., Lamprecht, I. 1978. Modern theories concerning the growth equations. In: Thermodynamics of Biological Processes. Berlin, de Gruyter, 143-162.

Wang, Y. K., Hioe, F. T. 1973. Phase transition in the Dicke model of superradiance. Phys. Rev. A 7, 831-836.

Wangsness, R. K. 1975. Minimum entropy production and logarithmic rate equations. Physica A 79, 543-557.

Watanabe, S. 1955 a. Symmetry of physical laws. I. Symmetry in space-time and balance theorems. Rev. mod. Phys. 27, 26-39.

Watanabe, S. 1955 b. Symmetry of physical laws. II. Q-number theory of space-time inversions and charge conjugation Rev. mod. Phys. 27, 40-76.

Watanabe, S. 1955 c. Symmetry of physical laws. III. Prediction and retrodiction. Rev. mod. Phys. 27, 179-186.

Watanabe, S. 1969. Knowing and Guessing. New York, Wiley.

Wax, N. 1954. Selected Topics in the Theory of Noise and Stochastic Processes. New York, Dover.

Webb, P. W. 1971. Swimming energetics of trout. II. Oxygen consumption and swimming efficiency. J. exp. Biol 55, 521-540.

Weis-Fogh, T. 1964. Biology and physics of locust flight. VIII. Life and metabolic rate of flying locust. J. exp. Biol. 41, 257-271.

Wigglesworth, V. B. 1974. Insect Physiology. London, Chapman & Hall.

Wigner, E. P. 1970. Symmetries and reflections. Bloomington, Indian Univ. Press.

Wilkie, D. R. 1959. The work output of animals: flight by birds and manpower. Nature 183, 1515-1516.

Wilson, K. G., Kogut, J. 1974. The Renormalization Group and the expansion. Amsterdam, North-Holland.

Wilson, K. J. 1974. The relationship of oxygen supply for activity to body temperature in four species of lizards. Copeia, No. 4, 920-934.

Wilson, K. J., Lee, A. K. 1970. Changes in oxygen consumption and heart-rate activity and body temperature in the tuatara *Spendon punctatum.* Comp. biochem. Physiol. 33, 311-322.

Winberg, G. G. 1950. Metabolism intensity and size of Crustacea. J. gen. Biol. 11, 367-380 (Russian).

Winberg, G. G. 1956 a. Metabolism Intensity and Nutrient Requirements of Fishes. Minsk, Belorussky Univ. Press (Russian).

Winberg, G. G. 1956 b. Dependence of metabolic rate on the body weight in Arthropoda. Proc. Belorussky Univ. 26, 243-254 (Russian).

Winberg, G. G. 1976. Dependence of energetic metabolism on the body weight in aquatic poikilothermic animals. J. gen. Biol. 37, 56-70 (Russian).

Winberg, G. G., Belyatskaya, Yu. S. 1959. Relation between metabolism intensity and the body weight in fresh water Gastropoda. Zool. J. 28, 1146-1151 (Russian).

Winfree, A. 1974. Rotating chemical reactions. Sci. Amer. 230, 82-95.

Woodcock, A. E. R., Poston, T. 1974. The geometrical study of the elementary catastrophe. Lecture Notes Mathemat., 373, Berlin, Springer.

Wolvekamp, H. P., Waterman, T. H. 1960. Respiration. In: The Physiology of Crustacea. v. I, 35-100, New York, Acad. Press.

Wynne-Edwards, V. C. 1962. Animal Dispersal in Relation to Social Behaviour. Edinburgh, Oliver & Boyd.

Yakobson, M. V. 1978. Topological and metric properties of one-dimensional endomorphisms. C. R. Acad. Sci. USSR 243, 866-869 (Russian).

Yamakoshi, K., Togawa, T., Kamiya, A. T. 1976. Optimal of mechanical porperty in structures of branching tree. Japan J. med. Electron. biol. Eng. 14, 296-302 (Japanese).

Yarzhombek, A. A., Klyashtorin, L. B., Podrazhanskaya, S. G. 1977. Active metabolism in fishes. Ecology 1, 65-75 (Russian).

Yoshida, K. 1974. Functional Analysis. Berlin, Springer.

Young, D. R., Mosher, R., Erue, P., Spector, H. 1959. Energy metabolism and gas exchange during treadmill running in dogs. J. appl. Physiol. 14, 834-838.

Yousef, M. K., Robertson, W. D., Dill, D. B. Johnson, H. D. 1970. Energy expenditure of running kangaroo rat *Dipodomys merriami*. Comp. biochem. Physiol. 36, 387-393.

Yousef, M. K. Robertson, W. D., Dill, D. B., Johnson, H. D. 1973. Energetic cost of running in the antelope ground squirrel *Ammospermophibus leucurus*. Physiol. Zool. 46, 139-147.

Yukhnovsky, I. R. 1978. Statistical sum of three-dimensional Ising model. Theoret. math. Phys. 36, 373-399 (Russian).

Zalevsky, K. 1973. Phenomenological and Statistical Thermodynamics. Moscow, Mir (Russian).

Zaslavsky, G. M. 1970. Statistical Irreversibility in Nonlinear Systems. Moscow, Nauka (Russian).

Zaslavsky, G. M. Chirikov, B. V. 1971. Stochastic non-stability nonlinear oscillations. Adv. phys. Sci. 105, 3-39 (Russian).

Zar, J. H. 1968. Standard metabolism comparisons between orders of Birds. Condor 70, 278-279.

Zarubin, A. M. Zharikov, G. G., Rubin, A. B. 1976. Some biological peculiarities and pattern of the growth of *Actinomyces levoris* 17-255 IBF. Sci. Report Higher School, Biol. Sci. No. 5, 100-106 (Russian).

Zebe, E. G. 1954. Über den Stoffwechsel der Lepidopteren. Z. vergl. Physiol. 36, 290-317.

Zeeman, E. C. 1973. Differential equations for the heartbeat and nerva impulse. In: Dynamical Systems. New York, Acad. Press, 683-741.

Zeeman, E. C. 1976. Catastrophe theory. Sci. Amer 234, 65-83.

Zeleznik, F. J. 1976. Thermodynamics. J. math. Phys. 17, 1572-1610.

Zeuthen, E. 1953. Oxygen uptake as related to body size in organisms. Quart. Rev. Biol. 28, 1-12.

Zeuthen, E. 1970. Rate of living as related to body size in organisms. Pol. Arch. Hydrobiol. 17, 21-30.

Zhukovsky, N. E. 1948. Selected Works. Moscow-Leningrad, Gostekhizdat (Russian).

Ziegler, H. 1963. Some extremum principles in irreversible thermodynamics with application to continuum mechanics. In: Progress in Solid Mechanics., v. 4, chap. 2, Amsterdam, North-Holland.

Ziegler, H. 1966. A possible generalization of Onsager's theory. In: Proc. IUTAM Symp. Vienna. Berlin, Springer.

Ziegler, H. 1972. Thermodynamics. Quart. appl. Math. 30, 91-107.

Ziegler, H. 1977. An Introduction to Thermodynamics. Amsterdam, North-Holland.

Zotin, A. I. 1972. Thermodynamic Aspects of Developmental Biology. Basel, Karger.

Zotin, A. I. 1974. Thermodynamical Approach to the Problems of Development, Growth and Aging. Moscow, Nauka (Russian).

Zotin, A. I. 1975. Heat production and respiration in oocytes of *Xenopus laevis* at different stages of oogenesis. Ontogenesis 6, 623-626 (Russian).

Zotin, A. I. 1979. Bound dissipation function - discrepancy between the data of the direct and indirect calorimetry in living organisms. J. gen. Biol. 40, 229-237 (Russian).

Zotin, A. I., Konoplev, V. A. 1978. Direction of the evolutionary progress of organisms. In: Thermodynamics of Biological Processes. Berlin, de Gruyter, 341-347.

Zotin, A. I., Konoplev, V. A., Grudnitzky, V. A. 1978. The questions of non-linearity for using criterion of orderliness. In: Thermodynamics of Biological Processes, Berlin, de Gruyter, 361-367.

Zotin, A. I., Zotina, R. S. 1967. Thermodynamic aspects of developmental biology. J. theor. Biol. 17, 57-75.

Zotin, A. I., Zotina, R. S. 1977. External dissipation function and evolution of the open system to the stationary state. In: Theor. exp. Biophys., Kaliningrad 7, 10-18 (Russian).

Zotin, A. I., Zotina, R. S. 1978. Experimental basis for qualitative phenomenological theory of development. In: Thermodynamics of Biological Processes. Berlin, de Gruyter, 61-84.

Zotin, A. I., Zotina, R. S., Konoplev, V. A. 1978. Theoretical basis for a qualitative phenomenological theory of development. In: Thermodynamics of Biological Processes. Berlin, de Gruyter. 85-98.

Zotin, A. I., Zotina, R. S., Prokofiev, E. A., Konoplev, V. A. 1978. Application of growth equations for the determination of maximal life-span of mammals and men. Proc. Acad. Sci. USSR, Ser. Biol., No. 1, 87-96 (Russian).

Zotina, R. S., Zotin, A. I. 1967. Quantitative relationship between the size, age and fertility in animals. J. gen. Biol. 28, 82-92 (Russian).

Zubarev, D. N. 1973. Nonequilibrium statistical operators and quasimeans in the theory of irreversible processes. In: Statistical Physics and the Quantum Field Theory. Moscow, Nauka, 81-95 (Russian).

Zubarev, D. N. 1974. Nonequilibrium Statistical Thermodynamics. New York, Plenum Press.

Zubarev, D. N., Kalashnikov, V. P. 1970. Plotting of statistical operators for nonequilibrium processes. Theor. math. Phys. 3, 126-134 (Russian).

Zubarev, D. N., Kalashnikov, V. P. 1971. The derivation of time-irreversible generalized master equation. Physica 56, 345-364.

Zubarev, D. N., Khazanov, A. M. 1978. Generalized Fokker-Planck equation and plotting of projection operators for different methods of brief description of nonequilibrium state. Theor. math. Phys. 34, 69-80 (Russian).

Zubarev, D. N., Novikov, M. Yu. 1972. Generalized formulating of boundary condition to Liouville equation and chains B-B-G-K-U. Theor. math. Phys. 13, 406-420 (Russian).

Zwanzig, R. 1960. Ensemble method in the theory of irreversibility. J. chem. Phys. 33, 1338-1341.

Zwanzig, R. W. 1961. Memory effects in irreversible thermodynamics. Phys. Rev. 124, 983-992.

Subject Index

Cited Organisms

Bei Fragen zur Produktsicherheit wenden Sie sich bitte an:
If you have any questions regarding product safety,
please contact:

Walter de Gruyter GmbH
Genthiner Straße 13
10785 Berlin
productsafety@degruyterbrill.com